Independent Component Analysis: Acoustic and Biomedical Aspects

Independent Component Analysis: Acoustic and Biomedical Aspects

Edited by **Howard Zea**

Clanrye International

New Jersey

Published by Clanrye International,
55 Van Reypen Street,
Jersey City, NJ 07306, USA
www.clanryeinternational.com

Independent Component Analysis: Acoustic and Biomedical Aspects
Edited by Howard Zea

International Standard Book Number: 978-1-63240-303-2 (Hardback)

Printed in the United States of America.

Contents

Preface

In my initial years as a student, I used to run to the library at every possible instance to grab a book and learn something new. Books were my primary source of knowledge and I would not have come such a long way without all that I learnt from them. Thus, when I was approached to edit this book; I became understandably nostalgic. It was an absolute honor to be considered worthy of guiding the current generation as well as those to come. I put all my knowledge and hard work into making this book most beneficial for its readers.

The acoustic and biomedical aspects of independent component analysis are elucidated in this insightful book. The signal-processing method used for drawing out individual sources from a provided analyzed data that is a blend of various unknown sources is called Independent Component Analysis (ICA). Blind Source Separation (BSS), a relatively new method, has gained significant importance in the past few years because of its capable signal-processing functions like speech enhancement systems, image processing, telecommunications, medical signal processing and several data mining issues. This book presents knowledge on the structure of some of the pivotal advanced discoveries of ICA, connected to Audio and Biomedical signal processing uses.

I wish to thank my publisher for supporting me at every step. I would also like to thank all the authors who have contributed their researches in this book. I hope this book will be a valuable contribution to the progress of the field.

Editor

Section 1

Introduction

1

Introduction: Independent Component Analysis

Ganesh R. Naik
RMIT University, Melbourne
Australia

1. Introduction

Consider a situation in which we have a number of sources emitting signals which are interfering with one another. Familiar situations in which this occurs are a crowded room with many people speaking at the same time, interfering electromagnetic waves from mobile phones or crosstalk from brain waves originating from different areas of the brain. In each of these situations the mixed signals are often incomprehensible and it is of interest to separate the individual signals. This is the goal of Blind Source Separation (BSS). A classic problem in BSS is the cocktail party problem. The objective is to sample a mixture of spoken voices, with a given number of microphones - the observations, and then separate each voice into a separate speaker channel -the sources. The BSS is unsupervised and thought of as a black box method. In this we encounter many problems, e.g. time delay between microphones, echo, amplitude difference, voice order in speaker and underdetermined mixture signal.

Herault and Jutten Herault, J. & Jutten, C. (1987) proposed that, in a artificial neural network like architecture the separation could be done by reducing redundancy between signals. This approach initially lead to what is known as independent component analysis today. The fundamental research involved only a handful of researchers up until 1995. It was not until then, when Bell and Sejnowski Bell & Sejnowski (1995) published a relatively simple approach to the problem named infomax, that many became aware of the potential of Independent component analysis (ICA). Since then a whole community has evolved around ICA, centralized around some large research groups and its own ongoing conference, International Conference on independent component analysis and blind signal separation. ICA is used today in many different applications, e.g. medical signal analysis, sound separation, image processing, dimension reduction, coding and text analysis Azzerboni et al. (2004); Bingham et al. (2002); Cichocki & Amari (2002); De Martino et al. (2007); Enderle et al. (2005); James & Hesse (2005); Kolenda (2000); Kumagai & Utsugi (2004); Pu & Yang (2006); Zhang et al. (2007); Zhu et al. (2006).

ICA is one of the most widely used BSS techniques for revealing hidden factors that underlie sets of random variables, measurements, or signals. ICA is essentially a method for extracting individual signals from mixtures. Its power resides in the physical assumptions that the different physical processes generate unrelated signals. The simple and generic nature of this assumption allows ICA to be successfully applied in diverse range of research fields. In ICA the general idea is to separate the signals, assuming that the original underlying source signals are mutually independently distributed. Due to the field's relatively young age, the distinction between BSS and ICA is not fully clear. When regarding ICA, the basic framework for most researchers has been to assume that the mixing is instantaneous and linear, as in infomax. ICA is often described as an extension to PCA, that uncorrelates the signals for higher order moments and produces a non-orthogonal basis. More complex models assume for example, noisy mixtures, Hansen (2000); Mackay (1996), nontrivial source distributions, Kab'an (2000); Sorenson (2002), convolutive mixtures Attias & Schreiner (1998); Lee (1997; 1998), time dependency, underdetermined sources Hyvarinen et al. (1999); Lewicki & Sejnowski (2000), mixture and classification of independent component Kolenda (2000); Lee et al. (1999). A general introduction and overview can be found in Hyvarinen et al. (2001).

1.1 ICA model

ICA is a statistical technique, perhaps the most widely used, for solving the blind source separation problem Hyvarinen et al. (2001); Stone (2004). In this section, we present the basic Independent Component Analysis model and show under which conditions its parameters can be estimated. The general model for ICA is that the sources are generated through a linear basis transformation, where additive noise can be present. Suppose we have N statistically independent signals, $s_i(t), i = 1, ..., N$. We assume that the sources themselves cannot be directly observed and that each signal, $s_i(t)$, is a realization of some fixed probability distribution at each time point t. Also, suppose we observe these signals using N sensors, then we obtain a set of N observation signals $x_i(t), i = 1, ..., N$ that are mixtures of the sources. A fundamental aspect of the mixing process is that the sensors must be spatially separated (e.g. microphones that are spatially distributed around a room) so that each sensor records a different mixture of the sources. With this spatial separation assumption in mind, we can model the mixing process with matrix multiplication as follows:

$$x(t) = As(t) \tag{1}$$

where A is an unknown matrix called the *mixing matrix* and $\boldsymbol{x}$(t), $\boldsymbol{s}$(t) are the two vectors representing the observed signals and source signals respectively. Incidentally, the justification for the description of this signal processing technique as *blind* is that we have no information on the mixing matrix, or even on the sources themselves.

The objective is to recover the original signals, $s_i(t)$, from only the observed vector $x_i(t)$. We obtain estimates for the sources by first obtaining the Şunmixing matrixŤ **W**, where, $W = A^{-1}$.

This enables an estimate, $\hat{s}(t)$, of the independent sources to be obtained:

$$\hat{s}(t) = Wx(t) \tag{2}$$

The diagram in Figure 1 illustrates both the mixing and unmixing process involved in BSS. The independent sources are mixed by the matrix **A** (which is unknown in this case). We seek to obtain a vector y that approximates s by estimating the unmixing matrix **W**. If the estimate of the unmixing matrix is accurate, we obtain a good approximation of the sources.

The above described ICA model is the simple model since it ignores all noise components and any time delay in the recordings.

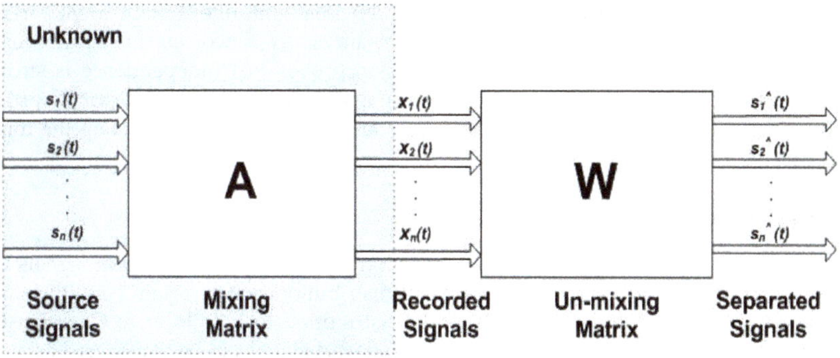

Fig. 1. Blind source separation (BSS) block diagram. $s(t)$ are the sources. $x(t)$ are the recordings, $\hat{s}(t)$ are the estimated sources A is mixing matrix and W is un-mixing matrix

1.2 Independence

A key concept that constitutes the foundation of independent component analysis is statistical independence. To simplify the above discussion consider the case of two different random variables s_1 and s_2. The random variable s_1 is independent of s_2, if the information about the value of s_1 does not provide any information about the value of s_2, and vice versa. Here s_1 and s_2 could be random signals originating from two different physical process that are not related to each other.

1.2.1 Independence definition

Mathematically, statistical independence is defined in terms of probability density of the signals. Consider the joint probability density function (pdf) of s_1 and s_2 be $p(s_1, s_2)$. Let the marginal pdf of s_1 and s_2 be denoted by $p_1(s_1)$ and $p_2(s_2)$ respectively. s_1 and s_2 are said to be independent if and only if the joint pdf can be expressed as;

$$p_{s_1,s_2}(s_1, s_2) = p_1(s_1)p_2(s_2) \tag{3}$$

Similarly, independence could be defined by replacing the pdf by the respective cumulative distributive functions as;

$$E\{p(s_1)p(s_2)\} = E\{g_1(s_1)\}E\{g_2(s_2)\} \tag{4}$$

where E{.} is the expectation operator. In the following section we use the above properties to explain the relationship between uncorrelated and independence.

1.2.2 Uncorrelatedness and Independence

Two random variables s_1 and s_2 are said to be uncorrelated if their covariance $C(s_1,s_1)$ is zero.

$$\begin{aligned} C(s_1, s_2) &= E\{(s_1 - m_{s1})(s_2 - m_{s2})\} \\ &= E\{s_1 s_2 - s_1 m_{s2} - s_2 m_{s1} + m_{s1} m_{s2}\} \\ &= E\{s_1 s_2\} - E\{s_1\}E\{s_2\} \\ &= 0 \end{aligned} \tag{5}$$

where m_{s1} is the mean of the signal. Equation 4 and 5 are identical for independent variables taking $g_1(s_1) = s_1$. Hence independent variables are always uncorrelated. How ever the opposite is not always true. The above discussion proves that independence is stronger than uncorrelatedness and hence independence is used as the basic principle for ICA source estimation process. However uncorrelatedness is also important for computing the mixing matrix in ICA.

1.2.3 Non-Gaussianity and independence

According to central limit theorem the distribution of a sum of independent signals with arbitrary distributions tends toward a Gaussian distribution under certain conditions. The sum of two independent signals usually has a distribution that is closer to Gaussian than distribution of the two original signals. Thus a gaussian signal can be considered as a liner combination of many independent signals. This furthermore elucidate that separation of independent signals from their mixtures can be accomplished by making the linear signal transformation as non-Gaussian as possible.

Non-Gaussianity is an important and essential principle in ICA estimation. To use non-Gaussianity in ICA estimation, there needs to be quantitative measure of non-Gaussianity of a signal. Before using any measures of non-Gaussianity, the signals should be normalised. Some of the commonly used measures are kurtosis and entropy measures, which are explained next.

- **Kurtosis**

Kurtosis is the classical method of measuring Non-Gaussianity. When data is preprocessed to have unit variance, kurtosis is equal to the fourth moment of the data.

The Kurtosis of signal (s), denoted by *kurt* (s), is defined by

$$kurt(s) = E\{s^4\} - 3(E\{s^4\})^2 \tag{6}$$

This is a basic definition of kurtosis using higher order (fourth order) cumulant, this simplification is based on the assumption that the signal has zero mean. To simplify things, we can further assume that (s) has been normalised so that its variance is equal to one: $E\{s^2\} = 1$.

Hence equation 6 can be further simplified to

$$kurt(s) = E\{s^4\} - 3 \tag{7}$$

Equation 7 illustrates that kurtosis is a nomralised form of the fourth moment $E\{s^4\} = 1$. For Gaussian signal, $E\{s^4\} = 3(E\{s^4\})^2$ and hence its kurtosis is zero. For most non-Gaussian signals, the kurtosis is nonzero. Kurtosis can be both positive or negative. Random variables that have positive kurtosis are called as *super Gaussian* or *platykurtotic*, and those with negative kurtosis are called as *sub Gaussian* or *leptokurtotic*. Non-Gaussianity is measured using the absolute value of kurtosis or the square of kurtosis.

Kurtosis has been widely used as measure of Non-Gaussianity in ICA and related fields because of its computational and theoretical and simplicity. Theoretically, it has a linearity property such that

$$kurt(s_1 \pm s_2) = kurt(s_1) \pm kurt(s_2) \tag{8}$$

and

$$kurt(\alpha s_1) = \alpha^4 kurt(s_1) \tag{9}$$

where α is a constant. Computationally kurtosis can be calculated using the fourth moment of the sample data, by keeping the variance of the signal constant.

In an intuitive sense, kurtosis measured how "spikiness" of a distribution or the size of the tails. Kurtosis is extremely simple to calculate, however, it is very sensitive to outliers in the data set. It values may be based on only a few values in the tails which means that its statistical significance is poor. Kurtosis is not robust enough for ICA. Hence a better measure of non-Gaussianity than kurtosis is required.

- **Entropy**

Entropy is a measure of the uniformity of the distribution of a bounded set of values, such that a complete uniformity corresponds to maximum entropy. From the information theory concept, entropy is considered as the measure of randomness of a signal. Entropy H of discrete-valued signal S is defined as

$$H(S) = -\sum P(S = a_i) log P(S = a_i) \tag{10}$$

This definition of entropy can be generalised for a continuous-valued signal (s), called differential entropy, and is defined as

$$H(S) = -\int p(s) log p(s) ds \tag{11}$$

One fundamental result of information theory is that Gaussian signal has the largest entropy among the other signal distributions of unit variance. entropy will be small for signals that have distribution concerned on certain values or have pdf that is very "spiky". Hence, entropy can be used as a measure of non-Gaussianity.

In ICA estimation, it is often desired to have a measure of non-Gaussianity which is zero for Gaussian signal and nonzero for non-Gaussian signal for computational simplicity. Entropy is closely related to the code length of the random vector. A normalised version of entropy is given by a new measure called Negentropy J which is defined as

$$J(S) = H(s_{gauss}) - H(s) \tag{12}$$

where s_{gauss} is the Gaussian signal of the same covariance matrix as (s). Equation 12 shows that Negentropy is always positive and is zero only if the signal is a pure gaussian signal. It is stable but difficult to calculate. Hence approximation must be used to estimate entropy values.

1.2.4 ICA assumptions

- *The sources being considered are statistically independent*

The first assumption is fundamental to ICA. As discussed in previous section, statistical independence is the key feature that enables estimation of the independent components $\hat{s}_(t)$ from the observations $x_i(t)$.

- *The independent components have non-Gaussian distribution*

The second assumption is necessary because of the close link between Gaussianity and independence. It is impossible to separate Gaussian sources using the ICA framework because the sum of two or more Gaussian random variables is itself Gaussian. That is, the sum of Gaussian sources is indistinguishable from a single Gaussian source in the ICA framework, and for this reason Gaussian sources are forbidden. This is not an overly restrictive assumption as in practice most sources of interest are non-Gaussian.

- *The mixing matrix is invertible*

The third assumption is straightforward. If the mixing matrix is not invertible then clearly the unmixing matrix we seek to estimate does not even exist.

If these three assumptions are satisfied, then it is possible to estimate the independent components modulo some trivial ambiguities. It is clear that these assumptions are not particularly restrictive and as a result we need only very little information about the mixing process and about the sources themselves.

1.2.5 ICA ambiguity

There are two inherent ambiguities in the ICA framework. These are (i) magnitude and scaling ambiguity and (ii) permutation ambiguity.

- *Magnitude and scaling ambiguity*

The true variance of the independent components cannot be determined. To explain, we can rewrite the mixing in equation 1 in the form

$$\begin{aligned} x &= As \\ &= \sum_{j=1}^{N} a_j s_j \end{aligned} \tag{13}$$

where a_j denotes the jth column of the mixing matrix A. Since both the coefficients a_j of the mixing matrix and the independent components s_j are unknown, we can transform Equation 13.

$$x = \sum_{j=1}^{N} (1/\alpha_j a_j)(\alpha_j s_j) \tag{14}$$

Fortunately, in most of the applications this ambiguity is insignificant. The natural solution for this is to use assumption that each source has unit variance: $E\{s_{j2}\} = 1$. Furthermore, the signs of the of the sources cannot be determined too. This is generally not a serious problem because the sources can be multiplied by -1 without affecting the model and the estimation

- *Permutation ambiguity*

The order of the estimated independent components is unspecified. Formally, introducing a permutation matrix P and its inverse into the mixing process in Equation 1.

$$\begin{aligned} x &= AP^{-1}Ps \\ &= A^{'}s^{'} \end{aligned} \tag{15}$$

Here the elements of P s are the original sources, except in a different order, and $A' = AP^{-1}$ is another unknown mixing matrix. Equation 15 is indistinguishable from Equation 1 within the ICA framework, demonstrating that the permutation ambiguity is inherent to Blind Source Separation. This ambiguity is to be expected Ű in separating the sources we do not seek to impose any restrictions on the order of the separated signals. Thus all permutations of the sources are equally valid.

1.3 Preprocessing

Before examining specific ICA algorithms, it is instructive to discuss preprocessing steps that are generally carried out before ICA.

1.3.1 Centering

A simple preprocessing step that is commonly performed is to Şcenterŧ the observation vector x by subtracting its mean vector $m = E\{x\}$. That is then we obtain the centered observation vector, x_c, as follows:

$$x_c = x - m \tag{16}$$

This step simplifies ICA algorithms by allowing us to assume a zero mean. Once the unmixing matrix has been estimated using the centered data, we can obtain the actual estimates of the independent components as follows:

$$\hat{s}(t) = A^{-1}(x_c + m) \tag{17}$$

From this point on, all observation vectors will be assumed centered. The mixing matrix, on the other hand, remains the same after this preprocessing, so we can always do this without affecting the estimation of the mixing matrix.

1.3.2 Whitening

Another step which is very useful in practice is to pre-whiten the observation vector x. Whitening involves linearly transforming the observation vector such that its components are uncorrelated and have unit variance [27]. Let x_w denote the whitened vector, then it satisfies the following equation:

$$E\{x_w x_w^T\} = I \tag{18}$$

where $E\{x_w x_w^T\}$ is the covariance matrix of x_w. Also, since the ICA framework is insensitive to the variances of the independent components, we can assume without loss of generality that the source vector, s, is white, i.e. $E\{ss^T\} = I$

A simple method to perform the whitening transformation is to use the eigenvalue decomposition (EVD) [27] of x. That is, we decompose the covariance matrix of x as follows:

$$E\{xx^T\} = VDV^T \tag{19}$$

where V is the matrix of eigenvectors of $E\{xx^T\}$, and D is the diagonal matrix of eigenvalues, i.e. $D = diag\{\lambda_1, \lambda_2, ..., \lambda_n\}$. The observation vector can be whitened by the following transformation:

$$x_w = VD^{-1/2}V^T x \tag{20}$$

where the matrix $D^{-1/2}$ is obtained by a simple component wise operation as $D^{-1/2} = diag\{\lambda_1^{-1/2}, \lambda_2^{-1/2}, ..., \lambda_n^{-1/2}\}$. Whitening transforms the mixing matrix into a new one, which is orthogonal

$$x_w = VD^{-1/2}V^T As = A_w s \tag{21}$$

hence,

$$\begin{aligned} E\{x_w x_w^T\} &= A_w E\{ss^T\} A_w^T \\ &= A_w A_w^T \\ &= I \end{aligned} \tag{22}$$

Whitening thus reduces the number of parameters to be estimated. Instead of having to estimate the n^2 elements of the original matrix A, we only need to estimate the new orthogonal mixing matrix, where An orthogonal matrix has $n(n-1)/2$ degrees of freedom. One can say that whitening solves half of the ICA problem. This is a very useful step as whitening is a simple and efficient process that significantly reduces the computational complexity of ICA. An illustration of the whitening process with simple ICA source separation process is explained in the following section.

1.4 Simple illustrations of ICA

To clarify the concepts discussed in the preceding sections two simple illustrations of ICA are presented here. The results presented below were obtained using the FastICA algorithm, but could equally well have been obtained from any of the numerous ICA algorithms that have been published in the literature (including the Bell and Sejnowsiki algorithm).

1.4.1 Separation of two signals

This section explains the simple ICA source separation process. In this illustration two independent signals, s_1 and s_2, are generated. These signals are shown in Figure2. The independent components are then mixed according to equation 1 using an arbitrarily chosen mixing matrix A, where

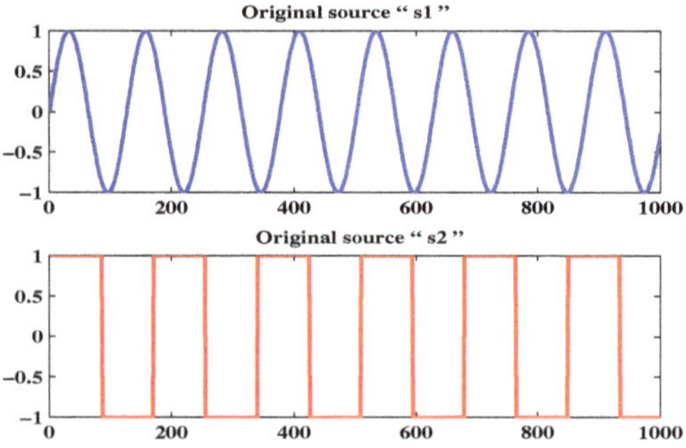

Fig. 2. Independent sources $s1$ and $s2$

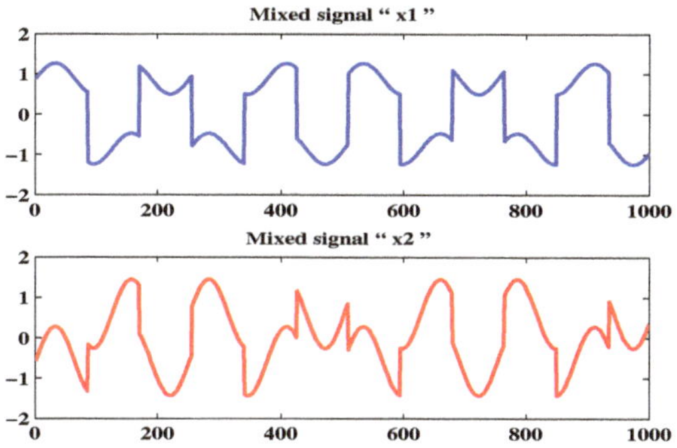

Fig. 3. Observed signals, $x1$ and $x2$, from an unknown linear mixture of unknown independent components

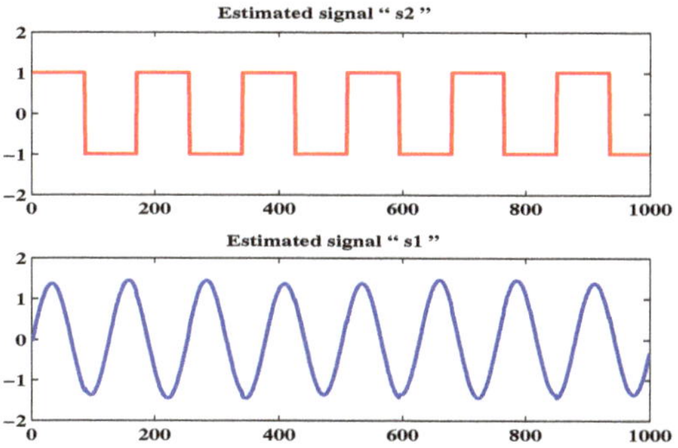

Fig. 4. Estimates of independent components

$$A = \begin{pmatrix} 0.3816 & 0.8678 \\ 0.8534 & -0.5853 \end{pmatrix}$$

The resulting signals from this mixing are shown in Figure 3. Finally, the mixtures x_1 and x_2 are separated using ICA to obtain s_1 and s_2, shown in Figure 4. By comparing Figure 4 to Figure 2 it is clear that the independent components have been estimated accurately and that the independent components have been estimated without any knowledge of the components themselves or the mixing process.

This example also provides a clear illustration of the scaling and permutation ambiguities discussed previously. The amplitudes of the corresponding waveforms in Figures 2 and 4 are different. Thus the estimates of the independent components are some multiple of the independent components of Figure 3, and in the case of $s1$, the scaling factor is negative. The permutation ambiguity is also demonstrated as the order of the independent components has been reversed between Figure 2 and Figure 4.

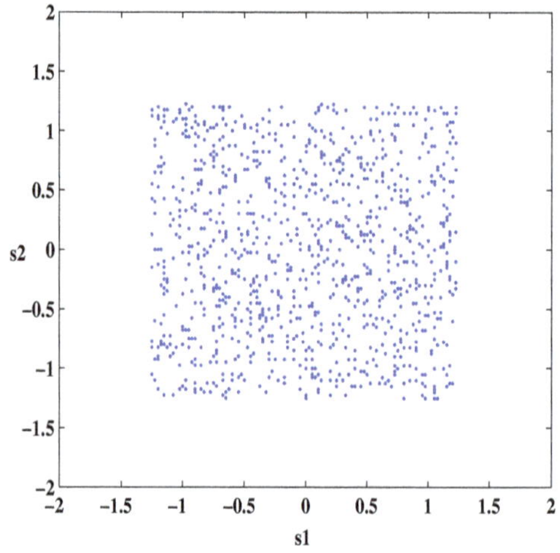

Fig. 5. Original sources

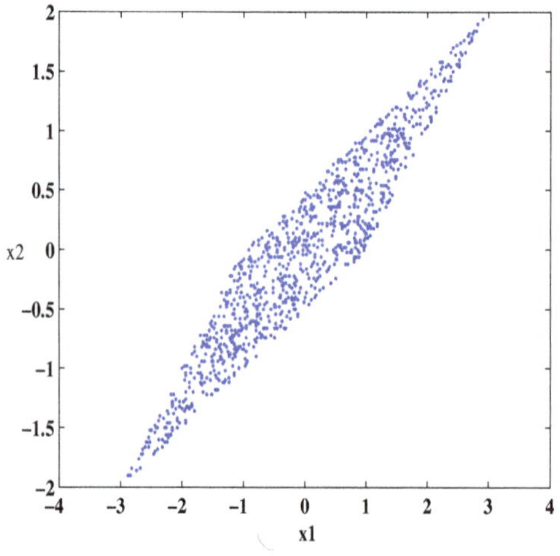

Fig. 6. Mixed sources

1.4.2 Illustration of statistical independence in ICA

The previous example was a simple illustration of how ICA is used; we start with mixtures of signals and use ICA to separate them. However, this gives no insight into the mechanics of ICA and the close link with statistical independence. We assume that the independent components can be modeled as realizations of some underlying statistical distribution at each time instant (e.g. a speech signal can be accurately modeled as having a Laplacian

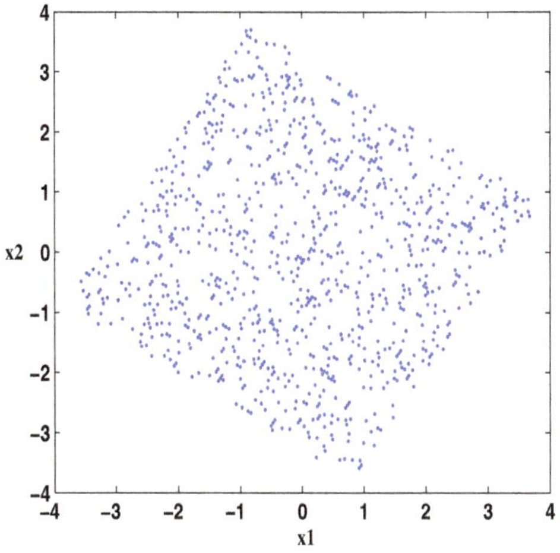

Fig. 7. Joint density of whitened signals obtained from whitening the mixed sources

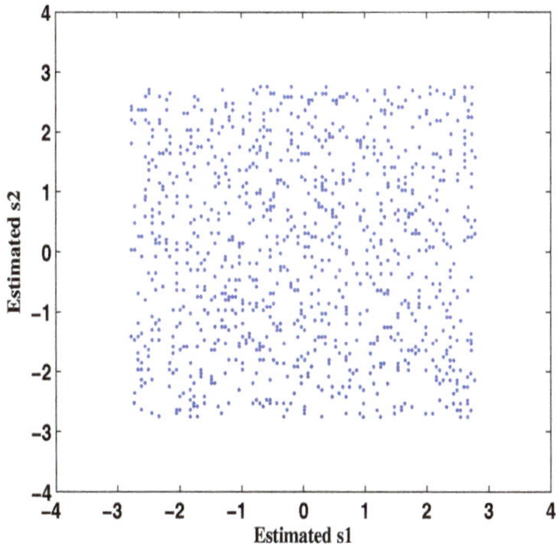

Fig. 8. ICA solution (Estimated sources)

distribution). One way of visualizing ICA is that it estimates the optimal linear transform to maximise the independence of the joint distribution of the signals X_i.

The statistical basis of ICA is illustrated more clearly in this example. Consider two random signals which are mixed using the following mixing process:

$$\begin{pmatrix} x_1 \\ x_2 \end{pmatrix} = \begin{pmatrix} 1 & 2 \\ 1 & 1 \end{pmatrix} \begin{pmatrix} s_1 \\ s_2 \end{pmatrix}$$

Figure 5 shows the scatter-plot for original sources s_1 and s_2. Figure 6 shows the scatter-plot of the mixtures. The distribution along the axis x_1 and x_2 are now dependent and the form of the density is stretched according to the mixing matrix. From the Figure 6 it is clear that the two signals are not statistically independent because, for example, if x_1 = -3 or 3 then x_2 is totally determined. Whitening is an intermediate step before ICA is applied. The joint distribution that results from whitening the signals of Figure 6 is shown in Figure 7. By applying ICA, we seek to transform the data such that we obtain two independent components.

The joint distribution resulting from applying ICA to x_1 and x_2 is shown in Figure 7. This is clearly the joint distribution of two independent, uniformly distributed random variables. Independence can be intuitively confirmed as each random variable is unconstrained regardless of the value of the other random variable (this is not the case for x_1 and x_2. The uniformly distributed random variables in Figure 8 take values between 3 and -3, but due to the scaling ambiguity, we do not know the range of the original independent components. By comparing the whitened data of Figure 7 with Figure 8, we can see that, in this case, pre-whitening reduces ICA to finding an appropriate rotation to yield independence. This is a simplification as a rotation is an orthogonal transformation which requires only one parameter.

The two examples in this section are simple but they illustrate both how ICA is used and the statistical underpinnings of the process. The power of ICA is that an identical approach can be used to address problems of much greater complexity.

2. ICA for different conditions

One of the important conditions of ICA is that the number of sensors should be equal to the number of sources. Unfortunately, the real source separation problem does not always satisfy this constraint. This section focusses on ICA source separation problem under different conditions where the number of sources are not equal to the number of recordings.

2.1 Overcomplete ICA

Overcomplete ICA is one of the ICA source separation problem where the number of sources are greater than the number of sensors, i.e $(n > m)$. The ideas used for overcomplete ICA originally stem from coding theory, where the task is to find a representation of some signals in a given set of generators which often are more numerous than the signals, hence the term overcomplete basis. Sometimes this representation is advantageous as it uses as few 'basis' elements as possible, referred to as sparse coding. Olshausen and Field Olshausen (1995) first put these ideas into an information theoretic context by decomposing natural images into an overcomplete basis. Later, Harpur and Prager Harpur & Prager (1996) and, independently, Olshausen Olshausen (1996) presented a connection between sparse coding and ICA in the square case. Lewicki and Sejnowski Lewicki & Sejnowski (2000) then were the first to apply these terms to overcomplete ICA, which was further studied and applied by Lee et al. Lee et al. (2000). De Lathauwer et al. Lathauwer et al. (1999) provided an interesting algebraic approach to overcomplete ICA of three sources and two mixtures by solving a system of linear equations in the third and fourth-order cumulants, and Bofill and Zibulevsky Bofill (2000) treated a special case ('delta-like' source distributions) of source signals after Fourier transformation. Overcomplete ICA has major applications in bio signal processing, due to the limited number of electrodes (recordings) compared to the number active muscles (sources) involved (in certain cases unlimited).

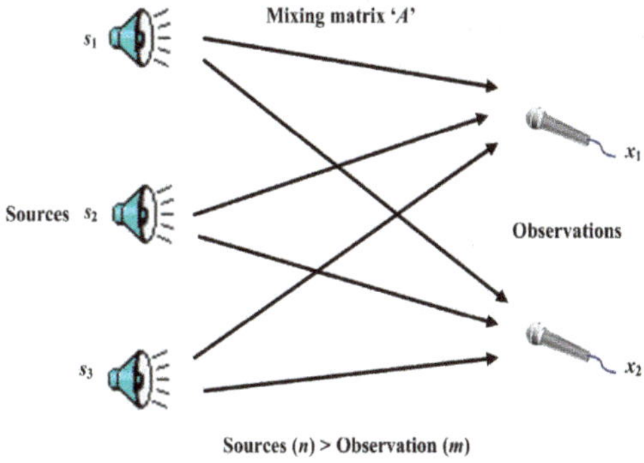

Fig. 9. Illustration of "overcomplete ICA"

In overcomplete ICA, the number of sources exceed number of recordings. To analyse this, consider two recordings $x_1(t)$ and $x_2(t)$ from three independent sources $s_1(t)$, $s_2(t)$ and $s_3(t)$. The $x_i(t)$ are then weighted sums of the $s_i(t)$, where the coefficients depend on the distances between the sources and the sensors (refer Figure 9):

$$\begin{aligned} x_1(t) &= a_{11}s_1(t) + a_{12}s_2(t) + a_{13}s_3(t) \\ x_2(t) &= a_{21}s_1(t) + a_{22}s_2(t) + a_{23}s_3(t) \end{aligned} \tag{23}$$

The a_{ij} are constant coefficients that give the mixing weights. The mixing process of these vectors can be represented in the matrix form as (refer Equation 1):

$$\begin{pmatrix} x_1 \\ x_2 \end{pmatrix} = \begin{pmatrix} a_{11} & a_{12} & a_{13} \\ a_{21} & a_{22} & a_{23} \end{pmatrix} \begin{pmatrix} s_1 \\ s_2 \\ s_3 \end{pmatrix}$$

The unmixing process and estimation of sources can be written as (refer Equation 2):

$$\begin{pmatrix} s_1 \\ s_2 \\ s_3 \end{pmatrix} = \begin{pmatrix} w_{11} & w_{12} \\ w_{21} & w_{22} \\ w_{31} & w_{32} \end{pmatrix} \begin{pmatrix} x_1 \\ x_2 \end{pmatrix}$$

In this example matrix A of size 2×3 matrix and unmixing matrix W is of size 3×2. Hence in overcomplete ICA it always results in pseudoinverse. Hence computation of sources in overcomplete ICA requires some estimation processes.

2.2 Undercomplete ICA

The mixture of unknown sources is referred to as under-complete when the numbers of recordings m, more than the number of sources n. In some applications, it is desired to have more recordings than sources to achieve better separation performance. It is generally believed that with more recordings than the sources, it is always possible to get better estimate of the sources. This is not correct unless prior to separation using ICA, dimensional reduction is conducted. This can be achieved by choosing the same number of principal recordings as the number of sources discarding the rest. To analyse this, consider three recordings $x_1(t)$, $x_2(t)$ and $x_3(t)$ from two independent sources $s_1(t)$ and $s_2(t)$. The $x_i(t)$ are then weighted sums of the $s_i(t)$, where the coefficients depend on the distances between the sources and the sensors (refer Figure 10):

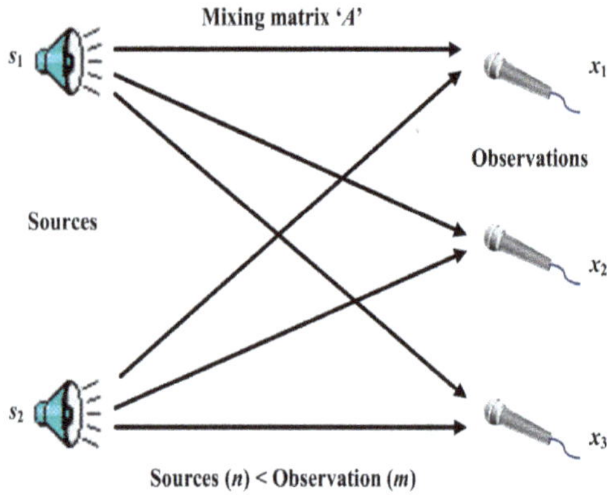

Fig. 10. Illustration of "undercomplete ICA"

$$\begin{aligned} x_1(t) &= a_{11}s_1(t) + a_{12}s_2(t) \\ x_2(t) &= a_{21}s_1(t) + a_{22}s_2(t) \\ x_3(t) &= a_{31}s_1(t) + a_{32}s_2(t) \end{aligned} \tag{24}$$

The a_{ij} are constant coefficients that gives the mixing weights. The mixing process of these vectors can be represented in the matrix form as:

$$\begin{pmatrix} x_1 \\ x_2 \\ x_3 \end{pmatrix} = \begin{pmatrix} a_{11} & a_{12} \\ a_{21} & a_{22} \\ a_{31} & a_{32} \end{pmatrix} \begin{pmatrix} s_1 \\ s_2 \end{pmatrix}$$

The unmixing process using the standard ICA requires a dimensional reduction approach so that, if one of the recordings is reduced then the square mixing matrix is obtained, which can use any standard ICA for the source estimation. For instance one of the recordings say x_3 is redundant then the above mixing process can be written as:

$$\begin{pmatrix} x_1 \\ x_2 \end{pmatrix} = \begin{pmatrix} a_{11} & a_{12} \\ a_{21} & a_{22} \end{pmatrix} \begin{pmatrix} s_1 \\ s_2 \end{pmatrix}$$

Hence unmixing process can use any standard ICA algorithm using the following:

$$\begin{pmatrix} s_1 \\ s_2 \end{pmatrix} = \begin{pmatrix} w_{11} & w_{12} \\ w_{21} & w_{22} \end{pmatrix} \begin{pmatrix} x_1 \\ x_2 \end{pmatrix}$$

The above process illustrates that, prior to source signal separation using undercomplete ICA, it is important to reduce the dimensionality of the mixing matrix and identify the required and discard the redundant recordings. Principal Component Analysis (PCA) is one of the powerful dimensional reduction method used in signal processing applications, which is explained next.

3. Applications of ICA

The success of ICA in source separation has resulted in a number of practical applications. Some of these includes,

- Machine fault detection Kano et al. (2003); Li et al. (2006); Ypma et al. (1999); Zhonghai et al. (2009)
- Seismic monitoring Acernese et al. (2004); de La et al. (2004)
- Reflection canceling Farid & Adelson (1999); Yamazaki et al. (2006)
- Finding hidden factors in financial data Cha & Chan (2000); Coli et al. (2005); Wu & Yu (2005)
- Text document analysis Bingham et al. (2002); Kolenda (2000); Pu & Yang (2006)
- Radio communications Cristescu et al. (2000); Huang & Mar (2004)
- Audio signal processing Cichocki & Amari (2002); Lee (1998)
- Image processing Cichocki & Amari (2002); Déniz et al. (2003); Fiori (2003); Karoui et al. (2009); Wang et al. (2008); Xiaochun & Jing (2004); Zhang et al. (2007)
- Data mining Lee et al. (2009)
- Time series forecasting Lu et al. (2009)
- Defect detection in patterned display surfaces Lu1 & Tsai (2008); Tsai et al. (2006)
- Bio medical signal processing Azzerboni et al. (2004); Castells et al. (2005); De Martino et al. (2007); Enderle et al. (2005); James & Hesse (2005); Kumagai & Utsugi (2004); Llinares & Igual (2009); Safavi et al. (2008); Zhu et al. (2006).

3.1 Audio and biomedical applications of ICA

Exemplary ICA applications in biomedical problems include the following:

- Fetal Electrocardiogram extraction, i.e removing/filtering maternal electrocardiogram signals and noise from fetal electrocardiogram signals Niedermeyer & Da Silva (1999); Rajapakse et al. (2002).
- Enhancement of low level Electrocardiogram components Niedermeyer & Da Silva (1999); Rajapakse et al. (2002)
- Separation of transplanted heart signals from residual original heart signals Wisbeck et al. (1998)
- Separation of low level myoelectric muscle activities to identify various gestures Calinon & Billard (2005); Kato et al. (2006); Naik et al. (2006; 2007)

One successful and promising application domain of blind signal processing includes those biomedical signals acquired using multi-electrode devices: Electrocardiography (ECG), Llinares & Igual (2009); Niedermeyer & Da Silva (1999); Oster et al. (2009); Phlypo et al. (2007); Rajapakse et al. (2002); Scherg & Von Cramon (1985); Wisbeck et al. (1998), Electroencephalography (EEG) Jervis et al. (2007); Niedermeyer & Da Silva (1999); Onton et al. (2006); Rajapakse et al. (2002); Vigário et al. (2000); Wisbeck et al. (1998), Magnetoencephalography (MEG) Hämäläinen et al. (1993); Mosher et al. (1992); Parra et al. (2004); Petersen et al. (2000); Tang & Pearlmutter (2003); Vigário et al. (2000).

One of the most practical uses for BSS is in the audio world. It has been used for noise removal without the need of filters or Fourier transforms, which leads to simpler processing methods.

There are various problems associated with noise removal in this way, but these can most likely be attributed to the relative infancy of the BSS field and such limitations will be reduced as research increases in this field Bell & Sejnowski (1997); Hyvarinen et al. (2001).

Audio source separation is the problem of automated separation of audio sources present in a room, using a set of differently placed microphones, capturing the auditory scene. The whole problem resembles the task a human listener can solve in a cocktail party situation, where using two sensors (ears), the brain can focus on a specific source of interest, suppressing all other sources present (also known as cocktail party problem) Hyvarinen et al. (2001); Lee (1998).

4. Conclusions

This chapter has introduced the fundamentals of BSS and ICA. The mathematical framework of the source mixing problem that BSS/ICA addresses was examined in some detail, as was the general approach to solving BSS/ICA. As part of this discussion, some inherent ambiguities of the BSS/ICA framework were examined as well as the two important preprocessing steps of centering and whitening. The application domains of this novel technique are presented. The material covered in this chapter is important not only to understand the algorithms used to perform BSS/ICA, but it also provides the necessary background to understand extensions to the framework of ICA for future researchers.

The other novel and recent advances of ICA, especially on Audio and Biosignal topics are covered in rest of the chapters in this book.

5. References

Acernese, F., Ciaramella, A., De Martino, S., Falanga, M., Godano, C. & Tagliaferri, R. (2004). Polarisation analysis of the independent components of low frequency events at stromboli volcano (eolian islands, italy), *Journal of Volcanology and Geothermal Research* 137(1-3): 153–168.

Attias, H. & Schreiner, C. E. (1998). Blind source separation and deconvolution: the dynamic component analysis algorithm, *Neural Comput.* 10(6): 1373–1424.

Azzerboni, B., Carpentieri, M., La Foresta, F. & Morabito, F. C. (2004). Neural-ica and wavelet transform for artifacts removal in surface emg, *Neural Networks, 2004. Proceedings. 2004 IEEE International Joint Conference on*, Vol. 4, pp. 3223–3228 vol.4.

Bell, A. J. & Sejnowski, T. J. (1995). An information-maximization approach to blind separation and blind deconvolution., *Neural Comput* 7(6): 1129–1159.

Bell, A. J. & Sejnowski, T. J. (1997). The "independent components" of natural scenes are edge filters., *Vision Res* 37(23): 3327–3338.

Bingham, E., Kuusisto, J. & Lagus, K. (2002). Ica and som in text document analysis, *SIGIR '02: Proceedings of the 25th annual international ACM SIGIR conference on Research and development in information retrieval*, ACM, pp. 361–362.

Bofill (2000). Blind separation of more sources than mixtures using sparsity of their short-time fourier transform, pp. 87–92.

Calinon, S. & Billard, A. (2005). Recognition and reproduction of gestures using a probabilistic framework combining pca, ica and hmm, *ICML '05: Proceedings of the 22nd international conference on Machine learning*, ACM, pp. 105–112.

Castells, F., Igual, J., Millet, J. & Rieta, J. J. (2005). Atrial activity extraction from atrial fibrillation episodes based on maximum likelihood source separation, *Signal Process.* 85(3): 523–535.

Cha, S.-M. & Chan, L.-W. (2000). Applying independent component analysis to factor model in finance, *IDEAL '00: Proceedings of the Second International Conference on Intelligent Data Engineering and Automated Learning, Data Mining, Financial Engineering, and Intelligent Agents*, Springer-Verlag, pp. 538–544.

Cichocki, A. & Amari, S.-I. (2002). *Adaptive Blind Signal and Image Processing: Learning Algorithms and Applications*, John Wiley & Sons, Inc.

Coli, M., Di Nisio, R. & Ippoliti, L. (2005). Exploratory analysis of financial time series using independent component analysis, *Information Technology Interfaces, 2005. 27th International Conference on*, pp. 169–174.

Cristescu, R., Ristaniemi, T., Joutsensalo, J. & Karhunen, J. (2000). Cdma delay estimation using fast ica algorithm, Vol. 2, pp. 1117–1120 vol.2.

de La, Puntonet, C. G., Górriz, J. M. & Lloret, I. (2004). An application of ica to identify vibratory low-level signals generated by termites, pp. 1126–1133.

De Martino, F., Gentile, F., Esposito, F., Balsi, M., Di Salle, F., Goebel, R. & Formisano, E. (2007). Classification of fmri independent components using ic-fingerprints and support vector machine classifiers, *NeuroImage* 34: 177–194.

Déniz, O., Castrillón, M. & Hernández, M. (2003). Face recognition using independent component analysis and support vector machines, *Pattern Recogn. Lett.* 24(13): 2153–2157.

Enderle, J., Blanchard, S. M. & Bronzino, J. (eds) (2005). *Introduction to Biomedical Engineering, Second Edition*, Academic Press.

Farid, H. & Adelson, E. H. (1999). Separating reflections and lighting using independent components analysis, *cvpr* 01.

Fiori, S. (2003). Overview of independent component analysis technique with an application to synthetic aperture radar (sar) imagery processing, *Neural Netw.* 16(3-4): 453–467.

Hämäläinen, M., Hari, R., Ilmoniemi, R. J., Knuutila, J. & Lounasmaa, O. V. (1993). Magnetoencephalography—theory, instrumentation, and applications to noninvasive studies of the working human brain, *Reviews of Modern Physics* 65(2): 413+.

Hansen (2000). *Blind separation of noicy image mixtures.*, Springer-Verlag, pp. 159–179.

Harpur, G. F. & Prager, R. W. (1996). Development of low entropy coding in a recurrent network., *Network (Bristol, England)* 7(2): 277–284.

Herault, J. & Jutten, C. (1987). Herault, J. and Jutten, C. (1987), Space or time adaptive signal processing by neural network models, in 'AIP Conference Proceedings 151 on Neural Networks for Computing', American Institute of Physics Inc., pp. 206-211.

Huang, J. P. & Mar, J. (2004). Combined ica and fca schemes for a hierarchical network, *Wirel. Pers. Commun.* 28(1): 35–58.

Hyvarinen, A., Cristescu, R. & Oja, E. (1999). A fast algorithm for estimating overcomplete ica bases for image windows, *Neural Networks, 1999. IJCNN '99. International Joint Conference on*, Vol. 2, pp. 894–899 vol.2.

Hyvarinen, A., Karhunen, J. & Oja, E. (2001). *Independent Component Analysis*, Wiley-Interscience.

James, C. J. & Hesse, C. W. (2005). Independent component analysis for biomedical signals, *Physiological Measurement* 26(1): R15+.

Jervis, B., Belal, S., Camilleri, K., Cassar, T., Bigan, C., Linden, D. E. J., Michalopoulos, K., Zervakis, M., Besleaga, M., Fabri, S. & Muscat, J. (2007). The independent components of auditory p300 and cnv evoked potentials derived from single-trial recordings, *Physiological Measurement* 28(8): 745–771.

Kab'an (2000). Clustering of text documents by skewness maximization, pp. 435–440.

Kano, M., Tanaka, S., Hasebe, S., Hashimoto, I. & Ohno, H. (2003). Monitoring independent components for fault detection, *AIChE Journal* 49(4): 969–976.

Karoui, M. S., Deville, Y., Hosseini, S., Ouamri, A. & Ducrot, D. (2009). Improvement of remote sensing multispectral image classification by using independent component analysis, *2009 First Workshop on Hyperspectral Image and Signal Processing: Evolution in Remote Sensing*, IEEE, pp. 1–4.

Kato, M., Chen, Y.-W. & Xu, G. (2006). Articulated hand tracking by pca-ica approach, *FGR '06: Proceedings of the 7th International Conference on Automatic Face and Gesture Recognition*, IEEE Computer Society, pp. 329–334.

Kolenda (2000). *Independent components in text*, Advances in Independent Component Analysis, Springer-Verlag, pp. 229–250.

Kumagai, T. & Utsugi, A. (2004). Removal of artifacts and fluctuations from meg data by clustering methods, *Neurocomputing* 62: 153–160.

Lathauwer, D., L. Comon, P., De Moor, B. & Vandewalle, J. (1999). Ica algorithms for 3 sources and 2 sensors, *Higher-Order Statistics, 1999. Proceedings of the IEEE Signal Processing Workshop on*, pp. 116–120.

Lee, J.-H. H., Oh, S., Jolesz, F. A., Park, H. & Yoo, S.-S. S. (2009). Application of independent component analysis for the data mining of simultaneous eeg-fmri: preliminary experience on sleep onset., *The International journal of neuroscience* 119(8): 1118–1136. URL: *http://view.ncbi.nlm.nih.gov/pubmed/19922343*

Lee, T. W. (1997). Blind separation of delayed and convolved sources, pp. 758–764.

Lee, T. W. (1998). *Independent component analysis: theory and applications*, Kluwer Academic Publishers.

Lee, T. W., Girolami, M., Lewicki, M. S. & Sejnowski, T. J. (2000). Blind source separation of more sources than mixtures using overcomplete representations, *Signal Processing Letters, IEEE* 6(4): 87–90.

Lee, T. W., Lewicki, M. S. & Sejnowski, T. J. (1999). Unsupervised classification with non-gaussian mixture models using ica, *Proceedings of the 1998 conference on Advances in neural information processing systems*, MIT Press, Cambridge, MA, USA, pp. 508–514.

Lewicki, M. S. & Sejnowski, T. J. (2000). Learning overcomplete representations., *Neural Comput* 12(2): 337–365.

Li, Z., He, Y., Chu, F., Han, J. & Hao, W. (2006). Fault recognition method for speed-up and speed-down process of rotating machinery based on independent component analysis and factorial hidden markov model, *Journal of Sound and Vibration* 291(1-2): 60–71.

Llinares, R. & Igual, J. (2009). Application of constrained independent component analysis algorithms in electrocardiogram arrhythmias, *Artif. Intell. Med.* 47(2): 121–133.

Lu, C.-J., Lee, T.-S. & Chiu, C.-C. (2009). Financial time series forecasting using independent component analysis and support vector regression, *Decis. Support Syst.* 47(2): 115–125.

Lu1, C.-J. & Tsai, D.-M. (2008). Independent component analysis-based defect detection in patterned liquid crystal display surfaces, *Image Vision Comput.* 26(7): 955–970.

Mackay, D. J. C. (1996). Maximum likelihood and covariant algorithms for independent component analysis, *Technical report*, University of Cambridge, London.

Mosher, J. C., Lewis, P. S. & Leahy, R. M. (1992). Multiple dipole modeling and localization from spatio-temporal meg data, *Biomedical Engineering, IEEE Transactions on* 39(6): 541–557.

Naik, G. R., Kumar, D. K., Singh, V. P. & Palaniswami, M. (2006). Hand gestures for hci using ica of emg, *VisHCI '06: Proceedings of the HCSNet workshop on Use of vision in human-computer interaction*, Australian Computer Society, Inc., pp. 67–72.

Naik, G. R., Kumar, D. K., Weghorn, H. & Palaniswami, M. (2007). Subtle hand gesture identification for hci using temporal decorrelation source separation bss of surface emg, *Digital Image Computing Techniques and Applications, 9th Biennial Conference of the Australian Pattern Recognition Society on*, pp. 30–37.

Niedermeyer, E. & Da Silva, F. L. (1999). *Electroencephalography: Basic Principles, Clinical Applications, and Related Fields*, Lippincott Williams and Wilkins; 4th edition .

Olshausen (1995). Sparse coding of natural images produces localized, oriented, bandpass receptive fields, *Technical report*, Department of Psychology, Cornell University.

Olshausen, B. A. (1996). Learning linear, sparse, factorial codes, *Technical report*.

Onton, J., Westerfield, M., Townsend, J. & Makeig, S. (2006). Imaging human eeg dynamics using independent component analysis, *Neuroscience & Biobehavioral Reviews* 30(6): 808–822.

Oster, J., Pietquin, O., Abächerli, R., Kraemer, M. & Felblinger, J. (2009). Independent component analysis-based artefact reduction: application to the electrocardiogram for improved magnetic resonance imaging triggering, *Physiological Measurement* 30(12): 1381–1397.
URL: *http://dx.doi.org/10.1088/0967-3334/30/12/007*

Parra, J., Kalitzin, S. N. & Lopes (2004). Magnetoencephalography: an investigational tool or a routine clinical technique?, *Epilepsy & Behavior* 5(3): 277–285.

Petersen, K., Hansen, L. K., Kolenda, T. & Rostrup, E. (2000). On the independent components of functional neuroimages, *Third International Conference on Independent Component Analysis and Blind Source Separation*, pp. 615–620.

Phlypo, R., Zarzoso, V., Comon, P., D'Asseler, Y. & Lemahieu, I. (2007). Extraction of atrial activity from the ecg by spectrally constrained ica based on kurtosis sign, *ICA'07: Proceedings of the 7th international conference on Independent component analysis and signal separation*, Springer-Verlag, Berlin, Heidelberg, pp. 641–648.

Pu, Q. & Yang, G.-W. (2006). Short-text classification based on ica and lsa, *Advances in Neural Networks - ISNN 2006* pp. 265–270.

Rajapakse, J. C., Cichocki, A. & Sanchez (2002). Independent component analysis and beyond in brain imaging: Eeg, meg, fmri, and pet, *Neural Information Processing, 2002. ICONIP '02. Proceedings of the 9th International Conference on*, Vol. 1, pp. 404–412 vol.1.

Safavi, H., Correa, N., Xiong, W., Roy, A., Adali, T., Korostyshevskiy, V. R., Whisnant, C. C. & Seillier-Moiseiwitsch, F. (2008). Independent component analysis of 2-d electrophoresis gels, *ELECTROPHORESIS* 29(19): 4017–4026.

Scherg, M. & Von Cramon, D. (1985). Two bilateral sources of the late aep as identified by a spatio-temporal dipole model., *Electroencephalogr Clin Neurophysiol* 62(1): 32–44.

Sorenson (2002). Mean field approaches to independent component analysis, *Neural Computation* 14: 889–918.

Stone, J. V. (2004). *Independent Component Analysis : A Tutorial Introduction (Bradford Books)*, The MIT Press.

Tang, A. C. & Pearlmutter, B. A. (2003). Independent components of magnetoencephalography: localization, pp. 129–162.

Tsai, D.-M., Lin, P.-C. & Lu, C.-J. (2006). An independent component analysis-based filter design for defect detection in low-contrast surface images, *Pattern Recogn.* 39(9): 1679–1694.

Vigário, R., Särelä, J., Jousmäki, V., Hämäläinen, M. & Oja, E. (2000). Independent component approach to the analysis of eeg and meg recordings., *IEEE transactions on bio-medical engineering* 47(5): 589–593.

Wang, H., Pi, Y., Liu, G. & Chen, H. (2008). Applications of ica for the enhancement and classification of polarimetric sar images, *Int. J. Remote Sens.* 29(6): 1649–1663.

Wisbeck, J., Barros, A. & Ojeda, R. (1998). Application of ica in the separation of breathing artifacts in ecg signals.

Wu, E. H. & Yu, P. L. (2005). Independent component analysis for clustering multivariate time series data, pp. 474–482.

Xiaochun, L. & Jing, C. (2004). An algorithm of image fusion based on ica and change detection, *Proceedings 7th International Conference on Signal Processing, 2004. Proceedings. ICSP '04. 2004.*, IEEE, pp. 1096–1098.

Yamazaki, M., Chen, Y.-W. & Xu, G. (2006). Separating reflections from images using kernel independent component analysis, *Pattern Recognition, 2006. ICPR 2006. 18th International Conference on*, Vol. 3, pp. 194–197.

Ypma, A., Tax, D. M. J. & Duin, R. P. W. (1999). Robust machine fault detection with independent component analysis and support vector data description, *Neural Networks for Signal Processing IX, 1999. Proceedings of the 1999 IEEE Signal Processing Society Workshop*, pp. 67–76.

Zhang, Q., Sun, J., Liu, J. & Sun, X. (2007). A novel ica-based image/video processing method, pp. 836–842.

Zhonghai, L., Yan, Z., Liying, J. & Xiaoguang, Q. (2009). Application of independent component analysis to the aero-engine fault diagnosis, *2009 Chinese Control and Decision Conference*, IEEE, pp. 5330–5333.

Zhu, Y., Chen, T. L., Zhang, W., Jung, T.-P., Duann, J.-R., Makeig, S. & Cheng, C.-K. (2006). Noninvasive study of the human heart using independent component analysis, *BIBE '06: Proceedings of the Sixth IEEE Symposium on BionInformatics and BioEngineering*, IEEE Computer Society, pp. 340–347.

Section 2

ICA: Audio Applications

2

Monaural Audio Separation Using Spectral Template and Isolated Note Information

Anil Lal and Wenwu Wang
Department of Electronic Engineering, University of Surrey, United Kingdom

1. Introduction

Musical sound separation systems attempt to separate individual musical sources from sound mixtures. The human auditory system gives us the extraordinary capability of identifying instruments being played (pitched and non-pitched) from a piece of music and also hearing the rhythm/melody of the individual instrument being played. This task appears 'automatic' to us but has proved to be very difficult to replicate in computational systems. Many methods have been developed recently for addressing this challenging source separation problem. They can be broadly classified into two categories, respectively, statistical learning based techniques such as independent component analysis (ICA) and non-negative matrix/tensor factorization (NMF/NTF), and computational auditory scene analysis (CASA) based techniques.

One of the popular methods for source separation was based on ICA [1-10], where the underlying unknown sources are assumed to be statistically independent, so that a criterion for measuring the statistical distance between the distribution of the sources can be formed and optimised either adaptively [5] [10] [11], or collectively (in block or batch processing mode) [2], given mixtures as the input signals. Both high-order statistics (HOS) [2] [4] [5] and second order statistics (SOS) [12] have been used for this purpose. The ICA techniques have been developed extensively since the pioneering contributions in early 1990s, made for example by Jutten [1], Comon [3], and Cardoso [2]. The early work of ICA concentrates on the instantaneous model which was soon found to be limited for real audio applications such as in a cocktail party environment, where the sound sources reach listeners (microphones) through multi-path propagations (with surface reflections). Convolutive ICA [13] was then proposed to deal with such situations (see [21] for a comprehensive survey). Using Fourier transform, the convolutive ICA problem can be converted to multiple instantaneous but complex valued ICA problems in the frequency domain [14-17] thanks to its computational efficiency, and the sources can be separated after permutation correction for all the frequency bins [18-20]. Most of the aforementioned methods consider (over-) determined cases where the number of sources is assumed to be no greater than the number of observed signals. In practical situations, however, an underdetermined separation problem is usually encountered. A widely used method for tackling this problem is based on sparse signal representations [22-29], where the sources are assumed to be sparse in either the time domain or a transform domain such that the overlap between the sources at each time instant (or time-frequency point) is minimal. Audio signals (such as music and speech) become sparser when transformed into the time-frequency domain, therefore, using such a representation, each source within the mixture can be identified based on the probability of each time-frequency point of the mixture that is dominated by a particular source, using either sparse coding [26], or time-frequency masking [30] [31] [33], based on the evaluation of various cues from the mixtures, including e.g. statistical cues [20], and binaural cues [30] [32]. Other methods for source separation include, for instance, the non-negative ICA method [34], the independent vector analysis (IVA) [35], and NMF/NTF [37-43]. Comprehensive review of ICA (and other statistical learning) methods is out of the scope of this chapter, and for more references, we refer the interested readers to the recent handbook on ICA edited by Comon and Jutten [36], and a book on NMF by Cichocki [44].

Many ICA methods discussed above can be broadly applied to different types of signals. In contrast, CASA is another important technique dealing specifically with audio signals, which is based on the principles of Auditory Scene Analysis (ASA). In [60], Bregman attempts to explain ASA principles by illustrating the ability of the human auditory system to identify and perceptually isolate several sources from acoustic mixtures by separating the sources into individual (perceptual) acoustic streams for each source, which suggests that the auditory system operates in two main stages, segmentation and grouping. The segmentation stage separates the mixture into (time-frequency) components that would relate to an individual source. The grouping stage then groups the components that are likely to be from the same source e.g. using information such as simultaneous onset/offset of particular frequency amplitudes or relationships of particular frequencies to source pitch [45-50]. It is well-known that the ICA technique is not effective in separating the underdetermined mixtures, for which, as mentioned above, one has to turn to, e.g. the technique of sparse representations, by sparsifying the underdetermined mixtures into a transform domain, and to reconstruct the sources using sparse recovery algorithms [51-53]. In contrast, CASA technique evaluates the temporal and frequency information of the sources directly from the mixtures, and therefore it has the advantage in dealing with underdetermined source separation problem, without having to assume explicitly the system to be (over-) determined, or the sources to be sparse. This is especially useful for addressing the monaural (single-channel) audio source separation problem, which is an extreme case of the underdetermined source separation problem. The task of computationally isolating acoustic sources from a mixture is extremely challenging, and recent efforts attempt to isolate speech/singing sources from monaural musical pieces or to isolate an individual's speech from a speech mixture [45] [54-59] [61] [62], and have achieved reasonable success. However, the task of separating musical sources from a monaural mixture has been, thus far, less successful in comparison.

The ability to isolate/extract individual musical components within an acoustic mixture would give an enormous amount of control over the sound. Musical pieces could be un-mixed and remixed for better musical fidelity. Signal processing, e.g. equalisation or compression, could be applied to individual instruments. Instruments could be removed from a mixture, possibly for musical students to accompany pieces of music for practice. Control over source location could be achieved in 3-D audio applications by placing the source in different locations within a 3D auditory scene.

Musical sources (instruments) have features in the frequency spectrum that are highly predictable due to the fact that they are typically constrained to specific notes (A to G# on the 12-tone musical scale) and so, frequencies are typically constrained to particular values. As such, harmonic frequencies are predictable as they can be derived from multiples of the fundamental frequency. If reliable pitch information for each source is available, harmonic frequencies for each source can be determined. With this information in hand, frequencies where harmonics from each source would overlap can be calculated. Non-overlapped harmonic frequencies in each source can therefore also be determined and non-overlapped and overlapped harmonic frequency regions in the mixture can be found, along with which particular source each non-overlapped harmonic would belong to. Existing systems [63-65] are successful in using this pitch information to identify non-overlapped harmonics and the source to which it belongs.

Polyphonic musical pieces typically have notes that complement each other (i.e. perfect 3rd, perfect 5th, minor 7th etc., explained by music theory) and so, result in a high, and regular, number of harmonics that overlap. For this reason, musical acoustic mixtures contain a larger number of overlapping harmonics in comparison to speech mixtures. Existing sound separation systems do not completely address the problem of resolving overlapping harmonics i.e. determining the contribution of each source to an overlapped harmonic. And so, because of typically higher numbers of overlapping harmonics in musical passages, musical sound separation is a difficult task and performance of existing source separation techniques has been limited. Therefore, the major challenge in musical sound separation is to effectively deal with overlapping harmonics.

A system proposed by Every and Szymanski [64] attempts to resolve overlapping harmonics by using adjacent non-overlapped harmonics to interpolate an estimate of the overlapped harmonic and so, 'fills out' the 'missing' harmonics for the spectrum of non-overlapped harmonics of each source. Nevertheless, this method relies heavily on the assumption that spectral envelopes are smooth and that amplitudes of any harmonic will have a 'middle value' of the amplitudes of the adjacent harmonics. In practice, however, spectral envelopes of real instruments rarely are smooth so this method produces varied results.

Hu [66] proposes a method of sound separation that uses onset/offset information (i.e. where performed notes start and end). Transient information in the amplitude envelope is used to determine onset/offset time by half-wave rectifying and low pass filtering the signals to obtain the amplitude envelope and the first order differential of the envelope highlights the time of sudden change in the envelope. This is a powerful cue as regions of isolated note performances can be determined. Li and Wang [63] also incorporate onset/offset information to separate sounds. However, the Li-Wang system uses the predetermined pitch information to find the onset/offset time; the time points where pitches change by at least a semi-tone are labelled appropriately as onset or offset times.

The Li-Woodruff-Wang system [67] incorporates a method utilizing common amplitude modulation (CAM) information to resolve overlapping harmonics. CAM suggests that all harmonics from a particular source have similar amplitude envelopes. The system uses the change in amplitude from the current time frame to the next of the strongest non-overlapped harmonic (in terms of a ratio), and the observed change in phase of the overlapped harmonic from the mixture to resolve the overlapped harmonic by means of least-squares estimation.

The focus of this chapter is to investigate the musical sound separation performance using pitch information and CAM principles described by Li-Woodruff-Wang [67] and proposing methods for the improvements of the system performance. The methods outlined by the pitch and CAM separation system have shown promising results, but only a small amount of research has been carried out that uses both pitch and CAM techniques together [67]. Preliminary work reveals that the pitch and CAM based system produces good results for mixtures containing long notes with considerable sustained portions e.g. a violin holding a note, but produces poor quality results for attack sections of notes, i.e. mixtures containing instruments with smaller, or no sustain sections (just attack and decay sections), e.g. a piano. Modern music typically has a high number of non-sustained note performances so the pitch and CAM method would fail with a vast number of musical pieces. In addition, the pitch and CAM method has difficulty in dealing with the overlapping harmonics, in particular, for audio sources playing similar notes.

This study aims to investigate more reliable methods of resolving harmonics for the pitch and CAM based technique of music separation which improves results, particularly for attack sections of note performances and overlapping harmonics. A method of using isolated (or relatively isolated) sections of performances in mixtures by obtaining onset/offset information is used to provide more reliable information to resolve harmonics. Such information is also used to generate a spectral template which is further used to improve the separation performance of overlapping spectral regions in the mixtures, based on the reliable information from non-overlapping regions. Implementation of the proposed methods is then attempted using a baseline pitch and CAM source separation algorithm, and system performance is evaluated.

2. Pitch and CAM system and its performance analysis

In general, the pitch and CAM system shows good performance for separating audio sources from single channel mixtures. However, according to our experimental evaluations briefly discussed below, its separation performance is limited for attack sections of notes and regions of same note performances.

We first evaluate the performance of the pitch and CAM system for separating the attack sections of music notes. To this end, we take the baseline pitch and CAM algorithm implemented in Matlab to test its performance. We use a sample database of real instrument recordings (available within ProTools music production software) to generate test files, so that the system performance on separating attack sections of notes could be evaluated. The audio file generated is a (monaural) single-channel mixture containing a melody played on a cello, and a different but complimentary melody played on a piano. The purpose of combining complimentary melodies from different sources is to generate a realistic amount of overlapping harmonics between sources, as would be found in typical musical pieces. Qualitative results show that the cello, which had long sustained portions of notes, is separated considerably well, while the attack sections of piano notes are in some cases lost as a result of the limited analysis time frame resolution. The piano has shorter notes with no sustain sections, only attacks and decays, but still contains considerable amount of harmonic content. As a result, the system performs less effectively in separating the piano source which highlights the difficulty the separation system has in isolating instruments playing short notes that are made up of regions of attack. Another experiment on the mixture of audio sources played by clarinet and cello again confirms that the pitch and CAM system has difficulty in separating the soft attack sections of the notes played by the clarinet.

We then evaluate the system performance for the regions of same notes in the mixture. We generated a mixture containing a piano and a cello performing the same note (C4). Using the pitch and CAM system, the cello was separated from the mixture but with some artefacts and distortions. However, the system was unsuccessful in separating the piano source, and only a low level signal could be heard that did not resemble the original piano signal. In another experiment, we generated a mixture with a cello playing the note C4 and a piano playing all notes in sequence from C4 to C5 (C4, C#4, D4, D#4... etc.) and ended on the note C4. The cello was separated well from the mixture as were all notes played by the piano except the notes C4 and C5 at the both ends of the sequence. Due to the slow attack of the cello, the C4 note played by the piano at the beginning of the piece was better separated than the C4 note at the end of the sequence, as the C4 note at the beginning is more isolated. In addition, we have examined the performance of the system for mixtures with the same note and varying octaves. To this end, we generated another mixture with a cello playing the note C3 and a piano playing notes C1 to C6 in sequence and then ending on note C2. The results again show that the cello was separated well but with high distortions in sections where the piano attacks occur. The piano notes C1 and C2 were separated with some distortions but notes C3 through to C6 were almost not separated at all.

In summary, the pitch and CAM system does not perform well for recovering the sharp transients of the amplitude envelope from mixtures due to the limited time frame resolution, and it also has difficulty in separating notes with same fundamental frequencies and harmonics, caused by insufficient data for resolving the overlapping harmonics and for extracting the CAM information. For example, if one source has a pitch frequency of 50 Hz, its harmonics would occur at 100 Hz, 150 Hz, etc. If the pitch frequency of a second source is an octave higher, i.e. 100 Hz, its harmonics would occur at 200 Hz, 300 Hz, etc. As a result, the harmonics of the second source will be overlapped with those of the first source. To address these problems, we suggest two methods to improve the pitch and CAM system, respectively, isolated note and spectral template methods, which attempt to better resolve the overlapping harmonics when the information used by the pitch and CAM system is considered to be unreliable, as described next in detail.

3. Isolated note method

The proposed isolated note system, shown in Figure 1, uses note onset/offset information to determine periods of isolated performance of an instrument so that the reliable spectral information from the isolated regions can be used to resolve overlapping harmonics in the remaining note performance regions. The proposed system is based on the pitch and CAM algorithm [67], with the addition of new processing stages shown in dotted lines in Figure 1. Same to the pitch and CAM system, the inputs to the proposed system are mixture signals and pitch information supplied by a pitch tracker. The details of each block in Figure 1 are explained below.

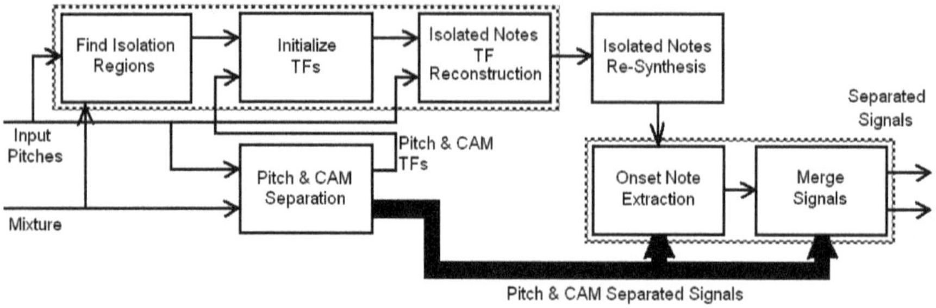

Fig. 1. Diagram of Isolated Note System.

The first processing stage is the *Pitch and CAM Separation* stage. The mixture signal is separated using the method described in [67] and by using the pitch information provided. The separated signals are used later in *Onset Note Extraction* and *Merge Signals* stages by the isolated note system. When the pitch and CAM separation is carried out the time-frequency (TF) representations of the mixture signal and the separated signals are generated which are then utilized later by the *Initialize TFs* processing stage.

The next processing stage is the *Find Isolated Regions* stage. Using input pitch information, we attempt to find time frames for each source where isolated performances of notes occur. Each time frame of each source is evaluated to determine if other sources contain pitch information (i.e. if other notes are performing during the same time frame). A list of time frames for each source is created and a flag is raised (the time frame is set to 1) if the note for the current frame and current source is isolated. Each occurrence of an isolated region (indicated by the flag) in each source is then numbered so that each region can be identified and processed independently at a later stage (achieved by simply searching through time frames and incrementing the region number at each encounter of a transition from 0 to 1 in the list of flagged time frames).

Next, we determine the non-isolated regions for the notes that contain a region of isolated note performance. For each numbered isolated region we find the corresponding non-isolated note performance and generate a new list where time frames for the non-isolated regions are numbered with the number relating to the corresponding isolated region. Note that we do not number the isolated time frames themselves in the newly generated list.

The new list is generated by searching back (previous frames) from the relevant isolated region and numbering all frames appropriately, and we then repeat by searching forward from the isolated region. Searches are terminated at endpoints of the note or at occurrences of another isolated region. Each isolated region that generates a set of corresponding non-isolated frames is saved in a new list separately, the list is then collapsed to form a final list where time frames for which we have non-isolated regions relating to two isolated regions are split halfway.

This is better illustrated by Fig. 2. Fig. 2(a) shows an occurrence of a note with three isolated regions for which information of time frames with isolated performance is determined. Fig. 2(b) illustrates that the non-isolated regions relating to each isolated region, are found by searching forwards and backwards and terminating at endpoints of notes or an occurrence of another isolated region. Each region is stored individually. Fig. 2(c) shows the final set of regions where time frames 'belonging' to two sets are split halfway.

The TF representation of each source is formed for the isolated notes in the *Initialize TFs* stage. We initialize the TF representation by starting with an empty set of frequency information for each time frame and then by searching through the list of isolated regions. For time frames that are identified as an isolated performance of a note (from the list), we copy all frequency information for those frames directly from the mixture to the corresponding TF representation of the sources. This is shown in Fig. 3 where the time frames for the isolated performances of the note C4 (in Fig. 3(a)) are copied directly to initialize the TF representation. Fig. 3(b) shows that all of the harmonic information is copied directly from the mixture; hence all harmonics are correctly present in the initialized isolated note TF representation.

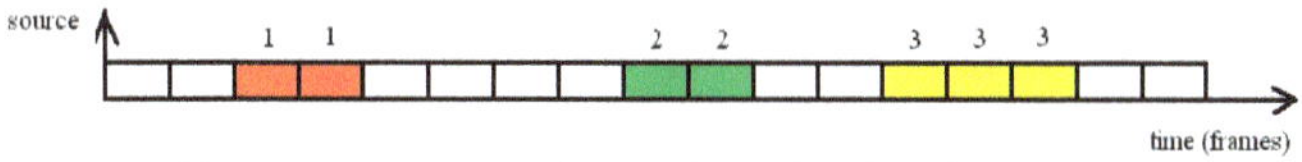

(a) Time Frames with Numbered Isolated Regions.

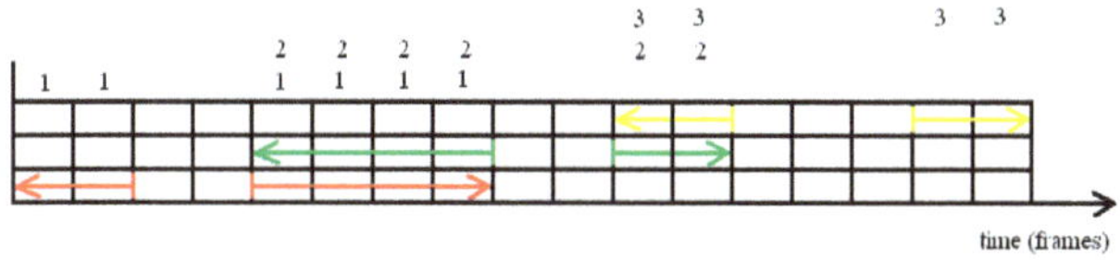

(b) Non-Isolated Regions Corresponding to Each Isolated Region.

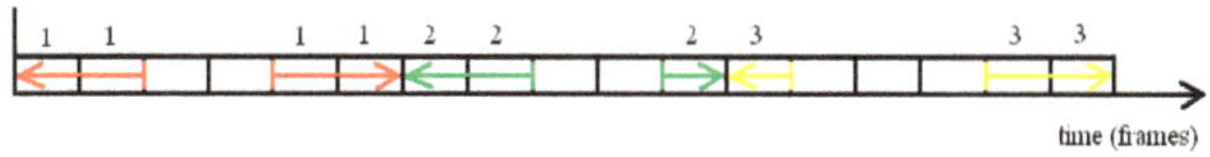

(c) Time Frames of Non-Isolated Regions Associated with Each Isolated Region.

Fig. 2. Method Used to Determine Non-Isolated Regions of Isolated Notes.

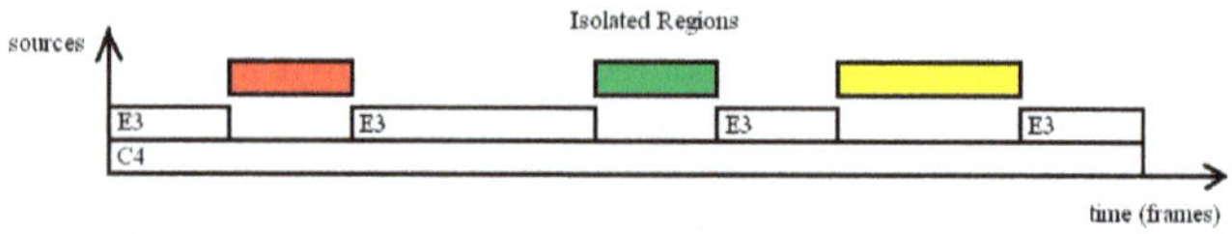

(a) Note Performance and Isolated Note Regions.

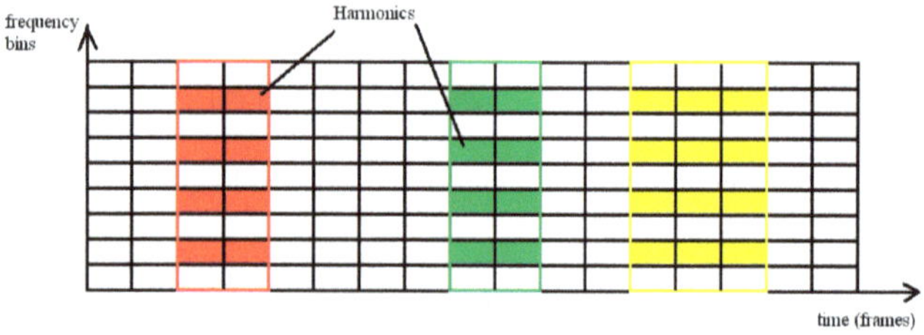

(b) Initialized TF Representations

Fig. 3. Method Used to Initialize TFs.

After the TF initialization, the *Isolated Notes TF Reconstruction* stage extends these isolated performance regions for the remaining parts of the note performances that contain isolated performance sections. Each region is evaluated in turn using information from the list of time frames for note performances which contain regions of isolated performance. The note for each time frame in the current region, and notes performed by other sources in the same time frame are determined so that a binary harmonic mask can be generated. This is then used to extract the non-overlapped harmonics for the note during sections of non-isolated performance (shown in Fig. 4(a)), which are then passed to the TF representation for relevant time frames.

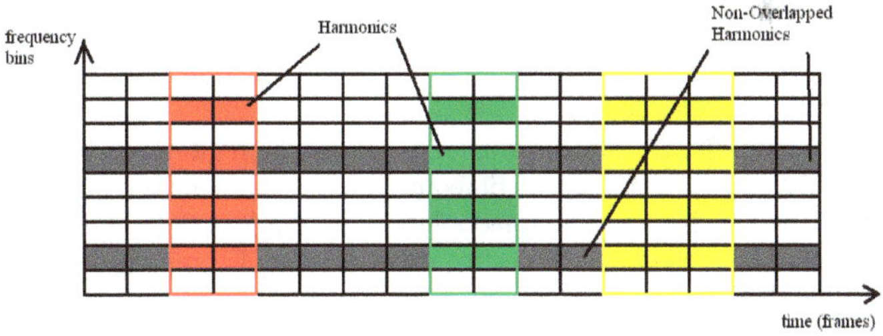

(a) TFs with Non-Overlapped Harmonics Added

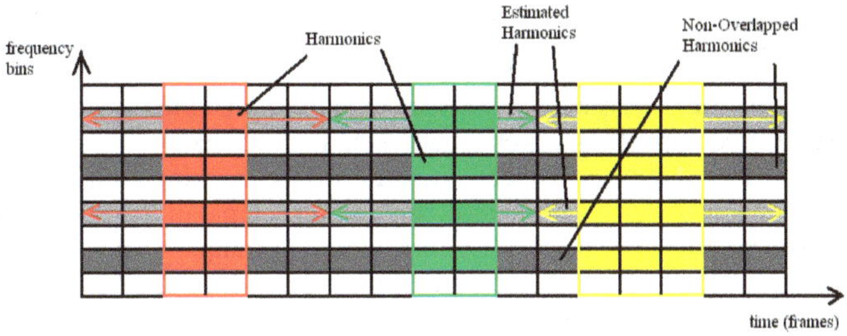

(b) TFs with Overlapped Harmonics Estimated Using Harmonic Information in Isolated Regions

Fig. 4. Method Used to Reconstruct TFs.

Having used non-overlapped harmonic information to update the isolated note TF representation, we can begin to estimate the overlapped harmonics for the relevant time frames. By using harmonic information available in isolated regions (for which information on all harmonics are available), amplitudes of overlapped harmonics can be estimated. Phase information for the overlapping harmonics is obtained from the corresponding harmonics in the separated TF representations found from the *Pitch and CAM Separation* stage.

As detailed earlier, each set of time frames for each source, relating to non-isolated notes containing an isolated section, are derived from time frames of corresponding isolated regions. Based on the boundary time frames, i.e. the first and last time frames of the isolated regions, we can estimate overlapped harmonic amplitudes (shown in figure 4(b)) by using the spectral information in these frames as templates. We use the first time frame frequency information in an isolated region to process previous time frames, and use the last time frame in the isolated region to process subsequent time frames. According to the CAM principle, amplitude envelopes are assumed to be the same for all harmonics. Hence, by following harmonic envelopes for the subsequent or previous time frames, we can determine the amplitude ratio $r_{t_0 \to t}$ between the template time frame t_0 and the time frame currently being processed B_t^h associated with harmonic h in time frame t

$$B_t^h = r_{t_0 \to t} B_{t_0}^h \tag{1}$$

Hence, by multiplying bins associated with an overlapped harmonic from the template frame with the ratio between frames, the amplitude for the corresponding bins in frame t can be found.

Once the TF information for notes with isolated performance regions has been constructed, it can be converted to the time domain as time-amplitude representation by the *Isolated Note Re-Synthesis* stage. The method is an adapted method used in [67]. Full frequency spectra are recreated from the half frequency spectra used in the TF representations, and the overlapped-add method is used to reconstruct the time-amplitude signals for each source. The system is designed to separate mixture signals comprising of two sources. Therefore, time domain signals of notes with isolated performance regions can be removed from the mixture signal to reveal the separated signal for the remaining source. We can simply subtract isolated note signal sample values from corresponding mixture signal sample values to generate the 'extracted' signals (performed by the *Onset Note Extraction* stage).

Finally, the *Merge Signals* stage uses isolated note signals, and the 'extracted' remaining signal, to update the separated signals obtained using the baseline pitch and CAM method. When isolated note information is available (determined by checking for a non-zero sample value) the final signal is updated with the corresponding sample in the isolated note signal for the current source being processed. The corresponding sample value is used to update the signal for the 'other' source, i.e. with the extracted signal. When isolated note information is unavailable (if a sample value of zero is encountered) the corresponding sample in the pitch and CAM separated signal, from the respective source, is used to update the final signal.

4. Spectral template method

This method aims to generate a database of spectral envelope templates of the sources from the mixtures, and then use the templates to resolve the overlapped harmonics when the pitch and CAM information is known to be unreliable. In this method, we generate a spectral envelope template for each note, using the information from the mixture. Eventually, it builds a database of spectral envelopes for all notes that are performed for each source, e.g. spectral envelopes for notes C4, E5, D# etc. The note information occurring in the mixture can be determined from the supplied pitch information. In particular, we use the non-overlapped harmonics from the most reliable sections of the mixture to fill in the spectral template for each note that appears in the mixture, where the most reliable section is regarded as the time section having the most non-overlapped harmonics for a particular instance of a note occurrence. The number of non-overlapped harmonics can vary, depending on the other notes being played simultaneously. Within this most reliable time section, the frequency spectrum at the time frame in which the largest harmonic occurs is used to train the template. Other occurrences of the note within the mixture are used to update the template for remaining unknown harmonics by analysing the ratio to adjacent non-overlapped harmonics (CAM information), based on the extraction of the 'exposed' non-overlapping harmonics. For example, when the note C5 from a source is being played together with another note G6, the 'exposed' non-overlapped harmonics of C5 can be used to train the C5 note template. Other occurrences of C5 from the same source, whilst the note A7 from the other source is being played, would 'expose' a different set of non-overlapping harmonics. These non-overlapped harmonics can be used to update the spectral template in order to 'fill out' the unknown harmonics by using the relative amplitudes of the harmonics. This provides a 'backup' set of information for the estimation of the overlapped harmonics and also enables us to better handle situations where other information for resolving overlapping harmonics is limited or unreliable e.g. concurrent same note occurrence. Figure 5 shows the diagram of the proposed system that uses the spectral envelope model, which uses the *Pitch and CAM Separation* algorithm (developed by Li, Woodruff and Wang [67]) as a basis and adds several components shown in large red blocks (implemented in Matlab), including *Find Reliable Time Frames, Template Generation, Refine Templates, Update TFs,* and *Envelope Correction,* discussed next.

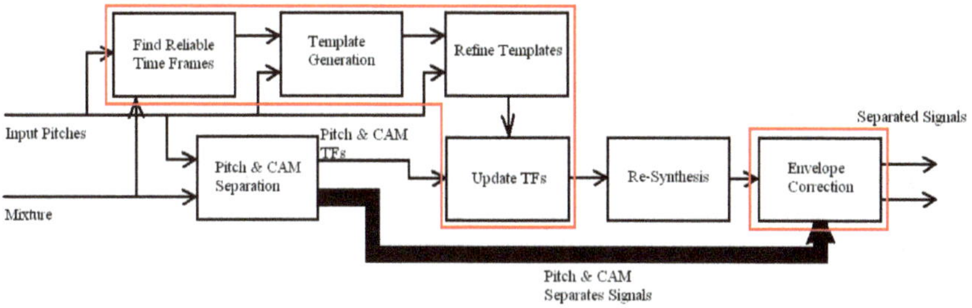

Fig. 5. Diagram of the spectral template method for audio mixture separation.

The proposed spectral template system has two inputs: the mixture signal and pitch information. The input signals are the audio mixtures we attempt to separate, which can be a time-domain representation. Pitch information of each source can be extracted from the time-frequency representation of the signals, using a pitch estimator or a pitch tracker, as done in the pitch and CAM system [67]. In our proposed system, however, we use the supplied pitch information as inputs, and this essentially eliminates the influence of pitch estimation process on the separation performance. The pitch information is needed by the pitch and CAM algorithm (shown in the *Pitch and CAM Separation* stage in Figure 5) for producing an initial estimate of the sources from the TF representations of the mixtures. It is also used in *Find Reliable Time Frames* stage to determine the time frames within the TF representations that would convey the most reliable harmonic information. These time frames are then passed onto the *Template Generation* stage, and the harmonic information from these frames is used to initialize the template. In the *Refine Templates* stage, the missing harmonics of each template are estimated from the templates of other notes, when limited information is available in the mixture. The *Update TFs* stage then uses the templates at time-frames with non-overlapped harmonics to resolve the overlapped harmonics (by the *Pitch and CAM Separation* stage). These modified TF representations are passed onto the *Re-Synthesis* stage for the reconstruction of the time domain signals of each source. The *Envelope Correction* stage obtains envelope information by subtracting all but the current source from the mixture, and then use it to correct the envelope for time regions of the sources where the template was used.

In the *Pitch and CAM Separation* stage, we use the baseline algorithm developed by Li, Woodruff and Wang [67] to separate the audio mixtures, using the additional pitch contour information. More specifically, the audio mixture is transformed to the TF domain using short-time Fourier transform (STFT) with overlaps between adjacent time frames. TF representations are generated for each separated source by the *pitch and CAM separation* algorithm, and are updated in later processing stages with improved information for time frames of unreliable information before being finally transformed back to the time domain. The separated time domain signals are also used in the *Envelope Correction* stage to obtain envelope information for the refinement of the separated signal.

In the *Find Reliable Time Frames* stage, we first find the time frames of the mixture that are most likely to yield the best set of harmonics, and we then use them to generate the spectral templates. Notes played by different instruments may have different harmonic structures, and many of them contain unreliable harmonic content. This is especially true for attack sections of many notes, due to the sharp transients and the noise content in the attack. For example, when a string is struck, its initial oscillations caused by the initial displacement will be non-periodic, and it takes a short amount of time for the string to settle into stable resonances of the instrument, and hence provide more reliable harmonic information. Some instruments may have long, slow and weak attack section, and in such a case, the harmonic content only becomes reliable at some time after the onset of the note. A similar problem also happens for notes of a short duration. In order to provide reliable frequency information for updating the note templates, we generate a list of time frames that does not include time frames containing short note performances and attack sections of note performances.

The pitch information is supplied to the *Find Reliable Time Frames* stage of the system in the form of fundamental frequencies for each source and for all time frames. The fundamental frequencies of the notes are converted into numbers representing the closest note in the 12 tone scale, i.e. C0 is 1, C#0 is 2, B0 is 12, C1 is 13 and so on up to 96 representing note B7. To find the corresponding note numbers for each frequency in the input pitch information we first determine which octave range the frequency is in by selecting an integer m such that $f_{\min} < \frac{f}{2^m} \le f_{\max}$, where the lower and upper frequency limits of the first octave (C0 to B0) are $f_{\min}$ and $f_{\max}$ respectively and f is the (fundamental) frequency value that we wish to convert to a note number. In practice, $f_{\min}$ is selected as the frequency value between C0 and the note that is one semi-tone lower (in theory, note B1), and $f_{\max}$ is selected as the frequency value between B0 and C1. The integer m can be determined by repeatedly halving the frequency until it falls within the first octave range. Once the octave range has been found, the note from A to G# on the 12-tone scale can then be found by further narrowing the searching range in terms of multiples of $f_{\min}$. In other words, we choose the integer n that satisfies the following inequality

$$\left(\sqrt[12]{2}\right)^{n-1} f_{\min} < \frac{f}{2^m} \le \left(\sqrt[12]{2}\right)^{n} f_{\min} \tag{2}$$

where m is the octave range value found previously. Once the octave range m and the note n are found, the list of note numbers at each time frame for each source can be easily calculated. From the list of notes selected, we further remove the invalid notes if they are from the attack sections of the notes, or their duration is too short. In our case, any notes whose duration is shorter than six time frames will be set to zero.

In the *Template Generation* stage, we update the spectral templates for each note with spectral information from the list of time frames that contains the valid notes obtained above. We search over each time frame in the list (also for each source in turn), and ignore the invalid time frames (with values of zero). We then determine the note performed by the current source and the notes by all other sources for the current time frame. Using such a particular note combination, we can generate binary harmonic masks to extract the non-overlapping harmonics from the TF representation of the mixtures for each of the frames. More specifically, the note performed by the current source is used to determine the frequencies of all harmonics. Notes performed simultaneously by all other sources are used to determine which of the current source harmonics are overlapped by all other sources thus, indicating the 'exposed' non-overlapping harmonics for the current note. Using such information, we can set frequency bins that are associated with non-overlapped harmonics to 1 and all other bins to 0. Firstly, the frequency of the note value for the current source must be found. According to the international standard (ISO 16 [68]), the note frequency for A4 is 440Hz, and note C0 is 57 semi-tones below A4. Hence, the frequency of C0 can be used as a basis to find the fundamental frequency of other notes such as A4 using $f = f_{C0}\left(\sqrt[12]{2}\right)^{p-1}$, where p is the note value and f_{C0} is the fundamental frequency of note C0. We then associate frequency bin b to harmonic h_i for the current source i using a similar method to that in [67], if it satisfies $|bf_a - h_i f| < \theta_1$, where θ_1 is a threshold and f_a is the frequency resolution of the TF representation (both θ_1 and f_a are determined previously in the *Pitch and CAM Separation* stage). We use a second threshold θ_2 to define the range in which the current source harmonic h_i is overlapped with any other source harmonic h_j, i.e. $\left|f_{h_j} - h_i f\right| < \theta_2$, where f_{h_j} is the frequency of harmonic h_j. Again, this is a similar method to that in [67], hence θ_2 can be chosen in the same way as used in the *Pitch and CAM Separation* stage. As a result, we can define a TF mask M_b which takes 1 if $|bf_a - h_i f| < \theta_1$, otherwise 0. This binary mask is then used to extract all non-overlapped harmonics for all time frames from the TF representation of the mixture. All the harmonic sets for the current note combination are evaluated to find the set that contains the largest amplitude harmonic, which is then used to update the note template (or simply stored if the template is empty and has not yet been initialized). We continue to go through the whole list of valid note regions, and when a new note combination is encountered, we update the note templates based on the new harmonic mask generated using the new set of 'exposed' non-overlapped harmonics. After all note combinations have been evaluated, the note templates may contain several sets of harmonics for each note combination. If this happens, we merge them to create a final set for the note template. Note that one may wish to apply scaling to each set of harmonic templates to ensure harmonics are of correct magnitude when merging the template.

As done in the *Refine Templates* stage, the spectral templates generated, are further refined and improved by using information from all the templates. The reason that the spectral templates need to be refined is because for some notes, there may be only a limited set of non-overlapped harmonics, as some harmonics may not be available in the mixture. To improve the templates, harmonic information from other note templates that are available within a specified range of notes is used. Spectra of other note templates are pitch shifted to match the note we intend to improve, so that information for correlating harmonics can be obtained (after harmonics are aligned). However, spectral quality tends to deteriorate as the degree of pitch shifting increases. Therefore we first use the templates of notes that are closest in frequency to the note template for which we wish to improve, and then continue with templates of decreasing quality. In addition, lower frequency note templates yield higher quality spectra when the pitch is shifted up to match the frequency of the note template we wish to improve, and vice versa. Hence, we limit the range of notes and also the number of note templates to be used for improving note templates. This essentially excludes note templates that have been excessively pitch shifted, and also improves computational efficiency of the proposed system.

In the *Update TFs* stage, we update the TF representations of the separated sources from the *Pitch and CAM Separation* stage, using the note templates. Pitch information is used to determine, for each source, the time frames where reliable non-overlapped harmonics are unavailable for separation. As already mentioned, if a source is playing a note which is one octave lower than a note played by another source, the former one would have every other harmonic overlapped whereas the harmonics of the latter one would be totally overlapped by those of the former one. As a consequence, no reliable information is available to resolve the overlapped harmonics of the latter source. However, there are many other note combinations, leading to unavailable non-overlapping harmonics to be used to resolve the overlapped harmonics, e.g. when one source performs a note 7 semi-tones higher (perfect fifth interval) than the other it would result in every third harmonic of the latter source being overlapped by the former one. Of course, it would be exhaustive to find all possible combinations of notes that result in all of the source harmonics being overlapped. Using pitch information is an efficient way to calculate the resulting number of overlapped harmonics at each time frame for each source. The number of overlapping harmonics $\varphi_i(t)$ for source i at time frame t can be determined by finding the number of harmonics in a complete set $H_{N_i}(t)$ that is not in the set of non-overlapped harmonics $\tilde{H}_{N_i}(t)$ based on the pitch information of note $N_i(t)$.

We use the same method discussed above to generate binary masks, using current note information and information on all other notes that are performed simultaneously. We also create a binary mask with a complete set of harmonics from which the mask with non-overlapped harmonics is subtracted. This gives a mask containing harmonics that are overlapped. Evaluating the magnitude at bins closest to the expected harmonic frequencies allows the number of overlapped harmonics present to be determined. For all t where $\varphi_i(t)=0$ i.e. time frames for source i that have no reliable information for source separation, frequency spectra for the respective note templates are used to replace the frequency spectra in the TF representation of the separated source.

The *Re-Synthesis* stage, adapted from [67], involves the reconstruction of the time domain signals from the TF representations for each source. Specifically, symmetric frequency spectra are created from the half spectra used in the TF representations and the overlap-add method is used to generate the time domain signals.

No amplitude envelope information has been conveyed in the note templates for refining the separated sources. Hence, in the *Envelope Correction* stage, for the time regions with unresolved overlapped harmonics, the amplitude envelopes of the separated sources will be corrected. All sources that have been separated (in the *Pitch and CAM Separation* stage) except the current source, for which the envelope is being corrected, are removed from the original mixture signal. The remaining signal would then be a crude representation of the source we are attempting to correct as most of the high energy components from all other sources are removed. The envelope of the remaining signal is found by finding peaks of absolute amplitude values. We detect peaks at time instances where the first order derivative of the absolute time-amplitude signal is zero. The envelopes of the separated sources are then adjusted by applying a certain amount of scaling determined by the desired envelope obtained above.

5. System evaluation

5.1 Evaluation method

The system is evaluated using test signals specifically designed to highlight differences between the proposed systems and the original pitch and CAM separation system. The proposed systems aim to address the weak points of the pitch and CAM system, i.e. the lack of time domain detail arising from poor separation of attack regions of notes, and its difficulty in resolving the overlapping harmonics due to similar note performances. Hence, tests were designed to evaluate differences in these particular points between the proposed systems and the original system, rather than an evaluation of overall performance of the system.

For the proposed isolated note system, test signals which were generated using real instrument recordings with different musical scores, contain isolated performances of notes in order to show the effectiveness of the proposed system. The isolated note system aims to better resolve attack sections of notes for which the pitch and CAM system performs poorly. Hence, instruments with fast attacks and relatively higher energy in the higher frequency range (of the attacks), e.g. instruments that are struck, or particular instruments that are plucked were selected for the test signals. Two test signals meeting these criteria were generated; the first signal (test signal 1) was a two-source mixture containing a cello and a piano performance, the cello was played throughout the signal and the piano had sections of performance interspersed with sections of silence giving the cello regions of isolated performance. The second signal (test signal 2) was also a two-source mixture containing a string section and a guitar performance, again, the string section was played throughout the test signal and the guitar had interspersed sections of silence. Both test mixtures were created by mixing clean source signals (16-bit, 44100Hz sample rate).

For the spectral template system, two test signals with the same musical score are generated containing sections with the same note performance and also sections with sufficient information to train the templates. The first piece was a two source mixture of a cello and a piano, the second piece was a two source mixture of a cello and a clarinet (both pieces approximately four seconds long at 16 bit, 44100 kHz sampling rate). All the test signals were created using ProTools music production software and instruments were selected to avoid synthesized replications to achieve performances as realistic as possible (this avoids signals being created with stable frequency spectra for note performances). A database of real recordings of instruments within the music production software was used to generate the test signals. Pitch and CAM separation was performed with default values.

System performance is evaluated by calculating the SNR for the pitch and CAM system and the proposed system with each test signal using

$$SNR(dB) = 10\log_{10} \frac{\sum_n (x[n])^2}{\sum_n (x[n] - \hat{x}[n])^2} \tag{3}$$

where $x[n]$ is the original signal and $\hat{x}[n]$ is the separated signal (n is the sample index). This allows us to quantify the sample-wise resemblance between the clean source signals and the separated signals generated by each of the systems.

For the evaluation of the isolated system, a direct comparison of SNR values for both systems would reveal the gains made by the isolated note system. However, differences in the attack sections only are difficult to quantify when evaluating the entire signal as they make up only a small proportion of the test signal. Hence, we expect the differences in perceptual quality to be more significant (i.e. differences would be heard, but are not represented as well in comparison using SNR measurements). Therefore, a listening test was also performed to observe the perceptual difference between the separated signals obtained using the pitch and CAM and the isolated note methods. Test signals for the listening test were generated by including the original clean source signal, followed by a one second silence, and then the separated signal allowing for a direct comparison to be made between the clean source and separated signals. 26 participants were asked to score the signals from 0 to 5, with 0 being extremely poor and 5 being perceptually transparent (with reference to the original signal). Scores were based on the details of attack sections as well as overall separation performance between the two systems (i.e. which system 'sounds better') all test signals were presented in a random order for each participant.

For the evaluation of the spectral template system, the separated signals are modified to remove the pitch and CAM sections so that the signals contain only same note performances, and the influence of the pitch and CAM results is ignored. Test signals are created by including the original signal at the start, followed by a one second silence, and then followed by the separated signal; this allows the listener to hear the original before hearing the separated signal so a direct comparison can be made. Test signals were generated for both pitch and CAM and note template systems. All test signals were played in a random order so that identification of each system remains unknown and cannot be anticipated. Signals were allowed to be repeated as many times as needed to assess signal quality.

5.2 Results

The results of the isolated note system are shown in Tables 1 and 2. When comparing results for test signal 1, source 1 (cello), we observe a reduction of -3.75 dB in SNR between the two systems. Nevertheless, this source contains sections of isolated performance which we use to better separate attack sections of source 2 (for which this study concerns). As can be seen for source 2 (piano), SNR of the proposed system is 15.08 dB higher than the pitch and CAM system, so a significant gain in separation performance is achieved. Looking at SNR results for test signal 2 source 1 (string section) we see a marginal increase of 0.34 dB in separation performance from the isolated note system, again, this source contains the isolated region of performance which is used to improve separation of source 2. For source 2 (guitar), we see a significant improvement in separation performance by the isolated note system with a SNR 8.44 dB higher than the pitch and CAM system.

Test Signal	Source	Pitch and CAM System	Isolated Note System
1	1	19.04	15.29
	2	5.87	20.95
2	1	15.78	16.09
	2	3.63	12.07

Table 1. SNR (dB) results for Isolated Note System as compared with the pitch and CAM system.

Test Signal	Source	Pitch and CAM Mean Score	Isolated Note Mean Score
1	1	4.88	4.73
	2	2.50	4.50
2	1	3.85	3.69
	2	1.65	3.54

Table 2. Listening test results for Isolated Note System as compared with the pitch and CAM system.

For test signal 1 we can see similar mean opinion scores for separation of source 1 by both systems suggesting a similar level of separation performance between the two systems. However, listening test results suggest a significant improvement of separation performance by the isolated note system for source 2. For test signal 1 and source 2, the pitch and CAM system achieved a mean score of 2.50 and the isolated note system achieved a mean score of 4.50. Again, the isolated note system achieved similar separation performance compared to the pitch and CAM system for test signal 2, source 1, while giving a significant improvement for source 2. The pitch and CAM system achieved a mean score of 1.65 whereas the isolated note achieved a higher mean score of 3.54. Both SNR and listening test results indicate that the note isolation separation system achieves better separation performance. We can see significant quantitative gains from the SNR results for signals with fast attacks (source 2 in both test signals 1 and 2). Qualitative results from the listening test also show significant perceptual gains obtained in the separation of attack sections in addition to overall separation.

The results of the spectral template system are summarised in Tables 3 and 4. Table 3 shows SNR results for the proposed note template separation system compared with the pitch and CAM separation system. We can see that for both test signals, we have the same separation performance for source 1 (cello). Sufficient harmonic information is available for source 1 to resolve overlapping harmonics so the note template system also uses the pitch and CAM method to separate the signal which is why the same performance result can be observed. However, for source 2 (piano), SNR results appear to be poor. For test signal 1 we see that the pitch and CAM system has a SNR of 0.79 dB whereas the note template system has a SNR of -

2.35 dB, suggesting that the level of noise introduced by the system is greater than the level of input signal. Likewise, test signal 2 shows poor SNR results for source 2, the pitch and CAM system has a SNR of 2.90 dB while the note template system has a SNR of -3.65 dB.

Test Signal	Source	Pitch and CAM System	Note Template System
1	1	2.62	2.62
	2	0.79	-2.35
2	1	7.79	7.79
	2	2.90	-3.65

Table 3. SNR (dB) results for Note Template System as compared with the pitch and CAM system.

Test Signal	Source	Pitch and CAM System	Note Template System
1	1	4.08	3.77
	2	1.96	0.92
2	1	4.77	4.81
	2	1.65	0.92

Table 4. Listening test results for Note Template System as compared with the pitch and CAM system.

Table 4 shows average results for the listening test for the pitch and CAM separation system and the note template separation system. For test signal 1, source 1, we see a mean score of 4.08 for the pitch and CAM separation system and a mean score of 3.77 for the note template system despite the same pitch and CAM separated signal being used by both systems, as explained earlier. For test signal 1, source 2, we see a mean score of 4.77 for the pitch and CAM system. We see a reduction of the score for the note template system, with a mean score of 0.92. Comparing scores for test signal 2, similar scores for source 1 can be seen for both systems, with the pitch and CAM system scoring a mean of 4.77 and the note template system scoring a mean of 4.81. Again, both systems use the pitch and CAM separated signals for source 1, as explained earlier. The score for the note template system is lower than the score for the pitch and CAM system, for test signal 2, source 2; we see a mean score of 1.65 for the pitch and CAM system and a mean score of 0.92 for the note template system. The spectral template system does not work as promising as we would have expected, due to the following possible reasons. The templates trained from mixtures may not be accurate enough to represent the sources, because of the limited number of non-overlapped harmonics and isolated notes within the mixture. Using clean music source data (instead of the monaural mixture) to train the templates may be able to mitigate this problem and further to improve the results. Also, in the proposed template systems, pitch shifting which was used to fill up the missing notes that are not available in the mixture, apparently introduces errors in harmonic estimation. These are interesting points for future investigation.

6. Conclusions

We have presented two new methods for music source separation from monaural mixture using the isolated note information and note spectral template, both evaluated from the sound mixture. The proposed methods were designed to improve the separation performance of the baseline pitch and CAM system especially for the separation of attack sections of notes, and overlapping time-frequency regions. In the pitch and CAM system, the fast attack sections are almost completely lost in the separated signals, resulting in poor separation results for the transient part of the signal. In the proposed isolate note system, accurate harmonic information available in the isolated regions is used to reconstruct harmonic content for the entire note performance, and so, the harmonic content can be removed from the mixture to reveal the remaining note performance (in a two-source case). The isolated note system has been shown to be successful in improving the separation performance of attack sections of notes, offering a large improvement in separation quality over the baseline system. In the proposed note template system, the overlapping time-frequency regions of the mixtures are resolved using the reliable information from the non-overlapping regions of the sources, based on the spectral template matching. Preliminary results show that the spectral templates evaluated from the mixtures can be noisy and may degrade the results. Using spectral template generated directly from clean training data (i.e. containing single signals, instead of mixtures) has the potential to improve the system performance which will be our future study.

7. Future directions

We have studied the potentials of using spectral template and isolated note information for music sound separation. A major challenge is however to identify the regions from which the note information can be regarded as reliable and thereby used to estimate the note information for the unreliable and overlapped regions. Under noisy and multiple source conditions, more ambiguous regions may be identified, and using such information may further distort the separation results. Pitch information is relatively reliable under noisy conditions and can be used to improve the system performance [81]. Another potential direction is to use the property of the sources and noise/interferences, such as sparseness, to facilitate the identification of the reliable regions within the mixture that can be used to estimate the sources [74-77]. This is mainly due to the following three reasons. Firstly, as mentioned earlier, music audio can be made sparser if it is transformed into another domain, such as the TF domain, using an analytically pre-defined dictionary such as discrete Fourier transform (DFT) or discrete cosine transform (DCT) [69] [70]. Recent studies show that signal dictionaries directly adapted from training data using machine learning techniques, based on some optimisation criterion (such as the reconstruction error regularised by a sparsity constraint), can offer better performance than the pre-defined dictionary [71] [72]. Secondly, the sparse techniques using learned dictionary have been shown to possess certain denoising capability for corrupted signals [72]. Thirdly, identification of reliable regions from sound mixtures, and estimation of the probability of each TF point dominated by a source can be potentially cast as an audio-inpainting [73] or matrix completion problem. This naturally links the two important areas: source separation and sparse coding. Hence, the emerging algorithms developed in the sparse coding area could be potentially used for the CASA based monaural separation system. Separating music sources from mixtures with uncertainties [78] [79], such as under the condition of unknown number of sources, is also a promising direction for future research, as required in many practical applications. In addition, online optimisation will be necessary when the separation algorithms operate on resource limited platforms [80].

8. References

[1] Jutten, C., & Herault, J. (1991). Blind Separation of Sources, Part I: An Adaptive Algorithm Based on Neuromimetic Architecture, *Signal Processing*, vol. 24, pp. 1-10.

[2] Cardoso, J.-F., & Souloumiac, A. (1993). Blind Beamforming for Non Gaussian Signals, *IEE Proc. F, Radar Signal Processing*, vol. 140, no. 6, pp. 362-370.

[3] Comon, P. (1994). Independent Component Analysis: a New Concept?, *Signal Processing*, vol. 36, no. 3, pp. 287-314.

[4] Bell, A. J., & Sejnowski, T. J. (1995). An Information Maximization Approach to Blind Separation and Blind Deconvolution, *Neural Computation*, vol. 7, no. 6, pp. 1129-1159.

[5] Amari, S.-I., Cichocki, A., & Yang, H. (1996). A New Learning Algorithm for Blind Signal Separation, *Advances Neural Information Processing System*, vol. 8, pp. 757-763.

[6] Cardoso, J.-F. (1998). Blind Signal Separation: Statistical Principles, *Proceedings of the IEEE*, vol. 9, no. 10, pp. 2009-2025.

[7] Lee, T.-W. (1998). *Independent Component Analysis: Theory and Applications*. Boston, MA: Kluwer Academic.

[8] Haykin, S. (2000). *Unsupervised Adaptive Filtering*, Volume 1, Blind Source Separation. New York: Wiley.

[9] Hyvärinen, A., Karhunen, J., & Oja, E. (2001). *Independent Component Analysis*. New York: Wiley.

[10] Cichocki, A., & Amari, S.-I. (2002). *Adaptive Blind Signal and Image Processing, Learning Algorithm and Applications*. New York: Wiley.

[11] Cardoso, J. F., & Laheld, B. (1996). Equivariant Adaptive Source Separation, *IEEE Transactions Signal Processing*, vol. 44, no. 12, pp. 3017-3030.

[12] Belouchrani, A., Abed-Meraim, K., Cardoso, J., & Moulines, E. (1997). A Blind Source Separation Technique Using Second-Order Statistics, *IEEE Transactions Signal Processing*, vol. 45, no. 2, pp. 434-444.

[13] Thi, H., & Jutten, C. (1995). Blind Source Separation for Convolutive Mixtures, *Signal Processing*, vol. 45, pp. 209-229.

[14] Smaragdis, P. (1998). Blind Separation of Convolved Mixtures in the Frequency Domain, *Neurocomputing*, vol. 22, pp. 21-34.

[15] Parra, L., & Spence, C. (2000). Convolutive Blind Source Separation of Nonstationary Sources, *IEEE Transactions Speech Audio Processing*, vol. 8, no. 3, pp. 320-327.

[16] Rahbar, K., & Reilly, J. (2001). Blind Source Separation of Convolved Sources by Joint Approximate Diagonalization of Cross-Spectral Density Matrices, in *Proc. IEEE International Conference Acoustics, Speech and Signal Processing*, Utah, USA.

[17] Davies, M. (2002). Audio Source Separation, in *Mathematics in Signal Separation V*. Oxford, U.K.: Oxford Univ. Press.

[18] Sawada, H., Mukai, R., Araki, S., & Makino, S. (2004). A Robust and Precise Method for Solving the Permutation Problem of Frequency-Domain Blind Source Separation, *IEEE Transactions Speech and Audio Processing*, vol.12, no. 5, pp. 530-538.

[19] Wang, W., Sanei, S., & Chambers, J.A. (2005). Penalty Function Based Joint Diagonalization Approach for Convolutive Blind Separation of Nonstationary Sources, *IEEE Transactions on Signal Processing*, vol. 53, no. 5, pp. 1654-1669.

[20] Sawada, H., Araki, S., & Makino, S. (2010). Underdetermined Convolutive Blind Source Separation via Frequency Bin-Wise Clustering and Permutation Alignment, *IEEE Transactions on Audio Speech and Language Processing*, vol. 18, pp. 516-527.

[21] Pedersen, M., Larsen, J., Kjems, U., & Parra, L. (2007). A Survey on Convolutive Blind Source Separation Methods, *Handbook on Speech Processing and Speech Communication*, Springer.

[22] Belouchrani, A., & Amin, M. G. (1998). Blind Source Separation Based on Time-Frequency Signal Representations, *IEEE Transactions Signal Processing*, vol. 46, no. 11, pp. 2888–2897, 1998.

[23] Chen, S., Donoho, D. L., & Saunders, M. A. (1998). Atomic Decomposition by Basis Pursuit, *SIAM Journal Scientific Computing*, vol. 20, no. 1, pp. 33-61.

[24] Lee, T., Lewicki, M., Girolami, M., & Sejnowski, T. (1998), Blind Source Separation of More Sources Than Mixtures Using Overcomplete Representations, *IEEE Signal Processing Letters*, vol. 6, no. 4, pp. 87-90.

[25] Lewicki, M. S., & Sejnowski, T. J. (1998). Learning Overcomplete Representations, *Neural Computation*, vol. 12, no. 2, pp. 337–365.

[26] Bofill, P., & Zibulevsky, M. (2001). Underdetermined Blind Source Separation Using Sparse Representation, *Signal Processing*, vol. 81, pp. 2253–2362.

[27] Zibulevsky, M., & Pearlmutter, B. A. (2001). Blind Source Separation by Sparse Decomposition in a Signal Dictionary, *Neural Computation*, vol. 13, no. 4, pp. 863–882.

[28] Li, Y., Amari, S., Cichocki, A., Ho, D. W. C., & Xie, S. (2006). Underdetermined Blind Source Separation Based on Sparse Representation. *IEEE Transactions on Signal Processing*, vol. 54, no. 2, pp. 423-437.

[29] He, Z., Cichocki, A., Li, Y., Xie, S., & Sanei, S. (2009). K-Hyperline Clustering Learning for Sparse Component Analysis. *Signal Processing*, vol. 89, no. 6, pp. 1011-1022.

[30] Yilmaz, O., & Richard, S. (2004). Blind Separation of Speech Mixtures via Time-Frequency Masking, *IEEE Transactions on Signal Processing*, vol. 52, no. 7, pp. 1830-1847.

[31] Wang, D.L. (2005). On Ideal Binary Mask as the Computational Goal of Auditory Scene Analysis. In Divenyi P. (ed.), *Speech Separation by Humans and Machines*, pp. 181-197, Kluwer Academic, Norwell MA.

[32] Mandel, M. I., Weiss, R. J., & Ellis, D. P. W. (2010). Model-based Expectation Maximisation Source Separation and Localisation, *IEEE Transactions on Audio Speech and Language Processing*, vol. 18, pp. 382-394.

[33] Duong, N. Q. K., Vicent, E., & Gribonval, R. (2010). Under-determined Reverberant Audio Source Separation Using a Full-Rank Spatial Covariance Model, *IEEE Transactions on Audio Speech and Language Processing*, vol. 18, pp. 1830-1840.

[34] Plumbley, M. D. (2003). Algorithms for Nonnegative Independent Component Analysis, *IEEE Transactions on Neural Networks*, vol. 14, no. 2, pp. 534–543.

[35] Kim, T., Attias, H., & Lee, T.-W. (2007). Blind Source Separation Exploiting Higher-Order Frequency Dependencies, *IEEE Transactions on Audio Speech and Language Processing*, vol. 15, pp. 70-79.

[36] Comon, P., & Jutten, C., eds. (2010). *Handbook of Blind Source Separation, Independent Component Analysis and Applications*, Academic Press.

[37] Smaragdis, P. (2004). Non-Negative Matrix Factor Deconvolution; Extraction of Multiple Sound Sources from Monophonic Inputs. In *Proceedings of the 5th International Conference on Independent Component Analysis and Blind Signal Separation*, Grenada, Spain.

[38] Schmidt, M. N., & Mørup, M. (2006). Nonnegative Matrix Factor 2-D Deconvolution for Blind Single Channel Source Separation, in *Proceedings of the International Conference on Independent Component Analysis and Signal Separation*, Charleston, USA.

[39] Wang, W., Cichocki, A., & Chambers, J. A. (2009). A Multiplicative Algorithm for Convolutive Non-negative Matrix Factorization Based on Squared Euclidean Distance, *IEEE Transactions on Signal Processing*, vol. 57, no. 7, pp. 2858-2864.

[40] Ozerov, A., & Févotte, C. (2010). Multichannel Nonnegative Matrix Factorization in Convolutive Mixtures for Audio Source Separation, *IEEE Transactions on Audio, Speech and Language Processing*.

[41] Mysore, G., Smaragdis, P., & Raj, B. (2010). Non-Negative Hidden Markov Modeling of Audio with Application to Source Separation. In *Proceedings of the 9th international conference on Latent Variable Analysis and Signal Separation (LCA/ICA)*. St. Malo, France.

[42] Ozerov, A., Févotte, C., Blouet, R., & Durrieu, J.L. (2011). Multichannel Nonnegative Tensor Factorization with Structured Constraints for User-guided Audio Source Separation, *Proceedings of the IEEE International Conference Acoustics, Speech and Signal Processing*, Prague, Czech Republic.

[43] Wang, W. & Mustafa, H. (2011). Single Channel Music Sound Separation Based on Spectrogram Decomposition and Note Classification, in *Computer Music Modelling and Retrieval*, Springer.

[44] Cichocki, A., Zdunek, R., Phan, A.H., & Amari, S. (2009). *Nonnegative Matrix and Tensor Factorizations: Applications to Exploratory Multi-way Data Analysis and Blind Source Separation*. Wiley.

[45] Brown, G.J., & Cooke, M.P. (1994). Computational Auditory Scene Analysis, *Computer Speech and Language*, vol. 8, pp. 297–336.

[46] Wrigley, S. N., Brown, G. J., Renals, S., & Wan, V. (2005). Speech and Crosstalk Detection in Multi-Channel Audio, *IEEE Transactions on Speech and Audio Processing*, vol. 13, no. 1, pp. 84-91.

[47] Palomäki, K. J., Brown, G. J., & Wang, D. L. (2004). A Binaural Processor for Missing Data Speech Recognition in the Presence of Noise and Small-Room Reverberation, *Speech Communication*, vol. 43, no. 4, pp. 361-378.

[48] Wang, D.L., & Brown, G. (2006). *Computational Auditory Scene Analysis: Principles, Algorithms, and Applications*, Wiley/IEEE.

[49] Shao, Y., & Wang, D.L. (2009). Sequential Organization of Speech in Computational Auditory Scene Analysis. *Speech Communication*, vol. 51, pp. 657-667.

[50] Hu, K., & Wang, D.L. (2011). Unvoiced Speech Segregation from Nonspeech Interference via CASA and Spectral Subtraction. *IEEE Transactions on Audio, Speech, and Language Processing*, vol. 19, pp. 1600-1609.

[51] Xu, T., & Wang, W. (2009). A Compressed Sensing Approach for Underdetermined Blind Audio Source Separation with Sparse Representations, in *Proceedings of the IEEE International Workshop on Statistical Signal Processing*, Cardiff, UK.
[52] Xu, T., & Wang, W. (2010). A Block-based Compressed Sensing Method for Underdetermined Blind Speech Separation Incorporating Binary Mask, in *Proceedings of the IEEE International Conference on Acoustics, Speech and Signal Processing*, Dallas, Texas, USA.
[53] Xu, T., & Wang, W. (2011). Methods for Learning Adaptive Dictionary for Underdetermined Speech Separation, in *Proceedings of the IEEE 21st International Workshop on Machine Learning for Signal Processing*, Beijing, China.
[54] Kim, M., & Choi, S. (2006). Monaural Music Source Separation: Nonnegativity, Sparseness, and Shift-Invariance, in *Proceedings of the IEEE International Conference on Independent Component Analysis and Blind Signal Separation*, Charleston, USA.
[55] Virtanen, T. (2006). *Sound Source Separation in Monaural Music Signals*, PhD Thesis, Tampere University of Technology.
[56] Virtanen, T. (2007). Monaural Sound Source Separation by Nonnegative Matrix Factorization With Temporal Continuity and Sparseness Criteria, *IEEE Transactions on Audio, Speech and Language Processing*, vol. 15, no. 3, pp. 1066-1074.
[57] Ozerov, A., Philippe, P., Bimbot, F., & Gribonval, R. (2007). Adaptation of Bayesian Models for Single-Channel Source Separation and its Application to Voice/Music Separation in Popular Songs, *Proceedings of the IEEE International Conference on Acoustics, Speech, and Signal Processing*, Hawaii, USA.
[58] Richard, G., & David, B. (2009). An Iterative Approach to Monaural Musical Mixture De-Soloing, in *Proceedings of the IEEE International Conference on Acoustics, Speech, and Signal Processing*, Taipei, Taiwan.
[59] Klapuri, A., Virtanen, T., & Heittola, T. (2010). Sound Source Separation in Monaural Music Signals Using Excitation-Filter Model and EM Algorithm, in *Proceedings of the IEEE International Conference on Acoustics, Speech, and Signal Processing*, Dallas, USA.
[60] Bregman, A. S. (1990). *Auditory Scene Analysis*, MIT Press.
[61] Li, Y. & Wang, D. L. (2007). Separation of Singing Voice From Music Accompaniment for Monaural Recordings, *IEEE Transactions on Audio, Speech and Language Processing*, vol. 15, no. 4, pp. 1475-1487.
[62] Parsons, T. W. (1976). Separation of Speech from Interfering Speech By Means of Harmonic Selection, *Journal of the Acoustical Society of America*, vol. 60, no. 4, pp. 911-918, 1976.
[63] Li, Y. & Wang, D. L. (2009). Musical Sound Separation Based on Binary Time-Frequency Masking, *EURASIP Journal on Audio, Speech and Music Processing*, article ID 130567.
[64] Every, M. R. & Szymanski, J. E. (2006). Separation of Synchronous Pitched Notes by Spectral Filtering of Harmonics, *IEEE Transactions on Audio, Speech and Language Processing*, vol. 14, no. 5, pp. 1845-1856.
[65] Virtanen, T. & Klapuri, A. (2001). Separation of Harmonic Sounds Using Multipitch Analysis and Iterative Parameter Estimation, in *Proceedings of the IEEE Workshop on Applications of Signal Processing in Audio and Acoustics*, pp. 83-86.
[66] Hu, G. (2006). *Monaural Speech Organization and Segregation*, Ph.D. Thesis, The Ohio State University, USA.

[67] Li, Y., Woodruff, J. & Wang, D. L. (2009). Monaural Musical Sound Separation Based on Pitch and Common Amplitude Modulation, *IEEE Transactions on Audio, Speech and Language Processing*, vol. 17, no. 7, pp. 1361-1371.

[68] ISO. *Acoustics – Standard Tuning Frequency (Standard Musical Pitch)*, ISO 16:1975, International Organization for Standardization, Geneva, 1975.

[69] Nesbit, A., Jafari, M. G., Vincent, E., & Plumbley, M. D. (2010). Audio Source Separation Using Sparse Representations. In W. Wang (Ed), *Machine Audition: Principles, Algorithms and Systems*. Chapter 10, pp.246-264. IGI Global.

[70] Plumbley, M. D., Blumensath, T., Daudet, L., Gribonval, R., & Davies, M. E. (2010). Sparse Representations in Audio and Music: from Coding to Source Separation, *Proceedings of the IEEE*, vol. 98, pp. 995-1005.

[71] Dai, W., Xu, T. & Wang, W. (2012). Dictionary Learning and Update based on Simultaneous Codeword Optimisation (SIMCO), *Proceedings of the IEEE International Conference on Acoustics, Speech and Signal Processing*, Kyoto, Japan.

[72] Dai, W., Xu, T., & Wang, W. (2011). Simultaneous Codeword Optimisation (SimCO) for Dictionary Update and Learning, *arXiv:1109.5302*.

[73] Adler, A., Emiya V., Jafari, M.G., Elad, M., Gribonval, G., & Plumbley, M.D. (2012). Audio Inpainting, *IEEE Transactions on Audio, Speech and Language Processing*, vol. 20, pp. 922-932.

[74] Wang, W. (2011). *Machine Audition: Principles, Algorithms and Systems*, IGI Global Press.

[75] Jan, T. & Wang, W. (2011). Cocktail Party Problem: Source Separation Issues and Computational Methods, in W. Wang (ed), *Machine Audition: Principles, Algorithms and Systems*, IGI Global Press, pp. 61-79.

[76] Jan, T., Wang, W., & Wang, D.L. (2011). A Multistage Approach to Blind Separation of Convolutive Speech Mixtures. *Speech Communication*, vol. 53, pp. 524-539.

[77] Luo, Y., Wang, W., Chambers, J. A., Lambotharan, S., & Prouder, I. (2006). Exploitation of Source Non-stationarity for Underdetermined Blind Source Separation With Advanced Clustering Techniques, *IEEE Transactions on Signal Processing*, vol. 54, no. 6, pp. 2198-2212.

[78] Adiloglu, K. & Vincent, E. (2011). An Uncertainty Estimation Approach for the Extraction of Source Features in Multisource Recordings, in *Proceedings of the European Signal Processing Conference*, Barcelona, Spain.

[79] Adiloglu, K., & Vincent, E. (2012). A General Variational Bayesian Framework for Robust Feature Extraction in Multisource Recordings, in *Proceedings of the IEEE International Conference on Acoustics, Speech and Signal Processing*, Kyoto, Japan.

[80] Simon, L. S. R., & Vincent, E. (2012). A General Framework for Online Audio Source Separation, in *Proceedings of the International conference on Latent Variable Analysis and Signal Separation*, Tel-Aviv, Israel.

[81] Hsu, C.-L., Wang, D.L., Jang J.-S.R., & Hu, K. (2012). A Tandem Algorithm for Singing Pitch Extraction and Voice Separation from Music Accompaniment, *IEEE Transactions on Audio, Speech, and Language Processing*, vol. 20, pp. 1482-1491.

3

On Temporomandibular Joint Sound Signal Analysis Using ICA

Feng Jin[1] and Farook Sattar[2]
[1]*Dept of Electrical & Computer Engineering, Ryerson University, Toronto, Ontario*
[2]*Dept of Electrical & Computer Engineering, University of Waterloo, Waterloo, Ontario*
Canada

1. Introduction

The Temporomandibular Joint (TMJ) is the joint which connects the lower jaw, called the mandible, to the temporal bone at the side of the head. The joint is very important with regard to speech, mastication and swallowing. Any problem that prevents this system from *functioning* properly may result in temporomandibular joint disorder (TMD). Symptoms include pain, limited movement of the jaw, radiating pain in the face, neck or shoulders, painful clicking, popping or grating sounds in the jaw joint during opening and/or closing of the mouth. TMD being the most common non-dental related chronic source of oral-facial pain(Gray et al., 1995)(Pankhurst C. L, 1997), affects over 75% of the United States population(Berman et al., 2006). TMJ sounds during jaw motion are important indication of dysfunction and are closely correlated with the joint pathology(Widmalm et al., 1992). The TMJ sounds are routinely recorded by auscultation and noted in dental examination protocols. However, stethoscopic auscultation is very subjective and difficult to document. The interpretations of the sounds often vary among different doctors. Early detection of TMD, before irreversible gross erosive changes take place, is extremely important.

Electronic recording of TMJ sounds therefore offers some advantages over stethoscopic auscultation recording by allowing the clinician to store the sound for further analysis and future reference. Secondly, the recording of TMJ sounds is also an objective and quantitative record of the TMJ sounds during the changes in joint pathology. The most important advantage is that electronic recording allows the use of advanced signal processing techniques to the automatic classification of the sounds. A cheap, efficient and reliable diagnostic tool for early detection of TMD can be developed using TMJ sounds recorded with a pair of microphones placed at the openings of the auditory canals. The analysis of these recorded TMJ vibrations offers a powerful non-invasive alternative to the old clinical methods such as auscultation and radiation.

In early studies, the temporal waveforms and power spectra of TMJ sounds were analyzed(Widmalm et al., 1991) to characterize signals based on their time behavior or their energy distribution over a frequency range. However, such approaches are not sufficient to fully characterize non-stationary signals like TMJ sounds. In other words, for non-stationary signals like TMJ vibrations, it is required to know how the frequency components of the signal change with time. This can be achieved by obtaining the distribution of signal energy over the TF plane(Cohen L., 1995). Several joint time-frequency analysis methods have then been applied to the analysis and classification of TMJ vibrations into different classes based on their time-frequency reduced interference distribution (RID)(Widmalm & Widmalm, 1996)(Akan et al., 2000). According to TF analysis, four distinct classes of defective TMJ sounds are defined: click, click with crepitation, soft crepitation, and hard crepitation(Watt, 1980) Here, clicks are identified as high amplitude peaks of very short duration, and crepitations are signals with multiple peaks of various amplitude and longer duration as well as a wide frequency range.

In this chapter, instead of discussing the classification of TMJ sounds into various types based on their TF characteristics, we address the problem of source separation of the stereo recordings of TMJ sounds. Statistical correlations between different type of sounds and the joint pathology have been explored by applying ICA based methods to present a potential diagnostic tool for temporomandibular joint disorder.

The chapter outline is as follows: The details for data acquisition are elaborated in Section 2, followed by the problem definition and the possible contribution of the independent component analysis (ICA) based approach. The proposed signal mixing and propagation models are then proposed in Section 3, with the theoretical background of ICA and the proposed ICA based solutions described in Sections 4 to 6. The illustrative results of the present method on both simulated and real TMJ signals are compared with other existing source separation methods in Section 7. The performance of the method has been further evaluated quantitatively in Section 8. Lastly, the chapter summary and discussion are presented in Section 9.

2. Data acquisition

The auditory canal is an ideal location for the non-invasive sensor (microphone) to come as close to the joint as possible. The microphones were held in place by earplugs made of a kneadable polysiloxane impression material (called the Reprosil putty and produced by Dentsply). A hole was punched through each earplug to hold the microphone in place and to reduce the interference of ambient noise in the recordings.

In this study, the TMJ sounds were recorded on a Digital Audio Tape (DAT) recorder. During recording session, the necessary equipments are two Sony ECM-77-B electret condenser microphones, Krohn-Hite 3944 multi-channel analog filter and TEAC RD-145T or TASCAM DA-P1 DAT recorder. The microphones have a frequency response ranges from 40–20,000 Hz and omni-directional. It acts as a transducer to capture the TMJ sounds. The signals were then passed through a lowpass filter to prevent aliasing effect of the digital signal. A Butterworth filter with a cut-off frequency of 20 KHz and attenuation slope of 24 dB/octave was set at the analog filter. There is an option to set the gain at the filter to boost up the energy level of the signal. The option was turned on when the TMJ sounds were too soft and the signals from the microphones were amplified to make full use of the dynamic range of the DAT recorder. Finally, the signals from the analog filter were sampled in the DAT recorder at the rate of 48 KHz and data were saved on a disc.

3. Problems and solution: The role of ICA

One common and major problem in both stethoscopic auscultation and digital recording is that the sound originating from one side will propagate to the other side, leading to misdiagnosis in some cases. It is shown in Fig. 1(a) that short duration TMJ sounds (less than 10ms) are frequently recorded in both channels very close in time. When the two channels show similar waveforms, with one lagging and attenuated to some degree, it can be concluded that the lagging signal is in fact the propagated version of the other signal(Widmalm et al., 1997).

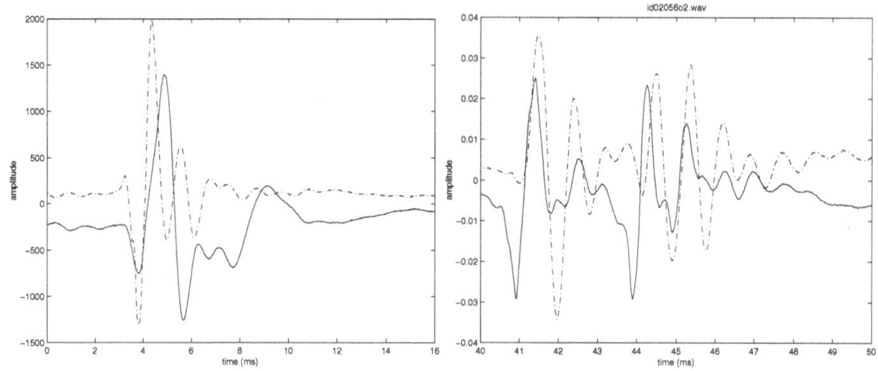

(a) (b)

Fig. 1. TMJ sounds of two channels.

This observation is very important. It means that a sound heard at auscultation on one side may have actually come from the other TMJ. This has great clinical significance because it is necessary to know the true source of the recorded sound, for example in diagnosing so called disk displacement with reduction(Widmalm et al., 1997). The TMJ sounds can be classified into two major classes: clicks and crepitations. A click is a distinct sound, of very limited duration, with a clear beginning and end. As the name suggests, it sounds like a "click". A crepitation has a longer duration. It sounds like a series of short but rapidly repeating sounds that occur close in time. Sometimes, it is described as "grinding of snow" or "sand falling". The duration of a click is very short (usually less than 10ms). It is possible to differentiate between the source and the propagated sound without much difficulty. This is due to the short delay (about 0.2ms) and the difference in amplitude between the signals of the two channels, especially if one TMJ is silent. However, it is sometimes very difficult to tell which is the source signals from the recordings. In Fig. 1(b), it seems that the dashed line is the source if we simply look at the amplitude. On the other hand, it might seem that the solid line is the source if we look at the time (it comes first). ICA could have vital role to solve this problem since both the sources (sounds from both TMJ) and the mixing process (the transfer function of the human head, bone and tissue) are unknown. If ICA is used, one output should be the original signal and the other channel should be silent with very low amplitude noise picked up by the microphone. Then it is very easy to tell which channel is the original sound. Furthermore, in the case of crepitation sounds, the duration of the signal is longer, and further complicated by the fact that both sides may crepitate at the same time. The ICA is then proposed as a means to recover the original sound for each channel.

4. Mixing model of TMJ sound signals

In this chapter, the study is not limited to patients with only one defective TMD joint. We thus consider the TMJ sounds recorded simultaneously from both sides of human head as a mixture of crepitations/clicks from the TMD affected joint and the noise produced by the other healthy TMJ or another crepitation/click. Instead of regarding the 'echo' recorded on the contra (i.e. opposite) side of the TMD joint as the lagged version of the TMD source(Widmalm et al., 2002), we consider here the possibility that this echo as a mixture of the TMD sources. Mathematically, the mixing model of the observed TMJ sound measurements is represented as

$$x_i(t) = \sum_{j=1}^{2} h_{ij} s_j(t - \delta_{ij}) + n_i(t) \tag{1}$$

with s_j being the jth source and x_i as the ith TMJ mixture signal with $i = 1, 2$. The additive white Gaussian noise at discrete time t is denoted by $n_i(t)$. Also, the attenuation coefficients, as well as the time delays associated with the transmission path between the jth source and the ith sensor (i.e. microphone) are denoted by h_{ij} and δ_{ij}, respectively.

Fig. 2 shows how the TMJ sounds are mixed. Sounds originating from a TMJ are picked up by the microphone in the auditory canal immediately behind the joint and also by the microphone in the other auditory canal as the sound travels through the human head.

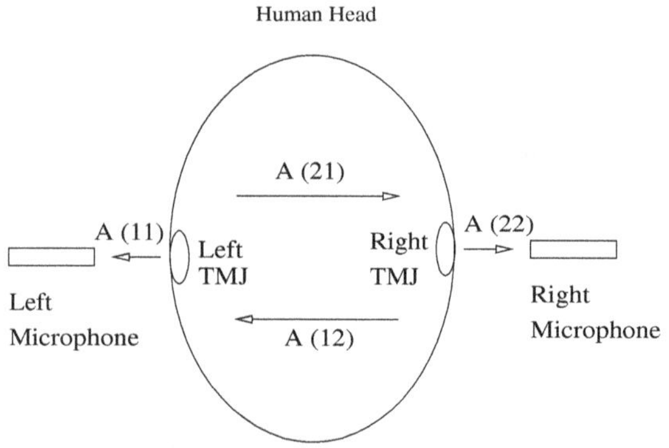

Fig. 2. Mixing model of TMJ sounds (A_{ij} refers to the acoustic path between the $j = 1$ (i.e. left side of human head) source and the $i = 2$ (right side of the human head) sensor.

The mixing matrix **H** could therefore be defined as below with z^{-1} indicating unit delay:

$$\mathbf{H} = \begin{pmatrix} h_{11} z^{-\delta_{11}} & h_{12} z^{-\delta_{12}} \\ h_{21} z^{-\delta_{21}} & h_{22} z^{-\delta_{22}} \end{pmatrix} \tag{2}$$

Please note that the time delay δ is not necessarily to be integer due to the uncertainty in sound transmission time in tissues.

The independency of the TMJ sound sources on both sides of the head might not hold as both joints operate synchronously during the opening and closing of mouth. Therefore, unlike the convolutive mixing model assumed in our previous paper(Guo et al., 1999), the instantaneous mixing model presented here does not depend on the assumption of statistical independence of the sources. In this work, the main assumptions made include the non-stationarity of all the source signals as well as anechoic head model. Here, the anechoic model is assumed due to the facts that:

1. TMJ sound made by the opposite side of the TMD joint has been reported as a delayed version of its ipsi(Widmalm et al., 2002).
2. The TMJ sounds has a wide bandwidth of [20, 3200]Hz. While it travels across the head, the high frequency components (>1200Hz) have been severely attenuated(OB́rien et al., 2005).

Single effective acoustic path from one side of the head to the other side is thus assumed. Also, the mixing model in Eq. (1) holds with mixing matrix presented in Eq. (2). However, due to the wide bandwidth of crepitation, source TMJ signals are not necessarily to be sparse in the time-frequency domain. This gives the proposed ICA based method better robustness as compared to the blind source separation algorithm proposed in(Took et al., 2008).

5. Theoretical background of ICA

There are three basic and intuitive principles for estimating the model of independent component analysis.

1) *ICA by minimization of mutual information*

This is based on information-theoretic concept, i.e. information maximization (InfoMax) as briefly explained here.

The differential entropy H of a random vector y with density $p(y)$ is defined as (Hyvärinen, 1999):

$$H(y) = -\int p(y) log p(y) dy \tag{3}$$

Basically, the mutual information I between m (scalar) random variables $y_i, i = 1 \cdots m$ is defined as follows:

$$I(y_1, y_2, \cdots, y_m) = \sum_{i=1}^{m} H(y_i) - H(y) \tag{4}$$

The mutual information is $I(y_1, y_2) = \sum_{i=1}^{2} H(y_i) - H(y_1, y_2)$, where $\sum_{i=1}^{2} H(y_i)$ is marginal entropy and $H(y_1, y_2)$ is joint entropy. The mutual information is a natural measure of the dependence between random variables. It is always nonnegative, and zero if and only if the variables are statistically independent. Therefore, we can use mutual information as the criterion for finding the ICA representation, i.e. to make the output "decorrelated". In any case, minimization of mutual information can be interpreted as giving the maximally independent components(Hyvärinen, 1999).

2) *ICA by maximization of non-Gaussianity*

Non-Gaussianity is actually most important in ICA estimation. In classic statistical theory, random variables are assumed to have Gaussian distributions. So we start by motivating the maximization of Non-Gaussianity by the central limit theorem. It has important consequences in independent component analysis and blind source separation. Even for a small number of sources the distribution of the mixture is usually close to Gaussian. We can simply explained the concept as follows:

Let us assume that the data vector x is distributed according to the ICA data model: $x = Hs$ is a mixture of independent source components s and H is the unknown full rank $(n \times m)$ mixing matrix for m mixed signals and n independent source components. Estimating the independent components can be accomplished by finding the right linear combinations of the mixture variables. We can invert the mixing model in vector form as: $s = H^{-1}x$, so the linear combination of x_i. In other words, we can denote this by $y = b^T x = \sum_{i=1}^{m} b_i x_i$. We could take b as a vector that maximizes the Non-Gaussianity of $b^T x$. This means that $y = b^T x$ equals one of the independent components. Therefore, maximizing the Non-Gaussianity of $b^T x$ gives us one of the independent components(Hyvärinen, 1999). To find several independent components, we need to find all these local maxima. This is not difficult, because the different independent components are uncorrelated. We can always constrain the search to the space that gives estimates uncorrelated with the previous ones(Hyvärinen, 2004).

3) *ICA by maximization of likelihood*

Maximization of likelihood is one of the popular approaches to estimate the independent components analysis model. Maximum likelihood (ML) estimator assumes that the unknown parameters are constants if there is no prior information available on them. It usually applies to large numbers of samples. One interpretation of ML estimation is calculating parameter values as estimates that give the highest probability for the observations. There are two algorithms to perform the maximum likelihood estimation:

- Gradient algorithm: this is the algorithms for maximizing likelihood obtained by the gradient based method(Hyvärinen, 1999).
- Fast fixed-point algorithm(Ella, 2000): the basic principle is to maximize the measures of Non-Gaussianity used for ICA estimation. Actually, the FastICA algorithm (gradient-based algorithm but converge very fast and reliably) can be directly applied to maximization of the likelihood.

6. The proposed ICA based TMJ analysis method

The proposed ICA based TMJ analysis method is based on the following considerations: i) asymmetric mixing, ii) non-sparse source conditions. Based on the above criterion we consider to apply the ICA technique based on information maximization as introduced above in Section 5 for the analysis of TMJ signals. We therefore propose an improved Infomax method based on its robustness against noise and general mixing properties. The present method has an adaptive contrast function (i.e. adaptive log-sigmoidal function) together with non-causal filters over the conventional Infomax method in Bell & Sejnowski (1995); Torkkola (1996) to get better performance for a pair of TMJ sources.

The nonlinear function, f, must be a monotonically increasing or decreasing function. In this paper, the nonlinear function proposed is defined as

$$y = f(u; b, m) = [1/(1 + e^{-bu})]^m. \tag{5}$$

Maximizing the output information can be then achieved by minimizing the mutual information between the outputs y_1 and y_2 of the above adaptive f function. In Eq. (5) the adaptation in the slope parameter b is equivalent to adaptive learning rate during our iterative process. This let us perform the iteration with a small learning rate followed by larger learning rate as the iteration proceeds. On the other hand, during iteration the exponent parameter m is kept as $m = 1$ in our case in order to make sure that the important 'click' signals are not skewed.

Moreover, algorithm in (Torkkola, 1996) performs well when there is stable inverse of the direct channel(i.e. ipsi side) which is not always feasible in real case. In the separation of TMJ sound signals, the direct channel is the path from the source (TMJ) through the head tissue to the skull bone, then to the air in the auditory canal directly behind the TMJ and finally to the ipsi microphone. The corresponding acoustic response would come from a very complex process, for which it is not guaranteed that there will a stable inverse for this transfer function.

However, even if a filter does not have a stable causal inverse, there still exists a stable non-causal inverse. Therefore, the algorithm of Torkkola can be modified and used even though there is no stable (causal) inverse filter for the direct channel. The relationships between the signals are now becomes:

$$\begin{aligned} u_1(t) &= \Sigma_{k=-M}^{M} w_k^{11} x_1(t-k) + \Sigma_{k=-M}^{M} w_k^{12} u_2(t-k) \\ u_2(t) &= \Sigma_{k=-M}^{M} w_k^{22} x_2(t-k) + \Sigma_{k=-M}^{M} w_k^{21} u_1(t-k) \end{aligned} \tag{6}$$

where M(even) is half of the (total filter length-1) and the zero lag of the filter is at $(M+1)$. In (6) there exist an initialization problem regarding filtering. To calculate the value of $u_1(t)$, the values of $u_2(t), u_2(t+1), \cdots, u_2(t+M)$ are required which are not initially available. Since learning is an iteration process, we have used some pre-assigned values to solve this filter initialization problem. For example, the value of $x_2(t)$ is used for $u_2(t)$ at the first iteration. The new values generated at the first iteration are then used for the second iteration. This process is repeated until its convergence to certain values. For each iteration, the value of the parameter b in the corresponding f function is updated based on its empirical initial/end values and total number of iterations. The expressions of b in Eq. (5) at the pth iteration is then defined as:

$$b(p) = b_o + (p-1)\Delta b \tag{7}$$

where $p = 1, 2, \cdots, iter$ and $iter$ is the total number of iterations. The Δb are obtained as $\Delta b = (b_e - b_o)/iter$ with $b \in [b_o, b_e]$. To avoid the saturation problem of the adaptive log-sigmoid function and for better use of nonlinearity, we restrict the b parameter within the interval $[1, 10]$.

The derivative of the learning rule can follow the same procedure as in Torkkola (1996). According to (6), only the unmixing coefficients of W_{12} and W_{21} have to be learned. The learning rule is the same in notation but different in nature because the values of k have changed:

$$\Delta w_k^{ij} \propto (1 - 2y_i) u_j(t-k) \tag{8}$$

where $k = -M, -M+1, \cdots, M$.

7. Results

7.1 Illustrative results on simulated signals

Simulated click signals are generated following the definition present in (Akan et al., 2000) as an impulse with very short duration (20 ms) and high amplitude peaks. Since normal TMJ is assumed to produce no sound, we have used a sine wave at 20 Hz with 1/10 click amplitude to represent normal TMJ sound. Fig. 3(a) shows an example simulation of TMJ sources with their corresponding mixtures captured by the sensors being simulated and illustrated in Fig. 3(b). The illustrated plots are generated to describe the signals being captured by sensors placed in each auditory canal of a patient with right TMD joint producing unilateral clicking. No power attenuation and delay for transmissions between sources and sensors on the same side of the head have thus been assumed.

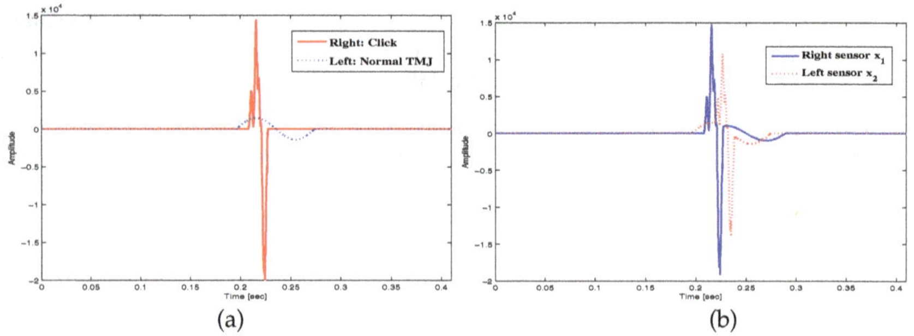

Fig. 3. Simulated (a) TMJ sources (click and normal TMJ sounds) and (b) the corresponding mixed signals.

The simulated mixing model parameters are therefore set as: $h_{11} = h_{22} = 1$ and $\delta_{11} = \delta_{22} = 0$. Also, a time delay of 1 ms between the two TMJ sounds has been observed in Fig. 6 which falls within the range of 0.2 to 1.2 ms reported in (Widmalm et al., 2002). Thus, $\delta_{12} = 1.1$ ms and $\delta_{21} = 1.5$ ms are set based on the above observations and the fact that signal with lower frequency travels slower than that of higher frequency. According to (Widmalm et al., 2003), frequencies greater than 1200 Hz of the ipsi TMJ sound were found to be severely attenuated when it propagates to the contra side, $h_{12} = h_{21}$ is thus set to have an impulse response of a lowpass filter with stopband at 1200 Hz and passband attenuation of 3 dB. The noise on both sides is considered as $n_1 = n_2 = 0$ for simplicity purpose. The illustrative results of ICA for the signals in Fig. 3 are presented in Figs. 4(a)-(b).

7.2 Illustrative results on real TMJ recordings

Fig. 5 shows a typical recording of TMJ crepitation sounds from the two microphones. Each recording is a mixture of the two TMJ sources. The sampling rate of the recorded TMJ signals is 48 KHz.

If we zoom in on Fig. 5 near 0.045s, the signals are shown in Fig. 6(a) for both channels (solid line for channel one and dashed line for channel two). It is difficult to tell how much of the signal in each channel comes from the ipsi (same side) TMJ and how much comes from the contra (opposite) TMJ(Widmalm et al., 1997). If we look at the signals near 0.35s (Fig. 6(b)), it is even more difficult to differentiate the source from the propagated component because the signals are almost 180° out of phase. It is almost impossible to determine the short time delay and difference in amplitudes between the two signals.

The results of ICA for the signals in Fig. 5 are presented in Fig. 7. In order to see the important role of ICA, let us look at the signals near 0.35s (Fig. 8). It clearly shows that the signal only comes from the first channel (solid line) and the second channel (dashed line) is basically silent. From Fig. 7, it is also clear that the source is now coming from channel two at the time near 0.045s.

8. Performance evaluation

For comparison purposes, the source estimates of Infomax(Guo et al., 1999), FastICA(Hyvärinen et al, 2000), and the original DUET(Widmalm et al., 2002) approach are also included in the simulation studies. The un-mixing filter length was set to be 161 for the Infomax approach, and the DUET approach assumes the W-disjoint orthogonality of the source signals, where at most one source is dominating over any particular TF interval. The illustrative results of the extracted sources from real TMJ recordings by these three reference methods are depicted in Fig 7. The source separation results on simulated mixture of click and normal TMJ sounds using various methods are denoted in Fig 4. The signal in the left column is evidently the click source, while the reduced traces of those prominent peaks in signal from right column suggests that it is in fact the sound produced by the healthy/normal joint. Although the FastICA method is able to estimate the normal source with the minimum

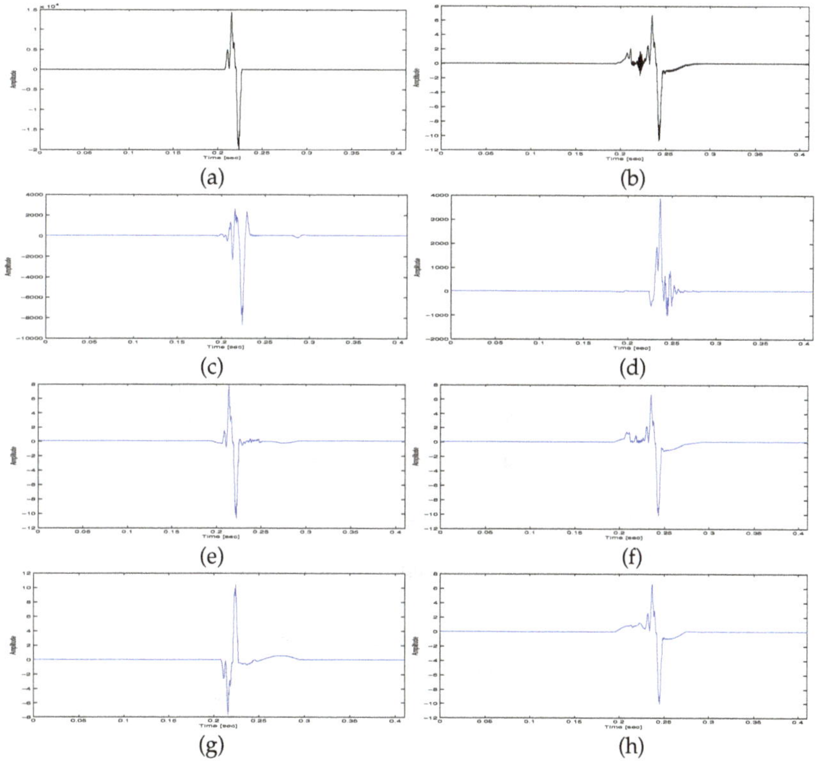

Fig. 4. The results of (a-b) the proposed ICA based source extraction method; (c-d) DUET method in (Widmalm et al., 2002); (e-f) Infomax approach in (Guo et al., 1999); (g-h) FastICA approach in (Hyvärinen et al, 2000) on simulated TMJ mixtures in Fig. 3.

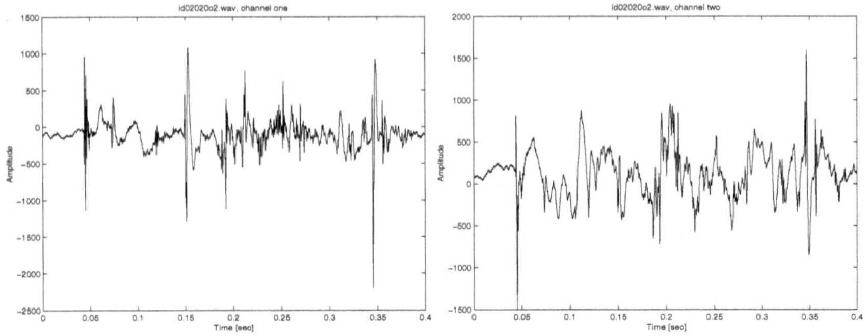

(a) (b)

Fig. 5. Typical TMJ signals (crepitation).

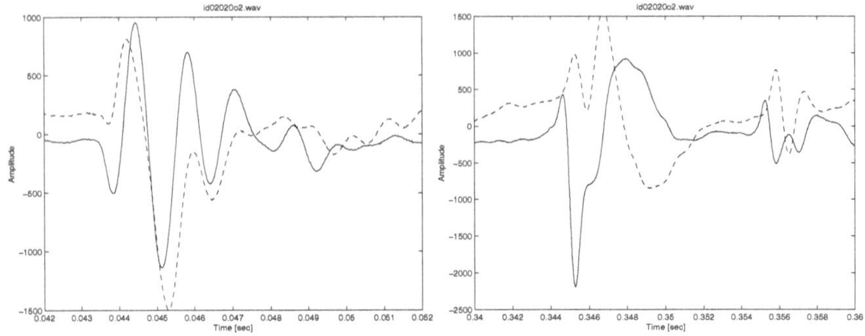

(a) (b)

Fig. 6. The zoomed version of the typical TMJ signals (crepitation).

click interference, it estimates the first source for click TMJ with a π phase shift, while the DUET method failed to estimate both source signals as shown in Fig. 4(d).

On the other hand, the proposed source separation method outperforms the existing ones on real TMJ mixtures as well. With the extracted crepitation labeled in circles, one could observe that both the FastICA and DUET approaches estimate sources with overlapping crepitations (i.e. the crepitations occur at the same time in both sources). This indicates the ineffectiveness of the separation scheme which does not happen in the estimated sources by the proposed method.

In order to quantitatively evaluate the separation performance, we have used mutual information of two separated TMJ signals. The mutual information of two random variables is a quantity that measures the mutual dependence of the two variables. Mutual information of two random variables y_1 and y_2 can be expressed as

$$I(y_1; y_2) = H(y_1) + H(y_2) - H(y_1, y_2) \tag{9}$$

where $H(y_1)$ and $H(y_2)$ are marginal entropies, and $H(y_1, y_2)$ is the joint entropy of y_1 and y_2. The value of mutual information (MI) for the pair of used mixture recording signals is

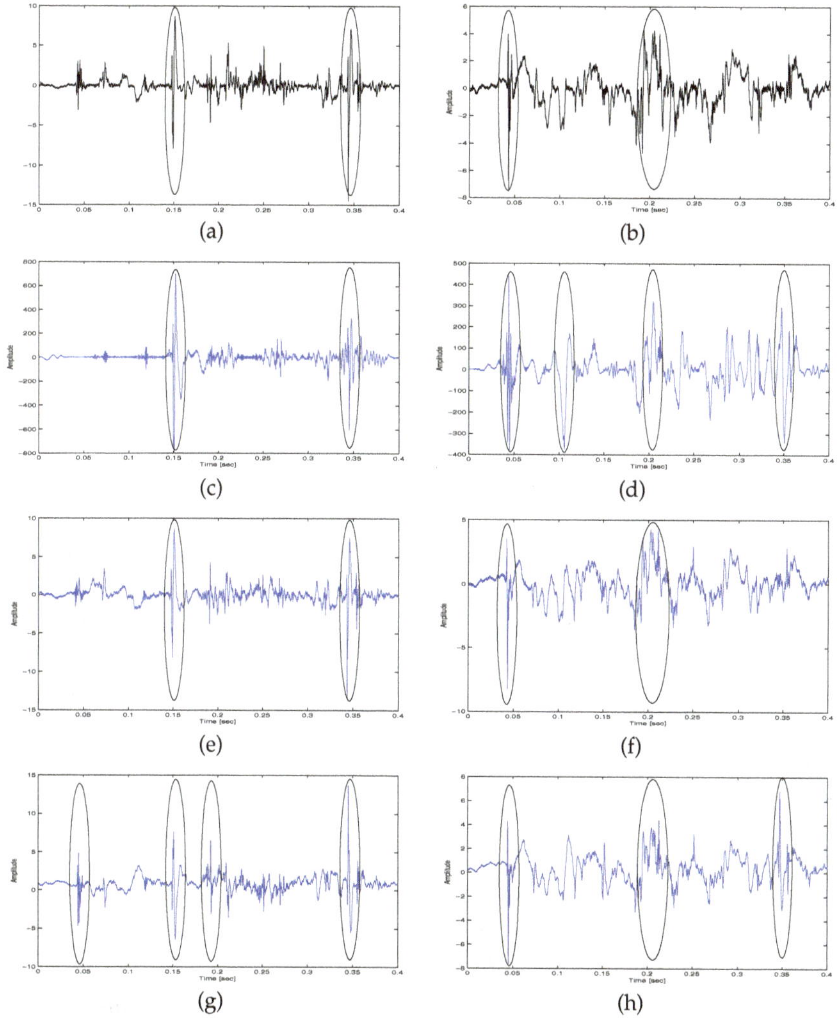

(a) (b) (c) (d) (e) (f) (g) (h)

Fig. 7. The results of (a)(b) the proposed ICA based source extraction method; (c-d) DUET method in (Widmalm et al., 2002); (e-f) Infomax approach in (Guo et al., 1999); (g-h) FastICA approach in Hyvärinen et al (2000) on real TMJ recordings in Fig. 5.

0.5853 for the simulated TMJ signals and the MI value for the real TMJ recordings are 0.2770, and the respective average mutual information for the pair of source signals estimated by various methods are summarized in Table 1. It can be seen that, as compared to the existing approaches, this value for the source signals estimated by the proposed method is much lower than the value of mutual information of the mixture signals. Also, for the real TMJ recordings, the lower values of mutual information between the pair of estimates as obtained

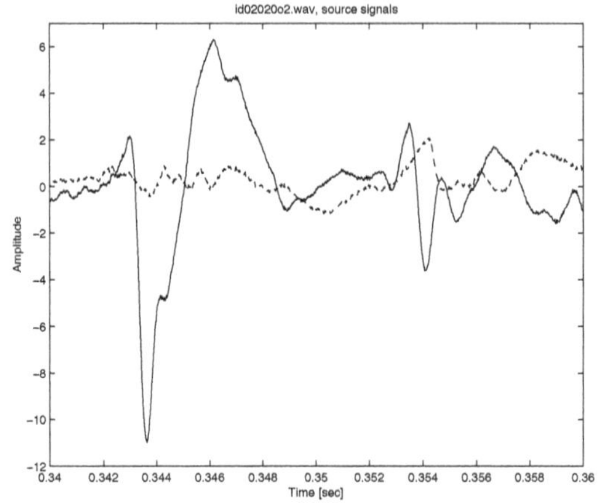

Fig. 8. The results of the proposed ICA based source extraction method (near 0.35s).

Method	Proposed ICA	DUET	Infomax	FastICA
Simulated TMJ Mixture	0.1275	0.4771	0.1296	0.4949
Real TMJ Recording	0.0812	0.1928	0.0875	0.2670

Table 1. The mutual information of the recovered/separated source signals by the proposed ICA based BSS and other existing methods in both simulated and real experiments

by the proposed method show that this ICA based source separation scheme achieves a better degree of statistical independence between their respective estimates than the Infomax and DUET estimates. Similar low values of mutual information has been observed between the pair of Infomax and the proposed ICA based estimates, which shows that both methods are able to achieve high degree of statistical independence between their respective estimates. Nevertheless, as compared to the Infomax estimates, the proposed method estimates the click source more accurately as depicted in Fig. 4. The averaged values of Pearson correlation between the simulated source and the estimated source obtained by the proposed ICA method, the DUET method, the Infomax method, and the FastICA method are 0.7968, 0.4524, 0.6686, 0.3355, respectively. The highest value by the proposed method indicates the high resemblance between the estimated and the actual sources.

To assess the robustness of the present scheme with respect to the noise effect, the performance of the proposed ICA based BSS method is also evaluated in the presence of additive noise. Pink noise and white Gaussian noise at various signal to noise ratio (SNR) have been added to the simulated TMJ mixtures.

The separation quality based on the estimated unmixing matrix $\hat{\mathbf{W}}$ obtained from simulated TMJ mixture and the actual mixing matrix $\mathbf{H}$ for the noisy simulated TMJ mixtures has been computed as follow (Choi et al, 2000):

$$PI = \frac{1}{2}\left[\sum_{i=1}^{i=2}\left(\frac{\sum_{j=1}^{j=2}|G_{ij}|^2}{\max_{\kappa}|G_{i\kappa}|^2}-1\right)+\sum_{j=1}^{j=2}\left(\frac{\sum_{i=1}^{i=2}|G_{ij}|^2}{\max_{\kappa}|G_{\kappa j}|^2}-1\right)\right] \tag{10}$$

with $\kappa \in \{1,2\}$ and

$$\mathbf{G} = \hat{\mathbf{W}}\mathbf{H} = \begin{pmatrix} G_{11} & G_{12} \\ G_{21} & G_{22} \end{pmatrix}, \; \hat{\mathbf{W}} = \begin{pmatrix} W^{11} & W^{12} \\ W^{21} & W^{22} \end{pmatrix} \tag{11}$$

Since the effect of additional noise has been estimated and ideally excluded from the estimated unmixing matrix $\hat{\mathbf{W}}$, the resulting index value PI should therefore give ideally consistent value of 0 (i.e. equivalent to negative large value in dB) with changing SNR.

The plot of performance index as defined in Eq. (10) vs SNR of the simulated noisy TMJ click signals with additive white Gaussian noise is illustrated in Fig. 9. We could observe that the PI remains at relatively low value as the level of the noise increases for the proposed method. In compare to other methods, the proposed method produces a relatively smaller values of PI. This result shows the robustness of the proposed scheme in the presence of noise.

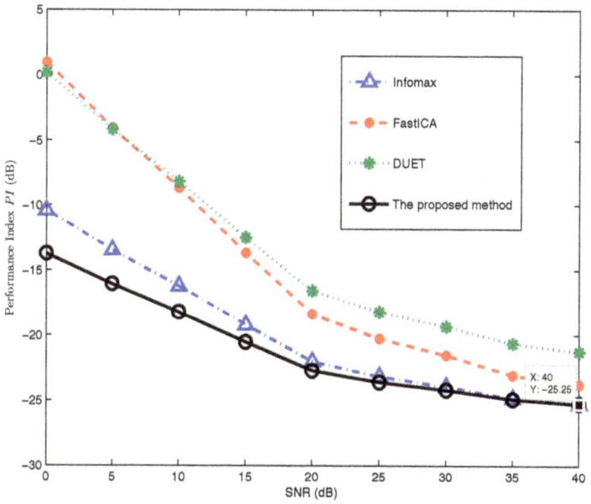

Fig. 9. Comparison of the noise resistance performance index PI by the proposed ICA based BSS method and other existing methods on simulated TMJ mixtures in the presence of white Gaussian noise.

The source separation performance is also evaluated by measuring the SNR gain of the extracted signals with respect to the original simulated source signals at various noise levels(Li et al, 2009). This SNR gain $\triangle SNR = SNR_{est} - SNR_{mix}$ of the estimated source signal is measured with

$$SNR_{est} = 10\log_{10}\frac{\sum_t s^2(t)}{\sum_t (s(t)-\hat{s}(t))^2} \tag{12}$$

and

$$SNR_{mix} = 10\log_{10}\frac{\sum_t s^2(t)}{\sum_t (s(t)-x(t))^2} \tag{13}$$

where $s(t)$ is the clean source signal prior to mixing, $\hat{s}(t)$ is the estimated source signal, and $x(t)$ is the mixture signal. The resulting gain is summarized in Fig. 10. Since the evaluation is performed on the two-source condition, the average value of $\triangle SNR$ for two sources is adopted. As compared to other methods, the consistent higher SNR_{est} with decreasing SNR produced by the proposed method verifies the noise resistance of the proposed source extraction scheme from another aspect. Furthermore, the proposed method is more resistant to white Gaussian noise than it to pink noise by providing higher SNR gain value at low SNR.

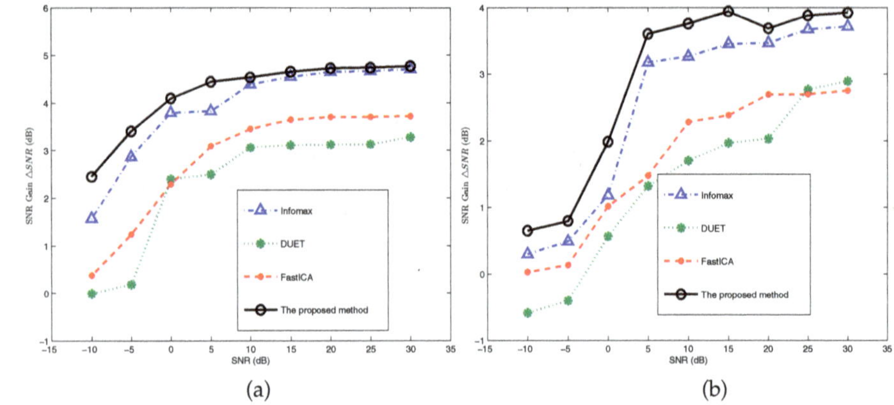

Fig. 10. The results of ICA based BSS method on simulated TMJ mixtures in the presence of (a) white Gaussian noise; and (b) pink noise.

9. Summary and discussion

In this article, it is shown that how ICA could play a vital role in order to develop a cheap, efficient and reliable diagnostic tool for the detection of temporomandibular joint disorders (TMD). The sounds from the temporomandibular joint (TMJ) are recorded using a pair of microphones inserted in the auditory canals. However, the TMJ sounds originating from one side of head can also be picked up by microphone at the other side. The presented ICA based method ensures that the signals used for the subsequent analysis are the actual source signals, and not contaminated by the sounds propagated from the contra side. The challenge of allocating the TMJ sound sources with respect to each side of the head has therefore been solved which provides an efficient non-invasive and non-intrusive procedure for TMD diagnosis.

The detailed technical issues together with elaborative results, quantitative evaluation as well as subsequent analysis have been presented in this chapter. In compared to the Infomax approach, the proposed ICA based method with adaptive parameters gives a better source separation performance and a higher noise resistance. Unlike any existing papers, the assumption that the two TMJ sound sources are non-overlapped in time-frequency (TF) domain has been removed here, and a more generalized mixing is considered by including the challenging source separation problem of two abnormal TMJ sounds which might be overlapped in TF domain. The proposed signal processing technique then allows for an enhanced clinical utility and an automated approach to the diagnosis of TMD.

10. Future directions

Since the anechoic head model is assumed in this paper, no reverberation of the sound sources is considered. A more generalized transmission models for the TMJ sounds would thus be discussed and compared together with the effects of the model parameters on the TMJ analysis in our future work. Although the general trend of blind source separation research makes use of linear models for source separation, it could also be possible to consider non-linear models for some complex systems (Ming et al, 2008) with additional constraints and assumptions. This topic therefore remains as a developing research area with a lot of potential for real-life.

Another interesting aspect of the problem could be the use of noise extracted from TMJ for the analysis of different types of non-stationarity to identify temporomandibular disorder (Ghodsi et al, 2009). Furthermore, it can be considered to extent ICA methods for 2D images in order to develop imaging methods in the diagnosis of temporomandibular joint disorders (Tvrdy, 2007). Besides, it could also be possible to further improve the TMD analysis performance assisted by a postprocessing scheme (Parikh, 2011). We would thus combine/fuse sound with the visual data such as those related to facial movement which are comfortable and safe to acquire, in order to further help the audio analysis to characterize TMD.

11. References

R. J. M. Gray, S. J. Davies, & A. A. Quayle (1995). Temporomandibular Disorders: A Clinical Approach, 1st ed. *London, U.K.: British Dental Association.*

C. L. Pankhurst(1997). Controversies in the aetiology of temporomandibular disorders, Part 1: Temporomandibular disorders all in the mind. *Primary Dental Care*, Vol. 3, 1997, 1–6.

S. A. Berman, A. Chaudhary, & J. Appelbaum (1997). Temporomandibular disorders. *Emedicine*. Available: *http://www.emedicine.com/neuro/topic366.htm*, Jun. 2006 [Online].

S. E. Widmalm, P. L. Westeesson, S. L. Brooks, M. P. Hatala, & D. Paesani (1992). Temporomandibular joint sounds: correlation to joint structure in fresh autopsy specimens. *American Journal of Orthodontics and Dentofacial Orthopedics*, Vol. 101, 1992, 60–69.

S. E. Widmalm, W. J. Williams, & C. Zheng (1991). Time frequency distributions of TMJ sounds. *J Oral Rehab*, Vol. 18, 1991, 403–412.

L. Cohen (1995) *Time-frequency analysis*. Englewood Cliffs, NJ: Prentice Hall, 1995.

S. E. Widmalm, W. J. Williams, R. L. Christiansen, S. M. Gunn, & D. K. Park (1996). Classification of temporomandibular joint sounds based upon there reduced interference distributions. *Journal of Oral Rehabilitation*, Vol. 23, 1996, 35–43.

A. Akan & R. B. Unsal (2000). Time-frequency analysis and classification of temporomandibular joint sounds. *J Franklin Inst Special issue on time-frequency signal analysis and applicatio*, Vol. 337, No. 4, 2000, 437–451.

W. D. OB́rien & Y. Liu (2005). Evaluation of acoustic propagation paths into the human head. *NATO Res. Technol. Organization: New Directions for Improving Audio Effectiveness*. Vol. 15, Apr. 2005, 1–24.

Y. Guo, F. Sattar, & C. Koh (1999). Blind separation of temporomandibular joint sound signals. *Proc. IEEE Int. Conf. Acoust. Speech Signal Process*. Vol. 2, 1999, 1069–1072.

C. C. Took, S. Sanei, S. Rickard, J. Chambers, & S. Dunne (2008). Fractional delay estimation for blind source separation and localization of Temporomandibular joint sounds. *IEEE Trans Biomed Eng*. Vol. 55, No. 3, 2008, 949–956.

A.J. Bell & T.J. Sejnowski(1995). An information maximisation approach to blind separation. *Neural Computation*, vol. 7, 1995, pp. 1129–1159.

K. Torkkola(1996). Blind separation of convolved sources based on information maximization. *IEEE Workshop Neural Networks for Signal Processing*, Kyoto, Japan, 1996.

S. E. Widmalm, W. J. Williams, K. H. Koh, E. T. Koh, E. K. Chua, K. B. C. Tan, A. U. J. Yap, & S. B. Keng (1997). Comparison of bilateral recording of TMJ joint sounds in TMJ disorders. *Int. Conf. Biomedical Engineering*, Singapore, 1997.

S. E. Widmalm, W. J. Williams, B. K. Ang, & D. C. Mckay (2002). Localization of TMJ sounds to side. *Oral Rehabil*, Vol. 29, 2002, 911–917.

A. Hyvärinen (1999). Survey on Independent Component Analysis. *Helsinki university of Technology Laboratory of Computer and Information Science*,1999.

A. Hyvärinen (2004). Principle of ICA estimation. *http://www.cis.hut.fi*, 2004.

S. E. Widmalm, W. J. Williams, D. Djurdjanovic, & D. C. Mckay (2003). The frequency range of TMJ sounds. *Oral Rehabil*, Vol. 30, 2003, 335–346.

D. M. Watt (1980). Temporomandibular joint sounds. *Journal of Dentistry*, Vol. 8, No. 2, 1980, 119–127.

B. Ella & H. Aapo (2002). A fast fixed-point algorithm for independent component analysis of complex valued signals. *Neural Networks Research Centre*, Helsinki University of Technology, 2000.

A. Hyvärinen & E. Oja (2000). Independent Component Analysis: Algorithms and Applications. *Neural Networks*, Vol. 13, No. 4-5, 2000, 411–430.

S. Choi & A. Cichocki (2000). Blind separation of non-stationary sources in noisy mixture. *Electronics Letters*, Vol. 36, No. 9, 2000, 848–849.

Y. Li, J. Woodruff, & D. L. Wang (2009). Monaural musical sound separation based on pitch and common amplitude modulation. *IEEE Trans Audio Speech Language Process*, Vol. 17, No. 7, 2009, 1361–1371.

M. Ming, Y. Y.-lin, G. W.-Juan, Z. Q.-Hai, & X. W.-Hao (2008). Blind speech separation based on homotopy nonlinear model. *Int. Conf. MultiMedia and Information Technology*, 2008, 369–372.

M. Ghodsi, H. Hassani, S. Sanei, & Y. Hicks (2009). The use of noise information for detection of temporomandibular disorder. *Biomedical Signal Processing and Control*, Vol. 4, 2009, 79–85.

P. Tvrdy (2007). Methods of imaging in the diagnosis of temporomandibular joint disorders. *Biomed Pap Med Fac Univ Palacky Olomouc Czech Repub*, 2007.

D. N. Parikh & D. V. Anderson (2011). Blind source separation with perceptual post processing. *IEEE 2011 DSP/SPE Workshop*, 2011, 321–325.

4

Non-Negative Matrix Factorization with Sparsity Learning for Single Channel Audio Source Separation

Bin Gao and W.L. Woo
School of Electrical and Electronic Engineering, Newcastle University, England, United Kingdom

1. Introduction

1.1 Single channel source separation (SCSS)

In this chapter, the special case of instantaneous underdetermined source separation problem termed as single channel source separation (SCSS) is focused. In general case and for many practical applications (e.g. audio processing) only one-channel recording is available and in such cases conventional source separation techniques are not appropriate. This leads to the SCSS research area where the problem can be simply treated as one observation instantaneous mixed with several unknown sources:

$$y(t)=\sum_{i=1}^{N_s}x_i(t) \tag{1}$$

where $i=1,\dots,N_s$ denotes number of sources and the goal is to estimate the sources $x_i(t)$ when only the observation signal $y(t)$ is available. This is an underdetermined system of equation problem. Recently, new advances have been achieved in SCSS and this can be categorized either as *supervised* SCSS methods or *unsupervised* SCSS methods. For *supervised* SCSS methods, the probabilistic models of the source are trained as a prior knowledge by using some or the entire source signals. The mixture is first transformed into an appropriate representation, in which the source separation is performed. The source models are either constructed directly based on knowledge of the signal sources, or by learning from training data (e.g. using Gaussian mixture model construct source models either directly based on knowledge of signal sources, or by learning from isolated training data). In the inference stage, the models and data are combined to yield estimates of the sources. This category predominantly includes the frequency model-based SCSS methods [1, 2] where the prior bases are modeled in time-frequency domain (e.g. spectrogram or power spectrogram), and the underdetermined-ICA time model-based SCSS method [3] which the prior bases are modeled in time domain. For *unsupervised* SCSS methods, this denotes the separation of completely unknown sources without using additional training information. These methods typically rely on the assumption that the sources are non-redundant, and the methods are based on, for example, decorrelation, statistical independence, or the minimum description length principle. This category includes several widely used methods: Firstly, the CASA-based *unsupervised* SCSS methods [4] whose goal is to replicate the process of human auditory system by exploiting signal processing approaches (e.g. notes in music recordings) and grouping them into auditory streams using psycho-acoustical cues. Secondly, the subspace technique based *unsupervised* SCSS methods using NMF [5, 6] or independent subspace analysis (ISA) [7] which usually factorizes the spectrogram of the input signal into elementary components. Of special interest, EMD [8] based *unsupervised* SCSS methods which can separate audio mixed signal in time domain and recover sources by combing other data analysis tools, e.g. independent component analysis (ICA) [9] or principle component analysis (PCA).

1.2 Unsupervised SCSS using NMF

In this book chapter, we propose a new NMF method for solving *unsupervised* SCSS problem. In a conventional NMF, given a data matrix $\mathbf{Y} = [\mathbf{y}_1, \ldots, \mathbf{y}_L] \in \Re_+^{K \times L}$ with $\mathbf{Y}_{k,l} > 0$, NMF factorizes this matrix into a product of two non-negative matrices:

$$\mathbf{Y} \approx \mathbf{DH} \tag{2}$$

where $\mathbf{D} \in \Re_+^{K \times \ell}$ and $\mathbf{H} \in \Re_+^{\ell \times L}$ where K and L represent the total number of rows and columns in matrix $\mathbf{Y}$, respectively. If ℓ is chosen to be $\ell = L$, no benefit is achieved at all. Thus the idea is to determine $\ell < L$ so that the matrix $\mathbf{D}$ can be compressed and reduced to its integral components such as $\mathbf{D}_{K \times \ell}$ is a matrix containing a set of dictionary vectors, and $\mathbf{H}_{\ell \times L}$ is an encoding matrix that describes the amplitude of each dictionary vector at each time point. A popular approach to solve the NMF optimization problem is the multiplicative update (MU) algorithm by Lee and Seung [10]. The MU update rule for Least square (LS) distance is given by:

$$\mathbf{D} \leftarrow \mathbf{D} \bullet \frac{\mathbf{YH}^{\mathrm{T}}}{\mathbf{DHH}^{\mathrm{T}}} \text{ and } \mathbf{H} \leftarrow \mathbf{H} \bullet \frac{\mathbf{D}^{\mathrm{T}}\mathbf{Y}}{\mathbf{D}^{\mathrm{T}}\mathbf{DH}} \tag{3}$$

Multiplicative update-based families of parameterized cost functions such as the Beta divergence [11], and Csiszar's divergences [12] have also been presented as well. A sparseness constraint [13, 14] can be added to the cost function, and this can be achieved by regularization using the L_1-norm. Here, 'sparseness' refers to a representational scheme where only a few units (out of a large population) are effectively used to represent typical data vectors [15]. In effect, this implies most units taking values close to zero while only few take significantly non-zero values. Several other types of prior over $\mathbf{D}$ and $\mathbf{H}$ can be defined e.g. in [16, 17], it is assumed that the prior of $\mathbf{D}$ and $\mathbf{H}$ satisfy the exponential density and the prior for the noise variance is chosen as an inverse gamma density. In [18], Gaussian distributions are chosen for both $\mathbf{D}$ and $\mathbf{H}$. The model parameters and hyperparameters are adapted by using the Markov chain Monte Carlo (MCMC) [19-21]. In all cases, a fully Bayesian treatment is applied to approximate inference for both model parameters and hyperparameters. While these approaches increase the accuracy of matrix factorization, it only works efficient when large sample dataset is available. Moreover, it consumes significantly high computational complexity at each iteration to adapt the parameters and its hyperparameters. Regardless of the cost function and sparseness constraint being used, the standard NMF or SNMF models [22] are only satisfactory for solving source separation provided that the spectral frequencies of the analyzed audio signal do not change over time. However, this is not the case for many realistic audio signals. As a result, the spectral dictionary obtained via the NMF or SNMF decomposition is not adequate to capture the temporal dependency of the frequency patterns within the signal. The recently developed two-dimensional sparse NMF deconvolution (SNMF2D) model [23, 24] extends the NMF model to be a two-dimensional convolution of $\mathbf{D}$ and $\mathbf{H}$ where the spectral dictionary and temporal code are optimized using the least square cost function with sparse penalty:

$$C_{LS}: \ \frac{1}{2}\sum_{k,l}(\mathbf{Y}_{k,l}-\tilde{\mathbf{Z}}_{k,l})^2+\lambda f(\mathbf{H}) \tag{4}$$

for $\forall k \in K, \forall l \in L$ where $\tilde{\mathbf{Z}}=\sum_{\tau,\phi}\overset{\downarrow\phi}{\tilde{\mathbf{D}}^{\tau}}\overset{\rightarrow\tau}{\mathbf{H}^{\phi}}$, $\tilde{\mathbf{D}}^{\tau}_{k,d}=\mathbf{D}^{\tau}_{k,d}\Big/\sqrt{\sum_{\tau,k}(\mathbf{D}^{\tau}_{k,d})^2}$ and $f(\mathbf{H})$ can be any function with positive derivative such as $L_{\alpha}-norm\,(\alpha>0)$ given by $f(\mathbf{H})=\|\mathbf{H}\|_{\alpha}=\left(\sum_{\phi,d,l}\left|\mathbf{H}^{\phi}_{d,l}\right|^{\alpha}\right)^{1/\alpha}$. Here $\overset{\downarrow\phi}{\tilde{\mathbf{D}}^{\tau}}$ denotes the downward shift which moves each element in the matrix $\tilde{\mathbf{D}}^{\tau}$ down by ϕ rows, and $\overset{\rightarrow\tau}{\mathbf{H}^{\phi}}$ denotes the right shift which moves each element in the matrix $\mathbf{H}^{\phi}$ to the right by τ columns. The SNMF2D is effective in single channel audio source separation (SCASS) because it is able to capture both the temporal structure and the pitch change of an audio source. However, the drawbacks of SNMF2D originate from its lack of a generalized criterion for controlling the sparsity of $\mathbf{H}$. In practice, the sparsity parameter is set manually. When SNMF2D imposes uniform sparsity on all temporal codes, this is equivalent to enforcing each temporal code to be identical to a fixed distribution according to the selected sparsity parameter. In addition, by assigning the fixed distribution onto each individual code, this is equivalent to constraining all codes to be stationary. However, audio signals are non-stationary in the TF domain and have different temporal structure and sparsity. Hence, they cannot be realistically enforced by a fixed probability distribution. These characteristics are even more pronounced between different types of audio signals. In addition, since the SNMF2D introduces many temporal shifts, this will result in more temporal codes to deviate from the fixed distribution. In such situation, the obtained factorization will invariably suffer from either under- or over-sparseness which subsequently lead to ambiguity in separating the audio mixture. Thus, the above suggests that the present form of SNMF2D is still technically lacking and is not readily suited for SCASS especially mixtures involving different types of audio signals.

In this chapter, an adaptive sparsity two-dimensional non-negative matrix factorization is proposed. The proposed model allows: (*i*) overcomplete representation by allowing many spectral and temporal shifts which are not inherent in the NMF and SNMF models. Thus, imposing sparseness is necessary to give unique and realistic representations of the non-stationary audio signals. Unlike the SNMF2D, our model imposes sparseness on $\mathbf{H}$ element-wise so that *each individual code* has its own distribution. Therefore, the sparsity parameter can be individually optimized for each code. This overcomes the problem of under- and over-sparse factorization. (*ii*) Each sparsity parameter in our model is learned and adapted as part of the matrix factorization. This bypasses the need of manual selection as in the case of SNMF2D. The proposed method is tested on the application of single channel music separation and the results show that our proposed method can give superior separation performance.

The chapter is organized as follows: In Section II, the new model is derived. Experimental results coupled with a series of performance comparison with other NMF techniques are presented in Section III. Finally, Section IV concludes the paper.

2. Adaptive sparsity two-dimensional non-negative matrix factorization

In this section, we derive a new factorization method termed as the *adaptive sparsity* two-dimensional non-negative matrix factorization. The model is given by

$$\mathbf{Y} = \sum_{\tau=0}^{\tau_{\max}} \sum_{\phi=0}^{\phi_{\max}} \overset{\downarrow\phi}{\mathbf{D}^{\tau}} \overset{\rightarrow\tau}{\mathbf{H}^{\phi}} + \mathbf{V} = \sum_{d=1}^{d_{\max}} \sum_{\tau=0}^{\tau_{\max}} \sum_{\phi=0}^{\phi_{\max}} \overset{\downarrow\phi}{\mathbf{D}_d^{\tau}} \overset{\rightarrow\tau}{\mathbf{H}_d^{\phi}} + \mathbf{V} \tag{5}$$

where $\mathbf{H}^{\phi} \sim p\left(\mathbf{H}^{\phi} \mid \boldsymbol{\lambda}^{\phi}\right) = \prod_{d=1}^{d_{\max}} \prod_{l=1}^{l_{\max}} \lambda_{d,l}^{\phi} \exp\left(-\lambda_{d,l}^{\phi} \mathbf{H}_{d,l}^{\phi}\right)$. In (5), it is worth pointing out that *each individual element* in $\mathbf{H}^{\phi}$ is constrained to an exponential distribution with independent decay parameter $\lambda_{d,l}^{\phi}$. Here, $\mathbf{D}_d^{\tau}$ is the dth column of $\mathbf{D}^{\tau}$, $\mathbf{H}_d^{\phi}$ is the dth row of $\mathbf{H}^{\phi}$ and $\mathbf{V}$ is assumed to be independently and identically distributed (i.i.d.) as Gaussian distribution with noise having variance σ^2. The terms $d_{\max}$, $\tau_{\max}$, $\phi_{\max}$ and $l_{\max}$ are the maximum number of columns in $\mathbf{D}^{\tau}$, τ shifts, ϕ shifts and column length in $\mathbf{Y}$, respectively. This is in contrast with the conventional SNMF2D where $\lambda_{d,l}^{\phi}$ is simply set to a fixed constant i.e. $\lambda_{d,l}^{\phi} = \lambda$ for all d,l,ϕ. Such setting imposes uniform constant sparsity on all temporal codes $\mathbf{H}^{\phi}$ which enforces each temporal code to be identical to a fixed distribution according to the selected constant sparsity parameter. The consequence of this uniform constant sparsity has already been discussed in Section I. In Section III, we will present the details of the sparsity analysis for source separation and evaluate its performance against with other existing methods.

2.1 Formulation of the proposed adaptive sparsity NMF2D

To facilitate such spectral dictionaries with adaptive sparse coding, we first define $\mathbf{D} = \left[\mathbf{D}^0 \ \ \mathbf{D}^1 \ \ \cdots \ \ \mathbf{D}^{\tau_{\max}}\right]$, $\mathbf{H} = \left[\mathbf{H}^0 \ \ \mathbf{H}^1 \ \ \cdots \ \ \mathbf{H}^{\phi_{\max}}\right]$ and $\boldsymbol{\lambda} = \left[\boldsymbol{\lambda}^1 \boldsymbol{\lambda}^2 \cdots \boldsymbol{\lambda}^{\phi_{\max}}\right]$, and then choose a prior distribution $p(\mathbf{D},\mathbf{H})$ over the factors $\{\mathbf{D},\mathbf{H}\}$ in the analysis equation. The posterior can be found by using Bayes' theorem as

$$p\left(\mathbf{D},\mathbf{H} \middle| \mathbf{Y},\sigma^2,\boldsymbol{\lambda}\right) = \frac{p\left(\mathbf{Y} \middle| \mathbf{D},\mathbf{H},\sigma^2\right) p\left(\mathbf{D},\mathbf{H} \middle| \boldsymbol{\lambda}\right)}{P(\mathbf{Y})} \tag{6}$$

where the denominator is constant and therefore, the log-posterior can be expressed as

$$\log p\left(\mathbf{D},\mathbf{H}\middle|\mathbf{Y},\sigma^2,\boldsymbol{\lambda}\right)=\log p\left(\mathbf{Y}\mid\mathbf{D},\mathbf{H},\sigma^2\right)+\log p\left(\mathbf{D},\mathbf{H}\middle|\boldsymbol{\lambda}\right)+\text{const} \tag{7}$$

where $'\text{const}'$ denotes constant. The likelihood of the observations given $\mathbf{D}$ and $\mathbf{H}$ can be written[1] as:

$$p\left(\mathbf{Y}\mid\mathbf{D},\mathbf{H},\sigma^2\right)=\frac{1}{\sqrt{2\pi\sigma^2}}\exp\left[-\left\|\mathbf{Y}-\sum_d\sum_\tau\sum_\phi \overset{\downarrow\phi}{\mathbf{D}}{}_d^\tau \overset{\rightarrow\tau}{\mathbf{H}}{}_d^\phi\right\|_F^2 \Big/ 2\sigma^2\right] \tag{8}$$

where $\|.\|_F$ denotes the Frobenius norm. The second term in (7) consists of the prior distribution of $\mathbf{D}$ and $\mathbf{H}$ where they are jointly independent. Each element of $\mathbf{H}$ is constrained to be exponential distributed with independent decay parameters, namely,

$$p(\mathbf{H}\mid\boldsymbol{\lambda})=\prod_\phi\prod_d\prod_l \lambda_{d,l}^\phi \exp\left(-\lambda_{d,l}^\phi \mathbf{H}_{d,l}^\phi\right) \text{ so that } f(\mathbf{H})=\sum_{\phi,d,l}\lambda_{d,l}^\phi \mathbf{H}_{d,l}^\phi \tag{9}$$

Hence, the negative log likelihood serves as the cost function defined as:

$$\begin{aligned} L &\propto \frac{1}{2\sigma^2}\left\|\mathbf{Y}-\sum_d\sum_\tau\sum_\phi \overset{\downarrow\phi}{\mathbf{D}}{}_d^\tau \overset{\rightarrow\tau}{\mathbf{H}}{}_d^\phi\right\|_F^2 + f(\mathbf{H}) \\ &= \frac{1}{2\sigma^2}\left\|\mathbf{Y}-\sum_d\sum_\tau\sum_\phi \overset{\downarrow\phi}{\mathbf{D}}{}_d^\tau \overset{\rightarrow\tau}{\mathbf{H}}{}_d^\phi\right\|_F^2 + \sum_{\phi,d,l}\lambda_{d,l}^\phi \mathbf{H}_{d,l}^\phi \end{aligned} \tag{10}$$

The sparsity term $f(\mathbf{H})$ forms the L_1-norm regularization which is used to resolve the ambiguity by forcing all structure in $\mathbf{H}$ onto $\mathbf{D}$. Therefore, the sparseness of the solution in (9) is highly dependent on the regularization parameter $\lambda_{d,l}^\phi$.

2.1.1 Estimation of the dictionary and temporal code

In (10), each spectral dictionary was constrained to unit length. This can be easily satisfied by normalizing each spectral dictionary according to $\tilde{\mathbf{D}}_{k,d}^\tau = \mathbf{D}_{k,d}^\tau \Big/ \sqrt{\sum_{\tau,k}(\mathbf{D}_{k,d}^\tau)^2}$ for all $d\in[1,\ldots,d_{\max}]$. With this normalization, the two-dimensional convolution of the spectral dictionary and temporal codes is now represented as $\tilde{\mathbf{Z}}=\sum_d\sum_\tau\sum_\phi \overset{\downarrow\phi}{\tilde{\mathbf{D}}}{}_d^\tau \overset{\rightarrow\tau}{\mathbf{H}}{}_d^\phi$. The derivatives of (10) corresponding to $\mathbf{D}^\tau$ and $\mathbf{H}^\phi$ of the adaptive sparsity factorization model are given by:

[1] To avoid cluttering the notation, we shall remove the upper limits from the summation terms. The upper limits can be inferred from (5).

$$\mathbf{H}^{\phi} \leftarrow \mathbf{H}^{\phi} \bullet \frac{\sum_{\tau} \overset{\downarrow\phi^{\mathrm{T}}}{\tilde{\mathbf{D}}^{\tau}} \overset{\leftarrow\tau}{\mathbf{Y}}}{\sum_{\tau} \overset{\downarrow\phi^{\mathrm{T}}}{\tilde{\mathbf{D}}^{\tau}} \overset{\leftarrow\tau}{\tilde{\mathbf{Z}}} + \boldsymbol{\lambda}^{\phi}} \tag{11}$$

$$\mathbf{D}^{\tau} \leftarrow \tilde{\mathbf{D}}^{\tau} \bullet \frac{\sum_{\phi} \overset{\uparrow\phi\rightarrow\tau^{\mathrm{T}}}{\mathbf{Y}\,\mathbf{H}^{\phi}} + \tilde{\mathbf{D}}^{\tau} diag\left(\sum_{\tau} \mathbf{1}\left(\left(\overset{\uparrow\phi\rightarrow\tau^{\mathrm{T}}}{\tilde{\mathbf{Z}}\,\mathbf{H}^{\phi}}\right) \bullet \tilde{\mathbf{D}}^{\tau}\right)\right)}{\sum_{\phi} \overset{\uparrow\phi\rightarrow\tau^{\mathrm{T}}}{\tilde{\mathbf{Z}}\,\mathbf{H}^{\phi}} + \tilde{\mathbf{D}}^{\tau} diag\left(\sum_{\tau} \mathbf{1}\left(\left(\overset{\uparrow\phi\rightarrow\tau^{\mathrm{T}}}{\mathbf{Y}\,\mathbf{H}^{\phi}}\right) \bullet \tilde{\mathbf{D}}^{\tau}\right)\right)} \quad \text{where } \tilde{\mathbf{D}}^{\tau}_{k,d} = \frac{\mathbf{D}^{\tau}_{k,d}}{\sqrt{\sum_{\tau,k}\left(\mathbf{D}^{\tau}_{k,d}\right)^{2}}} \tag{12}$$

In (11), superscript 'T' denotes matrix transpose, '$\bullet$' is the element wise product and $diag(\cdot)$ denotes a matrix with the argument on the diagonal. The column vectors of $\mathbf{D}^{\tau}$ will be factor-wise normalized to unit length.

2.1.2 Estimation of the adaptive sparsity parameter

Since $\overset{\rightarrow\tau}{\mathbf{H}^{\phi}}$ is obtained directly from the original sparse code matrix $\overset{\rightarrow 0}{\mathbf{H}^{\phi}}$, it suffices to compute just for the regularization parameters associated with $\overset{\rightarrow 0}{\mathbf{H}^{\phi}}$. Therefore, we can set the cost function in (10) with $\tau_{\max} = 0$ as

$$F(\mathbf{H}) = \frac{1}{2\sigma^{2}} \left\| Vec(\mathbf{Y}) - \sum_{\phi=0}^{\phi_{\max}} \left(\mathbf{I} \otimes \overset{\downarrow\phi}{\mathbf{D}}\right) Vec\left(\mathbf{H}^{\phi}\right) \right\|_{F}^{2} + \sum_{\phi=0}^{\phi_{\max}} \left(\underline{\boldsymbol{\lambda}}^{\phi}\right)^{\mathrm{T}} Vec\left(\mathbf{H}^{\phi}\right) \tag{13}$$

with $Vec(\cdot)$ represents the column vectorization, '$\otimes$' is the Kronecker product, and $\mathbf{I}$ is the identity matrix. Defining the following terms:

$$\underline{\mathbf{y}} = Vec(\mathbf{Y}) \;,\; \overline{\mathbf{D}} = \left[\mathbf{I} \otimes \overset{\downarrow 0}{\mathbf{D}}\; \mathbf{I} \otimes \overset{\downarrow 1}{\mathbf{D}} \cdots \mathbf{I} \otimes \overset{\downarrow \phi_{\max}}{\mathbf{D}}\right],$$
$$\underline{\mathbf{h}} = \begin{bmatrix} Vec(\mathbf{H}^{0}) \\ Vec(\mathbf{H}^{1}) \\ \vdots \\ Vec(\mathbf{H}^{\phi_{\max}}) \end{bmatrix} \;,\; \underline{\boldsymbol{\lambda}} = \begin{bmatrix} \underline{\boldsymbol{\lambda}}^{0} \\ \underline{\boldsymbol{\lambda}}^{1} \\ \vdots \\ \underline{\boldsymbol{\lambda}}^{\phi_{\max}} \end{bmatrix} \;,\; \underline{\boldsymbol{\lambda}}^{\phi} = \begin{bmatrix} \lambda^{\phi}_{1,1} \\ \lambda^{\phi}_{2,1} \\ \vdots \\ \lambda^{\phi}_{d_{\max}, l_{\max}} \end{bmatrix} \tag{14}$$

Thus, (13) can be rewritten in terms of $\underline{\mathbf{h}}$ as

$$F(\underline{\mathbf{h}}) = \frac{1}{2\sigma^{2}} \left\| \underline{\mathbf{y}} - \overline{\mathbf{D}}\underline{\mathbf{h}} \right\|_{F}^{2} + \underline{\boldsymbol{\lambda}}^{\mathrm{T}} \underline{\mathbf{h}} \tag{15}$$

Note that $\underline{\mathbf{h}}$ and $\underline{\boldsymbol{\lambda}}$ are vectors of dimension $R \times 1$ where $R = d_{\max} \times l_{\max} \times (\phi_{\max} + 1)$. To determine $\underline{\boldsymbol{\lambda}}$, we use the Expectation-Maximization (EM) algorithm and treat $\underline{\mathbf{h}}$ as the hidden variable where the log-likelihood function can be optimized with respect to $\underline{\boldsymbol{\lambda}}$. Using the Jensen's inequality, it can be shown that for any distribution $Q(\underline{\mathbf{h}})$, the log-likelihood function satisfies the following [25-27]:

$$\ln p\left(\underline{\mathbf{y}} \mid \underline{\boldsymbol{\lambda}}, \overline{\mathbf{D}}, \sigma^2\right) \geq \int Q(\underline{\mathbf{h}}) \ln\left(\frac{p\left(\underline{\mathbf{y}}, \underline{\mathbf{h}} \mid \underline{\boldsymbol{\lambda}}, \overline{\mathbf{D}}, \sigma^2\right)}{Q(\underline{\mathbf{h}})}\right) d\underline{\mathbf{h}} \tag{16}$$

One can easily check that the distribution that maximizes the right hand side of (16) is given by $Q(\underline{\mathbf{h}}) = p\left(\underline{\mathbf{h}} \mid \underline{\mathbf{y}}, \underline{\boldsymbol{\lambda}}, \overline{\mathbf{D}}, \sigma^2\right)$ which is the posterior distribution of $\underline{\mathbf{h}}$. In this paper, we represent the posterior distribution in the form of Gibbs distribution:

$$Q(\underline{\mathbf{h}}) = \frac{1}{Z_h} \exp\left[-F(\underline{\mathbf{h}})\right] \quad \text{where} \quad Z_h = \int \exp\left[-F(\underline{\mathbf{h}})\right] d\underline{\mathbf{h}} \tag{17}$$

The functional form of the Gibbs distribution in (17) is expressed in terms of $F(\underline{\mathbf{h}})$ and this is crucial as it will enable us to simplify the variational optimization of $\underline{\boldsymbol{\lambda}}$. The maximum likelihood estimation of $\underline{\boldsymbol{\lambda}}$ can be expressed by

$$\begin{aligned} \underline{\boldsymbol{\lambda}}^{ML} &= \arg\max_{\underline{\boldsymbol{\lambda}}} \ln p\left(\underline{\mathbf{y}} \mid \underline{\boldsymbol{\lambda}}, \overline{\mathbf{D}}, \sigma^2\right) \\ &= \arg\max_{\underline{\boldsymbol{\lambda}}} \int Q(\underline{\mathbf{h}}) \ln p(\underline{\mathbf{h}} \mid \underline{\boldsymbol{\lambda}}) d\underline{\mathbf{h}} \end{aligned} \tag{18}$$

Similarly,

$$\begin{aligned} \sigma^2_{ML} &= \arg\max_{\sigma^2} \int Q(\underline{\mathbf{h}}) \left(\ln p\left(\underline{\mathbf{y}} \mid \underline{\mathbf{h}}, \sigma^2, \overline{\mathbf{D}}\right) + \ln p(\underline{\mathbf{h}} \mid \underline{\boldsymbol{\lambda}})\right) d\underline{\mathbf{h}} \\ &= \arg\max_{\sigma^2} \int Q(\underline{\mathbf{h}}) \ln p\left(\underline{\mathbf{y}} \mid \underline{\mathbf{h}}, \sigma^2, \overline{\mathbf{D}}\right) d\underline{\mathbf{h}} \end{aligned} \tag{19}$$

Since each element of $\mathbf{H}$ is constrained to be exponential distributed with independent decay parameters, this gives $p(\underline{\mathbf{h}} \mid \underline{\boldsymbol{\lambda}}) = \prod_p \lambda_p \exp\left(-\lambda_p h_p\right)$ and therefore, (18) becomes:

$$\underline{\boldsymbol{\lambda}}^{ML} = \arg\max_{\underline{\boldsymbol{\lambda}}} \int Q(\underline{\mathbf{h}}) \left(\ln \lambda_p - \lambda_p h_p\right) d\underline{\mathbf{h}} \tag{20}$$

The Gibbs distribution $Q(\underline{\mathbf{h}})$ treats $\underline{\mathbf{h}}$ as the dependent variable while assuming all other parameters to be constant. As such, the functional optimization of $\underline{\boldsymbol{\lambda}}$ in (20) is obtained by differentiating the terms within the integral with respect to λ_p and the end result is given by

$$\lambda_p = \frac{1}{\int h_p Q(\underline{\mathbf{h}}) d\underline{\mathbf{h}}} \quad \text{for } p = 1,2,\ldots,R \tag{21}$$

where λ_p is the pth element of $\underline{\boldsymbol{\lambda}}$. Since $p\left(\underline{\mathbf{y}} \mid \underline{\mathbf{h}}, \overline{\mathbf{D}}, \sigma^2\right) = \frac{1}{\left(2\pi\sigma^2\right)^{N_0/2}} \exp\left(-\frac{1}{2\sigma^2}\left\|\underline{\mathbf{y}} - \overline{\mathbf{D}}\underline{\mathbf{h}}\right\|^2\right)$

where $N_o = K \times L$, the iterative update rule for σ_{ML}^2 is given by

$$\begin{aligned} \sigma_{ML}^2 &= \arg\max_{\sigma^2} \int Q(\underline{\mathbf{h}}) \left(-\frac{N_0}{2}\ln\left(2\pi\sigma^2\right) - \frac{1}{2\sigma^2}\left\|\underline{\mathbf{y}} - \overline{\mathbf{D}}\underline{\mathbf{h}}\right\|^2\right) d\underline{\mathbf{h}} \\ &= \frac{1}{N_0} \int Q(\underline{\mathbf{h}}) \left(\left\|\underline{\mathbf{y}} - \overline{\mathbf{D}}\underline{\mathbf{h}}\right\|^2\right) d\underline{\mathbf{h}} \end{aligned} \tag{22}$$

Despite the simple form of (21) and (22), the integral is difficult to compute analytically and therefore, we seek an approximation to $Q(\underline{\mathbf{h}})$. We note that the solution $\underline{\mathbf{h}}$ naturally partition its elements into distinct subsets $\underline{\mathbf{h}}_P$ and $\underline{\mathbf{h}}_M$ consisting of components $\forall p \in P$ such that $h_p = 0$, and components $\forall m \in M$ such that $h_m > 0$. Thus, the $F(\underline{\mathbf{h}})$ can be expressed as following:

$$\begin{aligned} F(\underline{\mathbf{h}}) &= \frac{1}{2\sigma^2}\left\|\underline{\mathbf{y}} - \overline{\mathbf{D}}_P\underline{\mathbf{h}}_P - \overline{\mathbf{D}}_M\underline{\mathbf{h}}_M\right\|_F^2 + \underline{\boldsymbol{\lambda}}_P^{\mathrm{T}}\underline{\mathbf{h}}_P + \underline{\boldsymbol{\lambda}}_M^{\mathrm{T}}\underline{\mathbf{h}}_M \\ &= \underbrace{\frac{1}{2\sigma^2}\left\|\underline{\mathbf{y}} - \overline{\mathbf{D}}_M\underline{\mathbf{h}}_M\right\|_F^2 + \underline{\boldsymbol{\lambda}}_M^{\mathrm{T}}\underline{\mathbf{h}}_M}_{F(\underline{\mathbf{h}}_M)} + \underbrace{\frac{1}{2\sigma^2}\left\|\underline{\mathbf{y}} - \overline{\mathbf{D}}_P\underline{\mathbf{h}}_P\right\|_F^2 + \underline{\boldsymbol{\lambda}}_P^{\mathrm{T}}\underline{\mathbf{h}}_P}_{F(\underline{\mathbf{h}}_P)} + \underbrace{\frac{1}{2\sigma^2}\left[2\left(\overline{\mathbf{D}}_M\underline{\mathbf{h}}_M\right)^{\mathrm{T}}\left(\overline{\mathbf{D}}_P\underline{\mathbf{h}}_P\right) - \left\|\underline{\mathbf{y}}\right\|^2\right]}_{G} \\ &= F(\underline{\mathbf{h}}_M) + F(\underline{\mathbf{h}}_P) + G \end{aligned} \tag{23}$$

In (23), the term $\left\|\underline{\mathbf{y}}\right\|^2$ in G is a constant and the cross-term $\left(\overline{\mathbf{D}}_M\underline{\mathbf{h}}_M\right)^{\mathrm{T}}\left(\overline{\mathbf{D}}_P\underline{\mathbf{h}}_P\right)$ measures the orthogonality between $\overline{\mathbf{D}}_M\underline{\mathbf{h}}_M$ and $\overline{\mathbf{D}}_P\underline{\mathbf{h}}_P$. where $\overline{\mathbf{D}}_P$ is the sub-matrix of $\overline{\mathbf{D}}$ that corresponds to $\underline{\mathbf{h}}_P$, $\overline{\mathbf{D}}_M$ is the sub-matrix of $\overline{\mathbf{D}}$ that corresponds to $\underline{\mathbf{h}}_M$. In this work, we intend to simply the expression in (23) by discounting the contribution from these terms and let $F(\underline{\mathbf{h}})$ be approximated as $F(\underline{\mathbf{h}}) \approx F(\underline{\mathbf{h}}_M) + F(\underline{\mathbf{h}}_P)$. Given this approximation, $Q(\underline{\mathbf{h}})$ can be decomposed as

$$\begin{aligned} Q(\underline{\mathbf{h}}) &= \frac{1}{Z_h}\exp\left[-F(\underline{\mathbf{h}})\right] \\ &\approx \frac{1}{Z_h}\exp\left[-\left(F(\underline{\mathbf{h}}_P) + F(\underline{\mathbf{h}}_M)\right)\right] \\ &= \frac{1}{Z_P}\exp\left[-F(\underline{\mathbf{h}}_P)\right]\frac{1}{Z_M}\exp\left[-F(\underline{\mathbf{h}}_M)\right] \\ &= Q_P(\underline{\mathbf{h}}_P)Q_M(\underline{\mathbf{h}}_M) \end{aligned} \tag{24}$$

with $Z_P = \int \exp\left[-F(\underline{\mathbf{h}}_P)\right] d\underline{\mathbf{h}}_P$ and $Z_M = \int \exp\left[-F(\underline{\mathbf{h}}_M)\right] d\underline{\mathbf{h}}_M$. Since $\underline{\mathbf{h}}_P = \underline{\mathbf{0}}$ is on the boundary of the distribution, this distribution is represented by using the Taylor expansion about the MAP estimate, $\underline{\mathbf{h}}^{MAP}$:

$$\begin{aligned} Q_P(\underline{\mathbf{h}}_P \geq 0) &\propto \exp\left\{-\left[\left(\frac{\partial F}{\partial \underline{\mathbf{h}}}\right)\Bigg|_{\underline{\mathbf{h}}^{MAP}}\right]_P^{\mathrm{T}} \underline{\mathbf{h}}_P - \frac{1}{2}\underline{\mathbf{h}}_P^{\mathrm{T}} \overline{\mathbf{\Lambda}}_P \underline{\mathbf{h}}_P\right\} \\ &= \exp\left[-\left(\overline{\mathbf{\Lambda}}\underline{\mathbf{h}}^{MAP} - \frac{1}{\sigma^2}\overline{\mathbf{D}}^{\mathrm{T}}\underline{\mathbf{y}} + \underline{\boldsymbol{\lambda}}\right)_P^{\mathrm{T}} \underline{\mathbf{h}}_P - \frac{1}{2}\underline{\mathbf{h}}_P^{\mathrm{T}} \overline{\mathbf{\Lambda}}_P \underline{\mathbf{h}}_P\right] \end{aligned} \tag{25}$$

where $\overline{\mathbf{\Lambda}}_P = \frac{1}{\sigma^2}\overline{\mathbf{D}}_P^{\mathrm{T}}\overline{\mathbf{D}}_P$, $\overline{\mathbf{\Lambda}} = \frac{1}{\sigma^2}\overline{\mathbf{D}}^{\mathrm{T}}\overline{\mathbf{D}}$. We perform variational approximation to $Q_P(\underline{\mathbf{h}}_P)$ by using the exponential distribution:

$$\hat{Q}_P(\underline{\mathbf{h}}_P \geq 0) = \prod_{p \in P} \frac{1}{u_p} \exp\left(-h_p / u_p\right) \tag{26}$$

The variational parameters $\underline{\mathbf{u}} = \{u_p\}$ for $\forall p \in P$ are obtained by minimizing the Kullback-Leibler divergence between Q_P and $\hat{Q}_P$:

$$\begin{aligned} \underline{\mathbf{u}} &= \arg\min_{\underline{\mathbf{u}}} \int \hat{Q}_P(\underline{\mathbf{h}}_P) \ln \frac{\hat{Q}_P(\underline{\mathbf{h}}_P)}{Q_P(\underline{\mathbf{h}}_P)} d\underline{\mathbf{h}}_P \\ &= \arg\min_{\underline{\mathbf{u}}} \int \hat{Q}_P(\underline{\mathbf{h}}_P)\left[\ln \hat{Q}_P(\underline{\mathbf{h}}_P) - \ln Q_P(\underline{\mathbf{h}}_P)\right] d\underline{\mathbf{h}}_P \end{aligned} \tag{27}$$

In Eqn. (27).

$$\begin{aligned} &\int \hat{Q}_P(\underline{\mathbf{h}}_P) \ln\left[\hat{Q}_P(\underline{\mathbf{h}}_P)\right] d\underline{\mathbf{h}}_P = \sum_{p \in P} \int \hat{Q}_P(h_p) \ln\left[\hat{Q}_P(h_p)\right] dh_p \\ &= \sum_{p \in P} \int_0^\infty dh_p \frac{1}{u_p} \exp\left(-h_p / u_p\right)\left(-\ln u_p - h_p / u_p\right) \\ &= -\sum_{p \in P} \ln u_p \int_0^\infty d\left(\frac{h_p}{u_p}\right) \exp\left(-h_p / u_p\right) - \sum_{p \in P} \int_0^\infty d\left(\frac{h_p}{u_p}\right) \frac{h_p}{u_p} \exp\left(-h_p / u_p\right) \\ &= -\sum_{p \in P} \ln u_p + 1 \end{aligned} \tag{28}$$

and

$$\begin{aligned}&\int \hat{Q}_P(\underline{\mathbf{h}}_P)\ln\left[Q_P(\underline{\mathbf{h}}_P)\right]d\underline{\mathbf{h}}_P\\&=-\int d\underline{\mathbf{h}}_P\left[\left(\overline{\boldsymbol{\Lambda}}\underline{\mathbf{h}}^{MAP}-\frac{1}{\sigma^2}\overline{\mathbf{D}}^{\mathrm{T}}\underline{\mathbf{y}}+\underline{\boldsymbol{\lambda}}\right)_P^{\mathrm{T}}\underline{\mathbf{h}}_P+\frac{1}{2}\underline{\mathbf{h}}_P^{\mathrm{T}}\overline{\boldsymbol{\Lambda}}_P\underline{\mathbf{h}}_P\right]\hat{Q}_P(\underline{\mathbf{h}}_P)\\&=-\sum_{p\in P,m\in M}\frac{1}{2}\left(\overline{\boldsymbol{\Lambda}}\right)_{pm}\left\langle h_p h_m\right\rangle-\sum_{p\in P}\left(\overline{\boldsymbol{\Lambda}}\underline{\mathbf{h}}^{MAP}-\frac{1}{\sigma^2}\overline{\mathbf{D}}^{\mathrm{T}}\underline{\mathbf{y}}+\underline{\boldsymbol{\lambda}}\right)_p\left\langle h_p\right\rangle\end{aligned} \tag{29}$$

with $\langle\bullet\rangle$ denotes the expectation under $\hat{Q}_P(\underline{\mathbf{h}}_P)$ distribution [28] such that $\left\langle h_p h_m\right\rangle = u_p u_m$ and $\left\langle h_p\right\rangle = u_p$ which leads to:

$$\min_{u_p}\hat{\underline{\mathbf{b}}}_P^{\mathrm{T}}\underline{\mathbf{u}}+\frac{1}{2}\underline{\mathbf{u}}^{\mathrm{T}}\hat{\boldsymbol{\Lambda}}\underline{\mathbf{u}}-\sum_{p\in P}\ln u_p \tag{30}$$

where $\hat{\underline{\mathbf{b}}}_P=\left(\overline{\boldsymbol{\Lambda}}\underline{\mathbf{h}}^{MAP}-\frac{1}{\sigma^2}\overline{\mathbf{D}}^{\mathrm{T}}\underline{\mathbf{y}}+\underline{\boldsymbol{\lambda}}\right)_P$ and $\hat{\boldsymbol{\Lambda}}=\overline{\boldsymbol{\Lambda}}_P+diag\left(\overline{\boldsymbol{\Lambda}}_P\right)$. The optimization of (30) can be accomplished be expanding (30) as follows:

$$G(\underline{\mathbf{u}},\underline{\tilde{\mathbf{u}}}) = \hat{\underline{\mathbf{b}}}_P^{\mathrm{T}}\underline{\mathbf{u}}+\frac{1}{2}\sum_{p\in P}\frac{\left(\hat{\boldsymbol{\Lambda}}\underline{\tilde{\mathbf{u}}}\right)_p}{\tilde{u}_p}u_p^2-\sum_{p\in P}\ln u_p \tag{31}$$

Taking the derivative of $G(\underline{\mathbf{u}},\underline{\tilde{\mathbf{u}}})$ in (31) with respect to $\underline{\mathbf{u}}$ and setting it to be zero, we have:

$$\frac{\left(\hat{\boldsymbol{\Lambda}}\underline{\tilde{\mathbf{u}}}\right)_p}{\tilde{u}_p}u_p+\hat{b}_p-\frac{1}{u_p}=0 \tag{32}$$

The above equation is equivalent to the following quadratic equations:

$$\frac{\left(\hat{\boldsymbol{\Lambda}}\underline{\tilde{\mathbf{u}}}\right)_p}{\tilde{u}_p}u_p^2+\hat{b}_p u_p-1=0 \tag{33}$$

Solving (33) for u_p leads to the following update:

$$u_p\leftarrow u_p\frac{-\hat{b}_p+\sqrt{\hat{b}_p^2+4\frac{\left(\hat{\boldsymbol{\Lambda}}\underline{\mathbf{u}}\right)_p}{\tilde{u}_p}}}{2\left(\hat{\boldsymbol{\Lambda}}\underline{\mathbf{u}}\right)_p} \tag{34}$$

As for components $\underline{\mathbf{h}}_M$, since none of the non-negative constraints are active, we approximate $Q_M(\underline{\mathbf{h}}_M)$ as unconstrained Gaussian with mean $\underline{\mathbf{h}}_M^{MAP}$. Thus using the factorized approximation $Q(\underline{\mathbf{h}})=\hat{Q}_P(\underline{\mathbf{h}}_P)Q_M(\underline{\mathbf{h}}_M)$ in (21), we obtain the following:

$$\lambda_p = \begin{cases} \dfrac{1}{\int h_p Q_M(\underline{\mathbf{h}}_M)\, d\underline{\mathbf{h}}_M} = \dfrac{1}{h_p^{MAP}} & \text{if } p \in M \\ \dfrac{1}{\int h_p \hat{Q}_P(\underline{\mathbf{h}}_P)\, d\underline{\mathbf{h}}_P} = \dfrac{1}{u_p} & \text{if } p \in P \end{cases} \tag{35}$$

for $p = 1, 2, \ldots, R$ and h_p^{MAP} is the pth element of sparse code $\underline{\mathbf{h}}_P$ computed from (11) and its covariance $\mathbf{C}$ is given by

$$C_{pm} = \begin{cases} \left(\overline{\boldsymbol{\Lambda}}_P^{-1}\right)_{pm} & \text{if } p, m \in M \\ u_p^2 \delta_{pm} & \text{Otherwise} \end{cases} \tag{36}$$

Thus, the update rule for σ^2 computed from (22) can be obtained as

$$\sigma^2 = \frac{1}{N_0}\left[\left(\underline{\mathbf{y}} - \overline{\mathbf{D}}\hat{\underline{\mathbf{h}}}\right)^{\mathrm{T}}\left(\underline{\mathbf{y}} - \overline{\mathbf{D}}\hat{\underline{\mathbf{h}}}\right) + \mathrm{Tr}\left(\overline{\mathbf{D}}^{\mathrm{T}}\overline{\mathbf{D}}\mathbf{C}\right)\right] \text{ where } \hat{h}_p = \begin{cases} h_p^{MAP} & \text{if } p \in M \\ u_p & \text{if } p \in P \end{cases} \tag{37}$$

The specific steps of the proposed method can be summarized as the following table:

1. Initialize $\mathbf{D}^{\tau}$ and $\mathbf{H}^{\phi}$ with nonnegative random values.
2. Define $\tilde{\mathbf{D}}_{k,d}^{\tau} = \mathbf{D}_{k,d}^{\tau} \Big/ \sqrt{\sum_{\tau,k} (\mathbf{D}_{k,d}^{\tau})^2}$ and Compute $\tilde{\mathbf{Z}} = \sum_d \sum_\tau \sum_\phi \overset{\downarrow\phi}{\tilde{\mathbf{D}}_d^{\tau}} \overset{\rightarrow\tau}{\mathbf{H}_d^{\phi}}$.
3. Assign $\lambda_p = \begin{cases} \dfrac{1}{h_p^{MAP}} & \text{if } p \in M \\ \dfrac{1}{u_p} & \text{if } p \in P \end{cases}$.
4. Assign $\sigma^2 = \dfrac{1}{N_0}\left[\left(\underline{\mathbf{y}} - \overline{\mathbf{D}}\hat{\underline{\mathbf{h}}}\right)^{\mathrm{T}}\left(\underline{\mathbf{y}} - \overline{\mathbf{D}}\hat{\underline{\mathbf{h}}}\right) + \mathrm{Tr}\left(\overline{\mathbf{D}}^{\mathrm{T}}\overline{\mathbf{D}}\mathbf{C}\right)\right]$.
5. Update $\mathbf{H}^{\phi} \leftarrow \mathbf{H}^{\phi} \bullet \dfrac{\sum_\tau \overset{\downarrow\phi^{\mathrm{T}}}{\tilde{\mathbf{D}}^{\tau}} \overset{\leftarrow\tau}{\mathbf{Y}}}{\sum_\tau \overset{\downarrow\phi^{\mathrm{T}}}{\tilde{\mathbf{D}}^{\tau}} \overset{\leftarrow\tau}{\tilde{\mathbf{Z}}} + \boldsymbol{\lambda}_p^{\phi}}$ and compute $\tilde{\mathbf{Z}} = \sum_d \sum_\tau \sum_\phi \overset{\downarrow\phi}{\tilde{\mathbf{D}}_d^{\tau}} \overset{\rightarrow\tau}{\mathbf{H}_d^{\phi}}$.
6. Update $\mathbf{D}^{\tau} \leftarrow \tilde{\mathbf{D}}^{\tau} \bullet \dfrac{\sum_\phi \overset{\uparrow\phi\rightarrow\tau^{\mathrm{T}}}{\mathbf{Y}\,\mathbf{H}^{\phi}} + \tilde{\mathbf{D}}^{\tau} diag\left(\sum_\tau \mathbf{1}\left(\left(\overset{\uparrow\phi\rightarrow\tau^{\mathrm{T}}}{\tilde{\mathbf{Z}}\,\mathbf{H}^{\phi}}\right) \bullet \tilde{\mathbf{D}}^{\tau}\right)\right)}{\sum_\phi \overset{\uparrow\phi\rightarrow\tau^{\mathrm{T}}}{\tilde{\mathbf{Z}}\,\mathbf{H}^{\phi}} + \tilde{\mathbf{D}}^{\tau} diag\left(\sum_\tau \mathbf{1}\left(\left(\overset{\uparrow\phi\rightarrow\tau^{\mathrm{T}}}{\mathbf{Y}\,\mathbf{H}^{\phi}}\right) \bullet \tilde{\mathbf{D}}^{\tau}\right)\right)}$.
7. Repeat steps 2 to 6 until convergence.

Table 1. Proposed Adaptive Sparsity NMF2D

3. Single channel audio source separation

3.1 TF representation

The classic spectrogram decomposes signals to components of linearly spaced frequencies. However, in western music, the typically used frequencies are geometrically spaced. Thus, obtaining an acceptable low-frequency resolution is absolutely necessary, while a resolution that is geometrically related to the frequency is desirable, although not critical. The constant Q transform as introduced in [29], tries to solve both issues. In general, the twelve-tone equal tempered scale which forms the basis of modern western music divides each octave into twelve half notes where the frequency ratio between each successive half note is equal [23]. The fundamental frequency of the note which is k_Q half note above can be expressed as $f^Q_{k_Q} = f_{\text{fund}} \cdot 2^{k_Q/24}$. Taking the logarithmic, this gives $\log f^Q_{k_Q} = \log f_{\text{fund}} + \frac{k_Q}{24}\log 2$. Thus, in a log-frequency representation the notes are linearly spaced. In our method, the frequency axis of the obtained spectrogram is logarithmically scaled and grouped into 175 frequency bins in the range of 50Hz to 8kHz (given $f_s = 16\text{kHz}$) with 24 bins per octave and the bandwidth follows the constant-Q rule. Figure 1 shows an example of the estimated spectral dictionary $\mathbf{D}$ and temporal code $\mathbf{H}$ based on SNMF2D method on the log-frequency spectrogram.

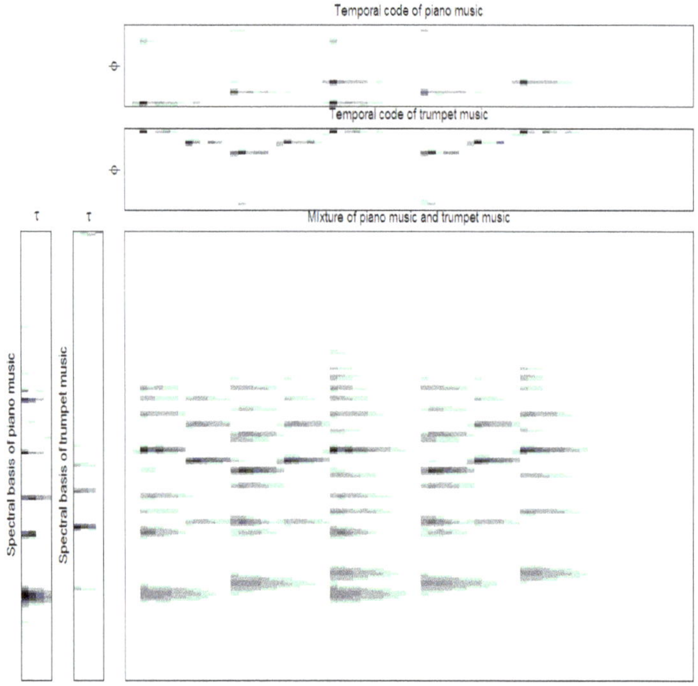

Fig. 1. The estimated spectral dictionary and temporal code of piano and trumpet mixture log-frequency spectrum using SNMF2D.

3.2 Source reconstruction

The Figure 2 shows the framework of the proposed *unsupervised* SCSS methods. The single channel audio mixture is constructed by several unknown sources, namely $y(t)=\sum_{d=1}^{d_{\max}} x_d(t)$. where $d=1,\ldots,d_{\max}$ denotes the sources number and $t=1,2,\ldots,T$ denotes the time index. The goal is to estimate the sources $x_d(t)$ when only the observation signal $y(t)$ is available. The mixture is then transformed into a suitable representation e.g. Time-Frequency (TF) representation. Thus the mixture $y(t)$ is given by $Y(f,t_s)=\sum_{d=1}^{d_{\max}} X_d(f,t_s)$ where $Y(f,t_s)$ and $X_d(f,t_s)$ denote the TF components obtained by applying the short time Fourier transform (STFT) on $y(t)$ and $x_d(t)$, respectively, e.g. $Y(f,t_s)=STFT\left(y(t)\right)$. The time slots are given by $t_s=1,2,\ldots,T_s$ while frequency bins by $f=1,2,\ldots,F$. Since each component is a function of t_s and f, we represent this as $\mathbf{Y}=\left[Y(f,t_s)\right]_{t_s=1,2,\ldots,T_s}^{f=1,2,\ldots,F}$ and $\mathbf{X}_d=\left[X_d(f,t_s)\right]_{t_s=1,2,\ldots,T_s}^{f=1,2,\ldots,F}$. The power spectrogram is defined as the squared magnitude STFT and hence, its matrix representation is given by $|\mathbf{Y}|^{.2}\approx\sum_{d=1}^{d_{\max}}|\mathbf{X}_d|^{.2}$ where the superscript '$\cdot$' represents element wise operation. The frequencies scale of power spectrogram $|\mathbf{Y}|^{.2}$ can be mapped into log-frequency scale which described in Section III A and this will result log-frequency power spectrogram $\left|\widehat{\mathbf{Y}}\right|^{.2}=\sum_{d=1}^{N_s}\left|\widehat{\mathbf{X}}_d\right|^{.2}$. The matrices we seek to determine are $\left\{\left|\widehat{\mathbf{X}}_d\right|^{.2}\right\}_{d=1}^{N_s}$ which will be obtained during the feature extraction process by using the proposed matrix factorization as $\left|\tilde{\mathbf{X}}_d\right|^{.2}=\sum_{\tau}\sum_{\phi}\overset{\downarrow\phi}{\mathbf{D}_d^{\tau}}\overset{\rightarrow\tau}{\mathbf{H}_d^{\phi}}$ where $\mathbf{D}_d^{\tau}$ and $\mathbf{H}_d^{\phi}$ are estimated using (11) and (12). Once these matrices are estimated, we form the dth binary mask according to $W_d(f,t_s)=1$ if $\left|\tilde{X}_d(f,t_s)\right|^{.2}>\left|\tilde{X}_j(f,t_s)\right|^{.2}$ $d\neq j$ and zero otherwise to approach source separation. Finally, the estimated time-domain sources are obtained as $\tilde{\mathbf{x}}_d=\xi^{-1}\left(\mathbf{W}_d\bullet\widehat{\mathbf{Y}}\right)$ where $\xi^{-1}(\bullet)$ denotes the inverse mapping of the log-frequency axis to the original frequency axis and followed by the inverse STFT back to the time domain. $\tilde{\mathbf{x}}_d=[\tilde{x}_d(1),\ldots,\tilde{x}_d(T)]^{\mathrm{T}}$ denotes the dth estimated audio sources in the time-domain.

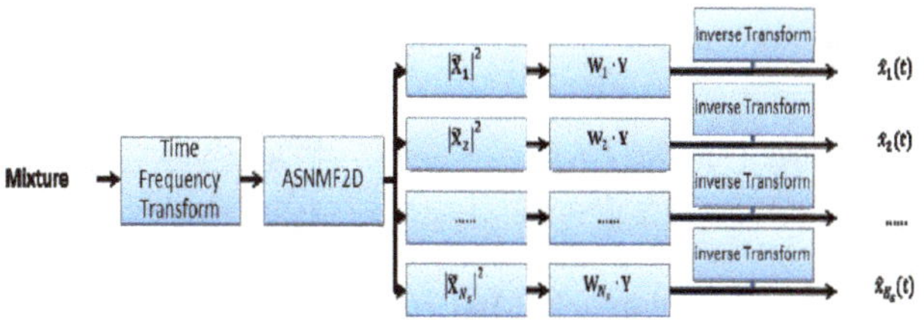

Fig. 2. A framework for the proposed *unsupervised* SCSS methods.

3.3 Efficiency of source extraction in TF domain

In this sub-section, we will analyze how different sparsity factorization methods impact on the source extraction performance in TF domain for SCASS. For separation, one generates the TF mask corresponding to each source and applies the generated mask to the mixture to obtain the estimated source TF representation. In particular, when the sources have no overlap in the TF domain, an optimum mask $W_d^{opt}(f,t_s)$ (optimal source extractor) exists which allows one to extract the dth original source from the mixture as

$$X_d(f,t_s) = W_d^{opt}(f,t_s)Y(f,t_s) \tag{38}$$

Given any TF mask $W_d(f,t_s)$ (source extractor) such that $0 \le W_d(f,t_s) \le 1$ for all (f,t_s), we define the efficiency of source extraction (ESE) in the TF domain for target source $x_d(t)$ in the presence of the interfering sources $\beta_d(t) = \sum_{j=1, j\neq d}^{d_{\max}} x_j(t)$ as

$$\psi(W_d) \triangleq \frac{\|W_d(f,t_s)X_d(f,t_s)\|_F^2}{\|X_d(f,t_s)\|_F^2} - \frac{\|W_d(f,t_s)B_d(f,t_s)\|_F^2}{\|X_d(f,t_s)\|_F^2} \tag{39}$$

where $X_d(f,t_s)$ and $B_d(f,t_s)$ are the TF representations of $x_d(t)$ and $\beta_d(t)$, respectively. The above represents the normalized energy difference between the extracted source and interferences. We also define the ESE of the mixture with respect to all the $d_{\max}$ sources as

$$\Omega = \frac{1}{d_{\max}} \sum_{d=1}^{d_{\max}} \psi(W_i) \tag{40}$$

Eqn. (39) is equivalent to measuring the ability of extracting the dth source $X_d(f,t_s)$ from the mixture $Y(f,t_s)$ given the TF mask $W_d(f,t_s)$. Eqn. (40) measures the ability of extracting all the $d_{\max}$ sources simultaneously from the mixture. To further study the ESE, we use the following two criteria [30]: (i) preserved signal ratio (PSR) which determines how well the mask preserves the source of interest and (ii) signal-to-interference ratio (SIR) which indicates how well the mask suppresses the interfering sources:

$$PSR_{W_d}^{X_d} \triangleq \frac{\|W_d(f,t_s)X_d(f,t_s)\|_F^2}{\|X_d(f,t_s)\|_F^2} \text{ and } SIR_{W_d}^{X_d} \triangleq \frac{\|W_d(f,t_s)X_d(f,t_s)\|_F^2}{\|W_d(f,t_s)B_d(f,t_s)\|_F^2} \tag{41}$$

Using (41), (39) can be expressed as $\psi(W_d) = PSR_{W_d}^{X_d} - PSR_{W_d}^{X_d} / SIR_{W_d}^{X_d}$. Analyzing the terms in (39), we have

$$PSR_{W_d}^{X_d} := \begin{cases} 1 & , \textit{if} \text{ supp } W_d^{opt} = \text{supp } W_d \\ <1 & , \textit{if} \text{ supp } W_d^{opt} \subset \text{supp } W_d \end{cases}$$
$$SIR_{W_d}^{X_d} := \begin{cases} \infty & , \textit{if} \text{ supp}[W_d X_d] \cap \text{supp} B_d = \varnothing \\ \textit{finite} & , \textit{if} \text{ supp}[W_d X_d] \cap \text{supp} B_d \neq \varnothing \end{cases} \tag{42}$$

where 'supp' denotes the support. When $\psi(W_d)=1$ (i.e. $PSR_{W_d}^{X_d}=1$ and $SIR_{W_d}^{X_d}=\infty$), this indicates that the mixture $y(t)$ is separable with respect to the dth source $x_d(t)$. In other words, $X_d(f,t_s)$ does not overlap with $B_d(f,t_s)$ and the TF mask $W_d(f,t_s)$ has perfectly separated the dth source $X_d(f,t_s)$ from the mixture $Y(f,t_s)$. This corresponds to $W_d(f,t_s)=W_d^{opt}(f,t_s)$ in (38). Hence, this is the maximum attainable $\psi(W_d)$ value. For other cases of $PSR_{W_d}^{X_d}$ and $SIR_{W_d}^{X_d}$, we have $\psi(W_d)<1$. Using the above concept, we can extend the analysis for the case of separating $d_{\max}$ sources. A mixture $y(t)$ is fully separable to all the N sources if and only if $\Omega=1$ in (40). For the case $\Omega<1$, this implies that some of the sources overlap with each other in the TF domain and therefore, they cannot be fully separated. Thus, Ω provides the quantitative performance measure to evaluate how separable the mixture is in the TF domain. In the following, we show the analysis of how different sparsity factorization methods affect the ESE of the mixture.

4. Results and analysis

4.1 Experiment set-up

The proposed method is tested by separating music sources. Several experimental simulations under different conditions have been designed to investigate the efficacy of the proposed method. All simulations and analyses are performed using a PC with Intel Core 2 CPU 6600 @ 2.4GHz and 2GB RAM. MATLAB is used as the programming platform. We have tested the proposed method in the wider types of music mixtures. All mixed signals are sampled at 16 kHz sampling rate. 30 music signals including 10 jazz, 10 piano and 10 trumpet signals are selected from the RWC [31] database. Three types of mixture have been generated: (i) jazz mixed with piano, (ii) jazz mixed with trumpet, (iii) piano mixed with trumpet. The sources are randomly chosen from the database and the mixed signal is generated by adding the chosen sources. In all cases, the sources are mixed with equal average power over the duration of the signals. The TF representation is computed by normalizing the time-domain signal to unit power and computing the STFT using 2048 point Hanning window FFT with 50% overlap. The frequency axis of the obtained spectrogram is then logarithmically scaled and grouped into 175 frequency bins in the range of 50Hz to 8kHz with 24 bins per octave. This corresponds to twice the resolution of the equal tempered musical scale. For the proposed adaptive sparsity factorization model, the convolutive components in time and frequency are selected to be (i) For piano and trumpet mixture $\tau=\{0,\ldots,3\}$ and $\phi=\{0,\ldots,31\}$, respectively; (ii) For piano and jazz mixture $\tau=\{0,\ldots,6\}$ and $\phi=\{0,\ldots,9\}$, respectively; (iii) For trumpet and jazz mixture $\tau=\{0,\ldots,6\}$ and $\phi=\{0,\ldots,9\}$, respectively. The corresponding sparse factor was determined by (35). We have evaluated our separation performance in terms of the signal-to-distortion ratio (SDR) which is one form of perceptual measure. This is a global measure that unifies source-to-interference ratio (SIR), source-to-artifacts ratio (SAR) and source-to-noise ratio (SNR). MATLAB routines for computing these criteria are obtained from the SiSEC'08 webpage [32, 33].

4.2 Impact of adaptive and fixed sparsity

In this implementation, we have conducted several experiments to compare the performance of the proposed method with SNMF2D under different sparsity regularization. In particular, Figures 3 and 4 show the separated sources by using the proposed method in terms of spectrogram and time-domain representation, respectively.

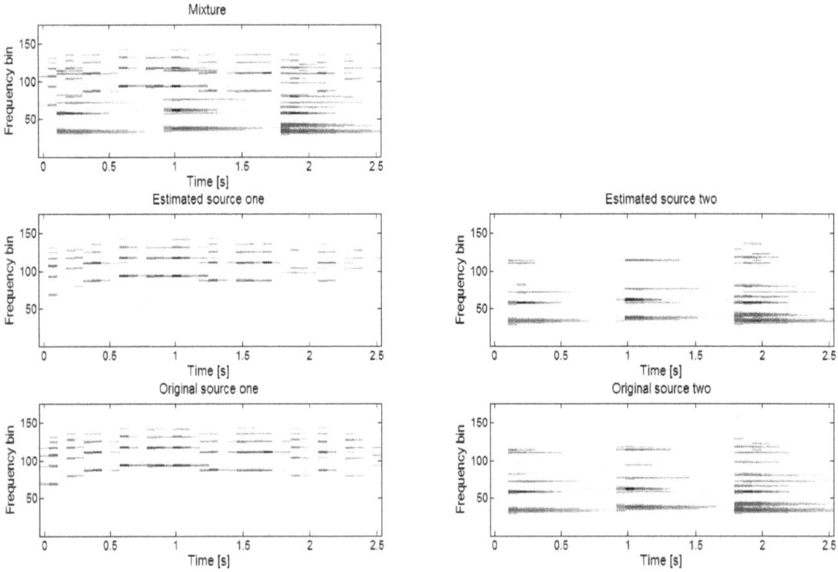

Fig. 3. Spectrogram of the mixed signal (top panel), the recovered trumpet music and piano music (middle panels) and original trumpet music and piano music (bottom panels).

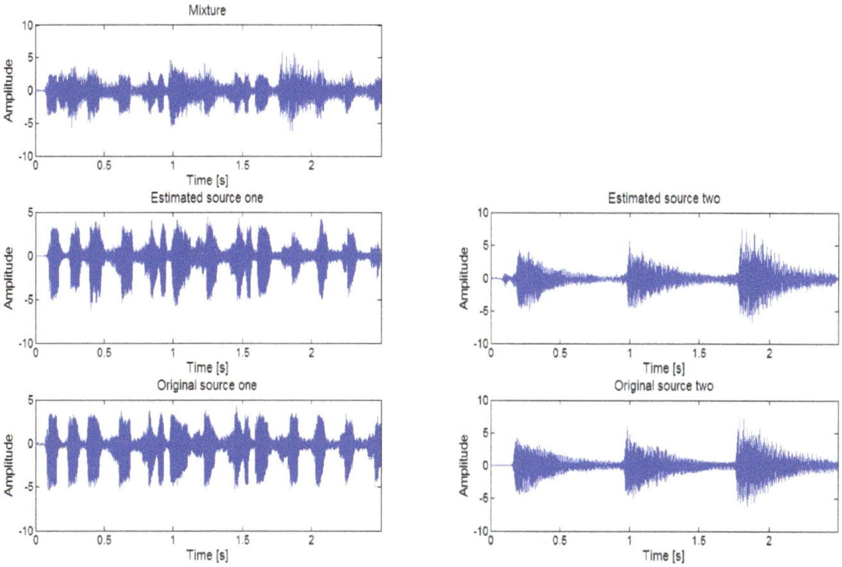

Fig. 4. Time domain of the mixed signal (top panel), the recovered trumpet music and piano music (middle panels) and original trumpet music and piano music (bottom panels).

To investigate this further, the impact of sparsity regularization on the separation results in terms of the SDR under different uniform regularization has been undertaken and the results are plotted in Figure 4. In this implementation, the uniform regularization is chosen as $c = 0, 0.5, \ldots, 10$ for all sparsity parameters i.e. $\lambda_{d,l}^{\phi} = \lambda = c$. The best result is retained and tabulated in Table I. In the case of the proposed method, it assigns a regularization parameter to each temporal code which is individually and adaptively tuned to yield the optimal number of times the spectral dictionary of a source recurs in the spectrogram. The sparsity on $\mathbf{H}_d^{\phi}$ is imposed *element-wise* in the proposed model so that each individual code in $\mathbf{H}_d^{\phi}$ is optimally sparse in the L_1-norm. In the conventional SNMF2D method, the sparsity is not fully controlled but is imposed uniformly on all the codes. The ensuing consequence is that the temporal codes are no longer optimal and this leads to 'under-sparse' or 'over-sparse' factorization which eventually results in inferior separation performance.

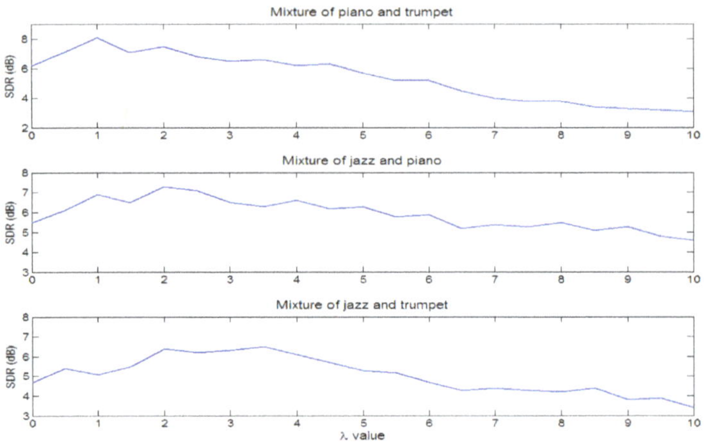

Fig. 5. Separation results of SNMF2D by using different uniform regularization.

In Figure 5, the results have clearly indicated that there are certain values of λ where the SNMF2D performs with exceptionally good results. In the case of piano and trumpet mixtures, the best performance is obtained when λ ranges from 0.5 to 2 where the highest SDR is 8.1dB. As for jazz and piano mixtures, the best performance is obtained when λ ranges from 1.0 to 2.5 where the highest SDR is 7.2dB and for jazz and trumpet mixtures, the best performance is obtained when λ ranges from 2 to 3.5 where the highest SDR is 8.6dB. On the contrary, when λ is set too high, the separation performance tends to degrade. It is also worth pointing out that the separation results are coarse when the factorization is non-regularized. Here, we see that (i) for piano and trumpet mixtures, the SDR is only 6.2dB, (ii) for jazz and piano mixtures, the SDR is only 5.6dB, (iii) for jazz and trumpet mixtures, the SDR is only 4.7dB. From above, it is evident that uniform sparsity scheme gives varying performance depending on the value of λ which in turn depends on the type of mixture. Hence, this poses a practical difficulty in selecting the appropriate level sparseness necessary for matrix factorization to resolve the ambiguity between the sources in the TF domain.

The overall comparison results between the adaptive and uniform sparsity methods have been summarized in Figure 6. According to the table, SNMF2D with adaptive sparsity tends to yield better result than the uniform sparsity-based methods. We may summarize the average performance improvement of our method against the uniform constant sparsity method: (i) For the piano and trumpet music, the improvement per source in terms of the SDR is 2dB (ii) For the piano and jazz music, the improvement per source in terms of SDR is 1.3dB. (iii) For the trumpet and jazz music, the improvement per source in terms of SDR is 1.1dB.

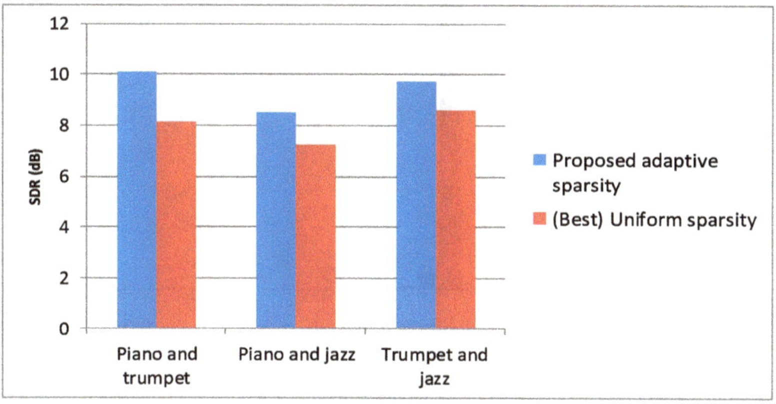

Fig. 6. SDR results comparison between adaptive and uniform sparsity methods.

4.2.1 Adaptive behavior of sparsity parameter

In this sub-section, the adaptive behavior of the sparsity parameters by using the proposed method will be demonstrated. Several sparsity parameters have been selected to illustrate its adaptive behavior. In the experiment, all sparsity parameters are initialized as $\lambda_{d,l}^{\phi}=5$ for all d,l,ϕ and are subsequently adapted according to (35). After 300 iterations, the sparsity parameters converge to their steady-states. We have plotted the histogram of the converged adaptive sparsity parameters in Figure 7. The figure suggests that the histogram can be represented as a bimodal distribution that each element code has its own sparseness. In addition, it is worth pointing out that in the case of piano and trumpet mixture the SDR result rises to 10dB when $\lambda_{d,l}^{\phi}$ is adaptive. This represents a 2dB per source improvement over the case of uniform constant sparsity (which is only 8.1dB in Figure 6). On the separate hand, when no sparsity is imposed onto the codes the SDR result immediately deteriorates to approximately 6dB. This represents a 4dB per source depreciation compared with the proposed adaptive sparsity method. From above, the results are ready to suggest that the performances of source separation have been undermined when the uniform constant sparsity scheme is used. On the contrary, improved performances can be obtained by allowing the sparsity parameters to be individually adapted for each element code. This is evident based on source separation performance as indicated in Figure 6.

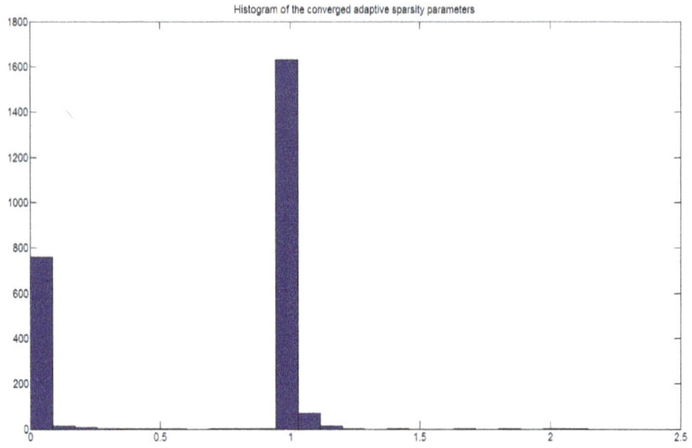

Fig. 7. The histogram of the converged adaptive sparsity parameter.

4.2.2 Efficiency of source extraction in TF domain

In this sub-section, we will analyze the efficiency of source extraction based on SNMF2D and the proposed method. Binary masks are constructed using the approach discussed in Section III B for each of the both methods. To ensure fair comparison, we generate the ideal binary mask (IBM) [34] from the original source which is used as a reference for comparison. The IBM for a target source is found for each TF unit by comparing the energy of the target source to the energy of all the interfering sources. Hence, the ideal binary mask produces the optimal signal-to-distortion ratio (SDR) gain of all binary masks and thus, it can be considered as an optimal source extractor in TF domain. The comparison results between IBM, uniform sparsity and proposed adaptive sparsity are tabulated in Table II.

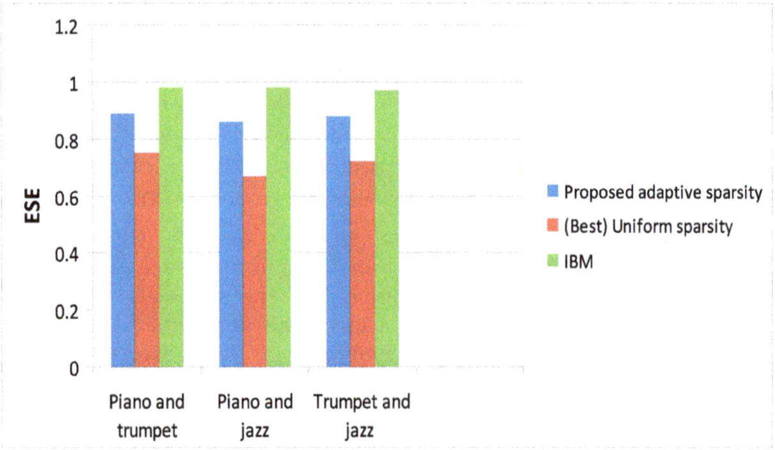

Fig. 8. Overall ESE performance

In Figure 8, the results of ESE for each mixture type are obtained by averaging over 100 realizations. From listening performance test, any $\psi(W_d) > 0.8$ indicates acceptable quality of source extraction performance in TF domain. Therefore, it is noted from the results in Figure 7 that both IBM and the proposed method satisfy this condition. In addition, the proposed method yields better ESE improvement against the uniform sparsity method. The average improvement results have been summarized as follows: (i) For the piano and trumpet music, 18.4%. (ii) For the piano and jazz music 26.5%. (iii) For the trumpet and jazz music, 20.6%. In addition, the average SIR of the proposed method exhibits much a higher value than the uniform sparsity SNMF2D. This clearly shows that the amount of interference between any two sources is lesser for the proposed method. Therefore, the above results unanimously indicate that the proposed adaptive sparsity method leads to higher ESE results than the uniform constant sparsity method.

4.3 Comparison with other sparse NMF-based SCASS methods

In Section IV B, analysis has been carried out to investigate effects between adaptive sparsity and uniform constant sparsity on source separation. In this evaluation, we compare the proposed method with other sparse NMF-based source separation methods. These consist of the followings:

- SNMF [13]. The uniform constant sparsity parameter is progressively varied from 0 to 10 with every increment of 0.1 (i.e. $\lambda = 0,0.1,0.2,\ldots,10$) and the best result is retained for comparison.
- Automatic Relevance Determination NMF (NMF-ARD) [35] exploits a hierarchical Bayesian framework SNMF that amounts to imposing an exponential prior for pruning and thereby enables estimation of the NMF model order. The NMF-ARD assumes prior on $\mathbf{H}$, namely, $p(\mathbf{H}|\lambda) = \prod_d \lambda_d^{l_{\max}} \exp-\left(\lambda_d \sum_l \mathbf{H}_{d,l}\right)$ and uses Automatic Relevance Determination (ARD) approach to determine the desirable number of components in $\mathbf{D}$.
- NMF with Temporal Continuity and Sparseness Criteria [14] (NMF-TCS) is based on factorizing the magnitude spectrogram of the mixed signal into a sum of components, which include the temporal continuity and sparseness criteria into the separation framework. In [14], the temporal continuity α is chosen as $[0,1,10,100,1000]$, sparseness weight β is chosen as $[0,1,10,100,1000]$. The best separation result is retained for comparison.

Figure 9 summarizes the SDR comparison results between our proposed method and the above three sparse NMF methods. From the results, it can be seen that the above methods fail to take into account the relative position of each spectrum and thereby discarding the temporal information. Better separation results will require a proper model that can represent both temporal structure and the pitch change which occurs when an instrument plays different notes simultaneously. If the temporal structure and the pitch change are not considered in the model, the mixing ambiguity is still contained in each separated source.

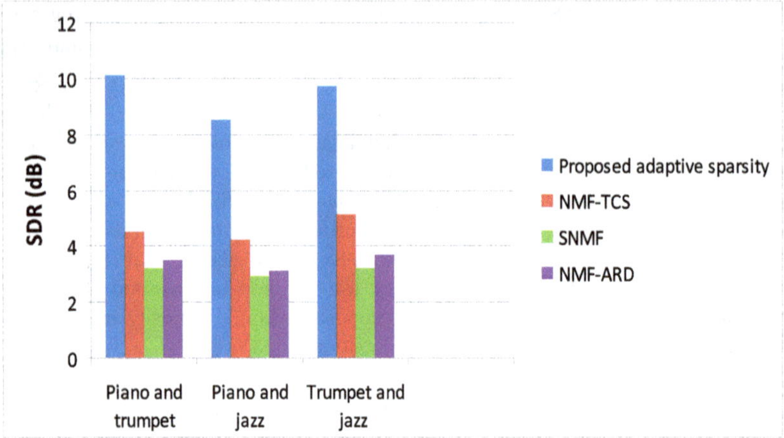

Fig. 9. Performance comparison between other NMF based SCASS methods and proposed method

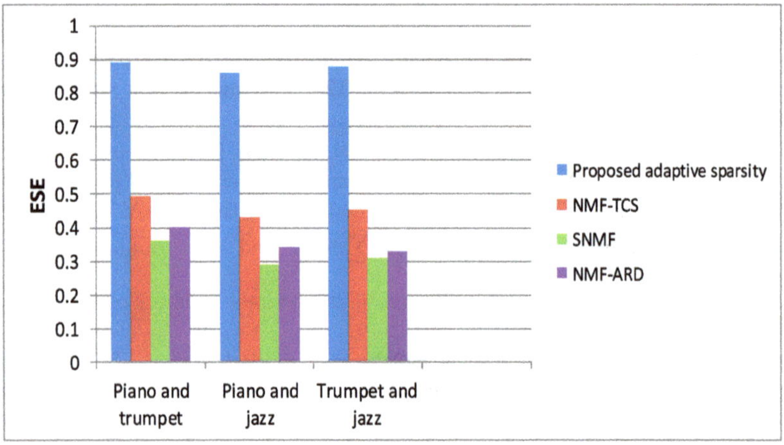

Fig. 10. ESE comparison between other NMF based SCASS methods and the proposed method

The improvement of our method compared with NMF-TCS, SNMF and NMF-ARD can be summarized as follows: (i) For the piano and trumpet music, the average improvement per source in terms of the SDR is 6.3dB. (ii) For the piano and jazz music, the average improvement per source in terms of SDR is 5dB. (iii) For the trumpet and jazz music, the average improvement per source in terms of SDR is 5.4dB. In the case of ESE (Figure 10), the proposed method exhibits much better average ESE of approximately 106.9%, 138.8% and 114.6% improvement with NMF-TCS, SNMF and NMF-ARD, respectively. Analyzing the separation results and ESE performance, the proposed method leads to the best separation performance for both recovered sources. The SNMF method performs with poorer results whereas the separation performance by the NMF-TCS method is slightly better than the NMF-ARD and SNMF methods. Our proposed method gives significantly better performance than the NMF-TCS, SNMF and NMF-ARD methods. The spectral dictionary obtained via NMF-TCS, SNMF and NMF-ARD methods are not adequate to capture the temporal dependency of the frequency patterns within the audio signal. In addition, the NMF-TCS, SNMF and NMF-ARD do not model notes but rather unique events only. Thus if two notes are always played simultaneously they will be modeled as one component. Also, some components might not correspond to notes but rather to the model e.g. background noise.

4.4 Comparison with underdetermined-based ICA SCSS method

In the underdetermined-ICA SCSS method [3], the key point is to exploit the prior knowledge of the sources such as the basis functions to generate the sparse codes. In this work, these basis functions are obtained in two stages: (i) Training stage: the basis functions are obtained by performing ICA on each concatenated sources. In this experiment, we derive a set of 64 basis functions for each type of source. These training data exclude the target sources which have been exclusively used to generate the mixture signals. (ii) Adaptation stage: the obtained ICA basis functions from the training stage are further adapted based on the current estimated sources during the separation process. In this method, both the estimated sources and the ICA basis functions are jointly optimized by maximizing the log-likelihood of the current mixture signal until it converges to the steady-state solution. If two sets of basis functions overlap significantly with each other, the underdetermined-ICA SCSS method is less efficient in resolving the mixing ambiguity between sources. The improvement of proposed method compared with underdetermined-ICA SCSS method can be summarized as follows: (i) For the piano and trumpet music, the average improvement per source in terms of the SDR is 4.3dB. (ii) For the piano and jazz music, the average improvement per source in terms of SDR is 4dB. (iii) For the trumpet and jazz music, the average improvement per source in terms of SDR is 4.2dB.

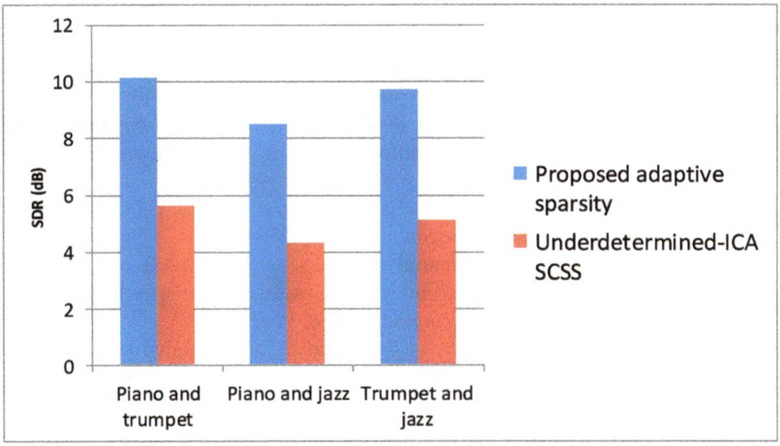

Fig. 11. Performance underdetermined-ICA SCSS method and proposed method

The performance of the underdetermined-ICA SCSS method relies on the ICA-derived time domain basis functions. High level performance can be achieved only when the basis functions of each source are sufficiently distinct. However, the result becomes considerably less robust in separating mixture where the original sources are of the same type e.g. mixture of music with music.

5. Conclusion

The chapter has presented an adaptive strategy to sparsifying the non-negative matrix factorization. The impetus behind this work is that the sparsity achieved by conventional SNMF and SNMF2D is not enough; in such situations it might be useful to control the degree of sparseness explicitly. In the proposed method, the regularization term is adaptively tuned using a variational Bayesian approach to yield desired sparse decomposition, thus enabling the spectral dictionary and temporal codes of non-stationary audio signals to be estimated more efficiently. This has been verified concretely based on our simulation results. In addition, the proposed method has yielded significant improvements in single channel music separation when compared with other sparse NMF-based source separation methods. Future work could investigate the extension of the proposed method to separate non-stationary (here non-stationary refers to the sources not located in the fixed places, e.g. the speakers are talking while on the move) and reverberant mixing model. For non-stationary reverberant mixing model, this gives

$$y(t) = \sum_{i=1}^{N_s} \sum_{\tau_r=0}^{L_r-1} m_i(\tau_r, t) x_i(t - \tau_r) + n(t)$$

where $m_i(\tau_r, t)$ is the finite impulse response of causal filter at t time and τ_r is the time delay. The expanded adaptive sparsity non-negative matrix factorization can then be developed to estimate mixing $\tilde{m}_i$ and sources $\tilde{x}_i$, respectively.

6. References

[1] Radfa M.H, Dansereau R.M. Single-channel speech separation using soft mask filtering. IEEE Trans. on Audio, Speech and Language Processing. 2007; 15: 2299-2310.
[2] Ellis D. Model-based scene analysis, in Computational Auditory Scene Analysis: Principles, Algorithms, and Applications, D. Wang and G. Brown, Eds. New York: Wiley/IEEE Press; 2006
[3] Jang G.J, Lee T.W. A maximum likelihood approach to single channel source separation. Journal of Machine Learning Research. 2003; 4: 1365–1392.
[4] Li P, Guan Y, Xu B, Liu W. Monaural speech separation based on computational auditory scene analysis and objective quality assessment of speech. IEEE Trans. on Audio, Speech and Language Processing. 2006; 14: 2014–2023.
[5] Paatero P, Tapper U. Positive matrix factorization: A non-negative factor model with optimal utilization of error estimates of data values. Environmetrics. 1994; 5: 111–126.

[6] Ozerov A, Févotte C. Multichannel nonnegative matrix factorization in convolutive mixtures for audio source separation. IEEE Transactions on Audio, Speech, and Language Processing 2010; 18: 550-563.
[7] Casey M. A, Westner A. Separation of mixed audio sources by independent subspace analysis. proceeding of. Int. Comput. Music Conf, 2000. 154-161; 2000
[8] Molla Md. K. I, Hirose K. Single-Mixture Audio Source Separation by Subspace Decomposition of Hilbert Spectrum. IEEE Trans. on Audio, Speech and Language Processing. 2007; 15: 893-900.
[9] Hyvarinen A, Karhunen J, Oja E, Independent component analysis and blind source separation, John Wiley & Sons 2005. p.20-60.
[10] Lee D, Seung H. Learning the parts of objects by nonnegative matrix factorisation. Nature. 1999; 401: 788-791.
[11] Kompass R. A generalized divergence measure for nonnegative matrix factorization. Neural Computation. 2007; 19: 780-791.
[12] Cichocki A, Zdunek R, Amari S.I. Csisz´ar's divergences for non-negative matrix factorization: family of new algorithms. Proceeding of. Intl. Conf. on Independent Component Analysis and Blind Signal Separation (ICABSS'06), Charleston, USA, March 2006, 3889: 32-39. 2006.
[13] Hoyer P. O. Non-negative matrix factorization with sparseness constraints. Journal of Machine Learning Research. 2004; 5: 1457-1469.
[14] Virtanen T (2007) Monaural sound source separation by non-negative matrix factorization with temporal continuity and sparseness criteria. IEEE Transactions on Audio, Speech, and Language Processing. 2007; 15: 1066-1074.
[15] Vincent E (2006) Musical source separation using time-frequency source priors. IEEE Trans. Audio, Speech and Language Processing. 2006; 14: 91-98.
[16] Ozerov A, Févotte C. Multichannel nonnegative matrix factorization in convolutive mixtures. With application to blind audio source separation. Proceeding of IEEE International Conference on Acoustics, Speech, and Signal Processing (ICASSP'09), 3137-3140. 2009.
[17] Mysore G, Smaragdis P, Raj B. Non-negative hidden Markov modeling of audio with application to source separation. Proceeding of 9th international conference on Latent Variable Analysis and Signal Separation (LCA/ICA). 2010.
[18] Nakano M, et al. Nonnegative Matrix Factorization with Markov-chained Bases for Modeling Time-varying in Music Spectrograms. Proceeding of 9th international conference on Latent Variable Analysis and Signal Separation (LCA/ICA). 2010
[19] Salakhutdinov R, Mnih A. Bayesian probabilistic matrix factorization using Markov chain Monte Carlo. Proceedings of the 25th international conference on Machine learning. 880-887. 2008.
[20] Cemgil A. T. Bayesian inference for nonnegative matrix factorisation models. Computational Intelligence and Neuroscience. 2009; doi: 10.1155/2009/785152.
[21] Moussaoui S, Brie D, Mohammad-Djafari A, Carteret C, Separation of non-negative mixture of non-negative sources using a Bayesian approach and MCMC sampling. IEEE Trans. on Signal Processing. 2006; 54: 4133-4145.
[22] Schmidt M. N, Winther O, Hansen L.K. Bayesian non-negative matrix factorization. Proceeding of Independent Component Analysis and Signal Separation, International Conference. 2009

[23] Morup M, Schmidt M.N. Sparse non-negative matrix factor 2-D deconvolution. Technical University of Denmark, Copenhagen, Denmark. 2006.

[24] Schmidt M.N, Morup M. Nonnegative matrix factor 2-D deconvolution for blind single channel source separation. Proceeding of Intl. Conf. Independent Component Analysis and Blind Signal Separation (ICABSS'06), Charleston, USA. 3889: 700-707. 2006.

[25] Lin Y. Q. l_1-norm sparse Bayesian learning: theory and applications. *Ph.D. Thesis, University of Pennsylvania*. 2008.

[26] Gao Bin, Woo W.L, Dlay S.S. Single Channel Source Separation Using EMD-Subband Variable Regularised Sparse Features. IEEE Trans. on Audio, Speech, and Language Processing. 2011; 19: 961-976.

[27] Gao Bin, Woo W.L, Dlay S.S. Adaptive Sparsity Non-negative Matrix Factorization for Single Channel Source Separation. IEEE the Journal of Selected Topics in Signal Processing. 2011; 5: 1932-4553.

[28] Sha F, Saul L.K, Lee D.D. Multiplicative updates for nonnegative quadratic programming in support vector machines. Proceeding of Advances in Neural Information Process. Systems. 15: 1041-1048. 2002

[29] Brown J. C. Calculation of a constant Q spectral transform. *J. Acoust. Soc. Am.* 1991; 89: 425-434.

[30] Yilmaz O, Rickard S. Blind separation of speech mixtures via time-frequency masking. IEEE Trans. Signal Processing. 2004; 52: 1830-1847.

[31] Goto M, Hashiguchi H, Nishimura T, Oka R. RWC music database: Music genre database and musical instrument sound database. in Proc. of Intl. Symp. on Music Information Retrieval (ISMIR), Baltimore, Maryland, USA. 229-230. 2003.

[32] Signal Separation Evaluation Campaign (SiSEC 2008), (2008). [Online]. Available: http://sisec.wiki.irisa.fr

[33] Vincent E, Gribonval R, Fevotte C. Performance measurement in blind audio source separation. IEEE Trans. Speech Audio Process. 2006; 14: 1462-1469.

[34] Wang D. L. On ideal binary mask as the computational goal of auditory scene analysis. in Speech Separation by Humans and Machines, P. Divenyi, Ed. Norwell, MA: Kluwer, pp. 181-197. 2005.

[35] Mørup M, Hansen K.L. Tuning Pruning in Sparse Non-negative Matrix Factorization. Proceeding of 17th European Signal Processing Conference (EUSIPCO'2009), Glasgow, Scotland. 2009.

5

Blind Source Separation for Speech Application Under Real Acoustic Environment

Hiroshi Saruwatari and Yu Takahashi
Nara Institute of Science and Technology
Japan

1. Introduction

A hands-free speech recognition system [1] is essential for the realization of an intuitive, unconstrained, and stress-free human-machine interface, where users can talk naturally because they require no microphone in their hands. In this system, however, since noise and reverberation always degrade speech quality, it is difficult to achieve high recognition performance, compared with the case of using a close-talk microphone such as a headset microphone. Therefore, we must suppress interference sounds to realize a noise-robust hands-free speech recognition system.

Source separation is one approach to removing interference sound source signals. Source separation for acoustic signals involves the estimation of original sound source signals from mixed signals observed in each input channel. Various methods have been presented for acoustic source signal separation. They can be classified into two groups: methods based on single-channel input, e.g., spectral subtraction (SS) [2], and those based on multichannel input, e.g., microphone array signal processing [3]. There have been various studies on microphone array signal processing; in particular, the delay-and-sum (DS) [4–6] array and adaptive beamformer (ABF) [7–9] are the most conventionally used microphone arrays for source separation and noise reduction. ABF can achieve higher performance than the DS array. However, ABF requires a priori information, e.g., the look direction and speech break interval. These requirements are due to the fact that conventional ABF is based on *supervised* adaptive filtering, which significantly limits its applicability to source separation in practical applications. Indeed, ABF cannot work well when the interfering signal is nonstationary noise.

Recently, alternative approaches have been proposed. Blind source separation (BSS) is an approach to estimating original source signals using only mixed signals observed in each input channel. In particular, BSS based on independent component analysis (ICA) [10], in which the independence among source signals is mainly used for the separation, has recently been studied actively [11–19]. Indeed, the conventional ICA could work, particularly in speech-speech mixing, i.e., all sources can be regarded as point sources, but such a mixing condition is very rare and unrealistic; real noises are often widespread sources. In this chapter, we mainly deal with generalized noise that cannot be regarded as a point source. Moreover, we assume this noise to be nonstationary noise that arises in many acoustical environments; however, ABF could not treat this noise well. Although ICA is not influenced by the nonstationarity of signals unlike ABF, this is still a very challenging task that can hardly be addressed by conventional ICA-based BSS because ICA cannot separate widespread sources.

To improve the performance of BSS, some techniques combining conventional ICA and beamforming have been proposed [18, 20]. However, these studies dealt with the separation of point sources, and the behavior of such methods under a non-point-source condition was not explicitly analyzed to our knowledge. Therefore, in this chapter, first, we analyze ICA under a non-point-source noise condition and point out that ICA is proficient in noise estimation rather than in speech estimation under such a noise condition. This analysis implies that we can still utilize ICA as an accurate noise estimator.

Next, we review blind spatial subtraction array (BSSA) [21], an improved BSS algorithm recently proposed in order to deal with real acoustic sounds. BSSA consists of an ICA-based noise estimator, and noise reduction in the proposed BSSA is achieved by subtracting the power spectrum of the estimated noise via ICA from the power spectrum of the noisy observations. This "power-spectrum-domain subtraction" procedure provides better noise reduction than conventional ICA with estimation-error robustness. The efficacy of BSSA can be determined in various experiments, including computer-simulation-based and real-recording-based experiments. This chapter shows strong evidence of BSSA providing promising speech enhancement results in a railway-station environment.

Finally, the real-time implementation issue of BSS is discussed. Several recent studies have dealt with the real-time implementation of ICA, but they still required high-speed personal computers. Consequently, BSS implementation on a small LSI still receives much attention in industrial applications. In this chapter, an example of hardware implementation of BSSA is introduced, which has yielded commercially available microphones adopted by the Japanese National Police Agency.

The rest of this chapter is organized as follows. In Sect. 2, the sound mixing model and conventional ICA are discussed. In Sect. 3, the analysis of ICA under a non-point-source condition is described in detail. In Sect. 4, BSSA is reviewed in detail. In Sect. 5, the experimental results are shown and compared with those of conventional methods. In Sect. 6, an example of hardware implementation of BSSA is introduced. Following the example, the chaper conclusions are given in Sect. 7.

2. Data model and conventional BSS method

2.1 Sound mixing model of microphone array

In this chapter, a straight-line array is assumed. The coordinates of the elements are designated $d_j (j = 1, \ldots, J)$, and the direction-of-arrivals (DOAs) of multiple sound sources are designated $\theta_k (k = 1, \ldots, K)$ (see Fig. 1). Then, we consider that only one target speech signal, some interference signals that can be regarded as point sources, and additive noise exist. This additive noise represents noises that cannot be regarded as point sources, e.g., spatially uncorrelated noises, background noises, and leakage of reverberation components outside the frame analysis. Multiple mixed signals are observed at microphone array elements, and a short-time analysis of the observed signals is conducted by frame-by-frame discrete Fourier transform (DFT). The observed signals are given by

$$\boldsymbol{x}(f,\tau) = \boldsymbol{A}(f)\left\{\boldsymbol{s}(f,\tau) + \boldsymbol{n}(f,\tau)\right\} + \boldsymbol{n}_a(f,\tau), \tag{1}$$

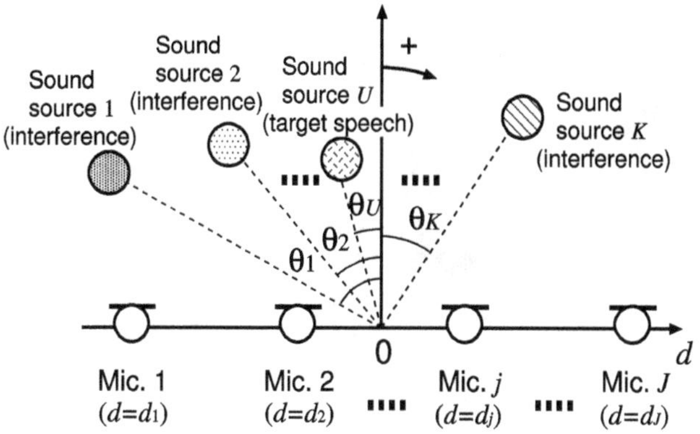

Fig. 1. Configurations of microphone array and signals.

where f is the frequency bin and τ is the time index of DFT analysis. Also, $\boldsymbol{x}(f,\tau)$ is the observed signal vector, $\boldsymbol{A}(f)$ is the mixing matrix, $\boldsymbol{s}(f,\tau)$ is the target speech signal vector in which only the Uth entry contains the signal component $s_U(f,\tau)$ (U is the target source number), $\boldsymbol{n}(f,\tau)$ is the interference signal vector that contains the signal components except the Uth component, and $\boldsymbol{n}_a(f,\tau)$ is the nonstationary additive noise signal term that generally represents non-point-source noises. These are defined as

$$\boldsymbol{x}(f,\tau) = [x_1(f,\tau),\ldots,x_J(f,\tau)]^{\mathrm{T}}, \tag{2}$$

$$\boldsymbol{s}(f,\tau) = [\underbrace{0,\ldots,0}_{U-1},s_U(f,\tau),\underbrace{0,\ldots,0}_{K-U}]^{\mathrm{T}}, \tag{3}$$

$$\boldsymbol{n}(f,\tau) = [n_1(f,\tau),\ldots,n_{U-1}(f,\tau),0,n_{U+1},\ldots,n_K(f,\tau)]^{\mathrm{T}}, \tag{4}$$

$$\boldsymbol{n}_a(f,\tau) = [n_1^{(a)}(f,\tau),\ldots,n_J^{(a)}(f,\tau)]^{\mathrm{T}}, \tag{5}$$

$$\boldsymbol{A}(f) = \begin{bmatrix} A_{11}(f) & \cdots & A_{1K}(f) \\ \vdots & & \vdots \\ A_{J1}(f) & \cdots & A_{JK}(f) \end{bmatrix}. \tag{6}$$

2.2 Conventional frequency-domain ICA

Here, we consider a case where the number of sound sources, K, equals the number of microphones, J, i.e., $J = K$. In addition, similarly to that in the case of the conventional ICA contexts, we assume that the additive noise $\boldsymbol{n}_a(f,\tau)$ is negligible in (1). In frequency-domain ICA (FDICA), signal separation is expressed as

$$\boldsymbol{o}(f,\tau) = [o_1(f,\tau),\ldots,o_K(f,\tau)]^{\mathrm{T}} = \boldsymbol{W}_{\mathrm{ICA}}(f)\boldsymbol{x}(f,\tau), \tag{7}$$

$$\boldsymbol{W}_{\mathrm{ICA}}(f) = \begin{bmatrix} W_{11}^{(\mathrm{ICA})}(f) & \cdots & W_{1J}^{(\mathrm{ICA})}(f) \\ \vdots & & \vdots \\ W_{K1}^{(\mathrm{ICA})}(f) & \cdots & W_{KJ}^{(\mathrm{ICA})}(f) \end{bmatrix}, \tag{8}$$

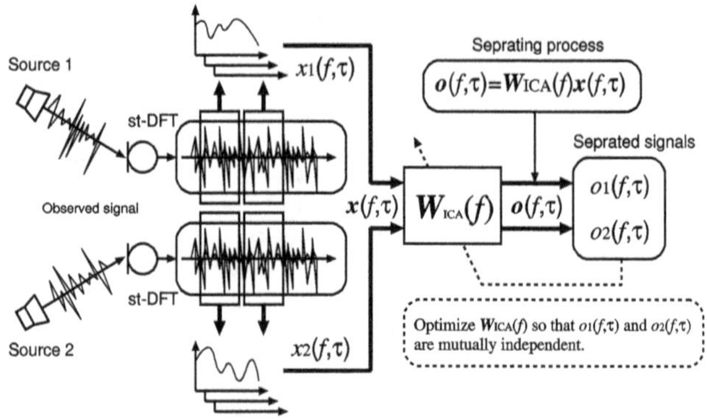

Fig. 2. Blind source separation procedure in FDICA in case of $J = K = 2$.

where $\boldsymbol{o}(f,\tau)$ is the resultant output of the separation and $\boldsymbol{W}_{\mathrm{ICA}}(f)$ is the complex-valued unmixing matrix (see Fig. 2).

The unmixing matrix $\boldsymbol{W}_{\mathrm{ICA}}(f)$ is optimized by ICA so that the output entries of $\boldsymbol{o}(f,\tau)$ become mutually independent. Indeed, many kinds of ICA algorithm have been proposed. In the second-order ICA (SO-ICA) [15, 17], the separation filter is optimized by the joint diagonalization of co-spectra matrices using the nonstationarity and coloration of the signal. For instance, the following iterative updating equation based on SO-ICA has been proposed by Parra and Spence [15]:

$$\boldsymbol{W}_{\mathrm{ICA}}^{[p+1]}(f) = -\mu \sum_{\tau_b} \chi(f)\,\text{off-diag}\left(\boldsymbol{R}_{oo}\left(f,\tau_b\right)\right) \boldsymbol{W}_{\mathrm{ICA}}^{[p]}(f) \boldsymbol{R}_{xx}(f,\tau_b) + \boldsymbol{W}_{\mathrm{ICA}}^{[p]}(f), \tag{9}$$

where μ is the step-size parameter, $[p]$ is used to express the value of the pth step in iterations, off-diag$[\boldsymbol{X}]$ is the operation for setting every diagonal element of matrix $\boldsymbol{X}$ to zero, and $\chi(f) = (\sum_{\tau_b} \|\boldsymbol{R}_{xx}(f,\tau_b)\|^2)^{-1}$ is a normalization factor ($\|\cdot\|$ represents the Frobenius norm). $\boldsymbol{R}_{xx}(f,\tau_b)$ and $\boldsymbol{R}_{oo}(f,\tau_b)$ are the cross-power spectra of the input $\boldsymbol{x}(f,\tau)$ and output $\boldsymbol{o}(f,\tau)$, respectively, which are calculated around multiple time blocks τ_b. Also, Pham et al. have proposed the following improved criterion for SO-ICA [17]:

$$\sum_{\tau_b} \left\{ \frac{1}{2} \log\det \mathrm{diag}[\boldsymbol{W}_{\mathrm{ICA}}(f) \boldsymbol{R}_{oo}(f,\tau_b) \boldsymbol{W}_{\mathrm{ICA}}(f)^{\mathrm{H}}] - \log\det[\boldsymbol{W}_{\mathrm{ICA}}(f)] \right\}, \tag{10}$$

where the superscript H denotes Hermitian transposition. This criterion is to be minimized with respect to $\boldsymbol{W}_{\mathrm{ICA}}(f)$.

On the other hand, a higher-order-statistics-based approach exists. In higher-order ICA (HO-ICA), the separation filter is optimized on the basis of the non-Gaussianity of the signal. The optimal $\boldsymbol{W}_{\mathrm{ICA}}(f)$ in HO-ICA is obtained using the iterative equation

$$\boldsymbol{W}_{\mathrm{ICA}}^{[p+1]}(f) = \mu[\boldsymbol{I} - \langle \boldsymbol{\varphi}(\boldsymbol{o}(f,\tau)) \boldsymbol{o}^{\mathrm{H}}(f,\tau) \rangle_\tau] \boldsymbol{W}_{\mathrm{ICA}}^{[p]}(f) + \boldsymbol{W}_{\mathrm{ICA}}^{[p]}(f), \tag{11}$$

where $\boldsymbol{I}$ is the identity matrix, $\langle\cdot\rangle_\tau$ denotes the time-averaging operator, and $\boldsymbol{\varphi}(\cdot)$ is the nonlinear vector function. Many kinds of nonlinear function $\boldsymbol{\varphi}(f,\tau)$ have been proposed.

Considering a batch algorithm of ICA, it is well-known that $\tanh(\cdot)$ or the sigmoid function is appropriate for super-Gaussian sources such as speech signals [22]. In this study, we define the nonlinear vector function $\boldsymbol{\varphi}(\cdot)$ as

$$\boldsymbol{\varphi}(\boldsymbol{o}(f,\tau)) \equiv [\varphi(o_1(f,\tau)),\ldots,\varphi(o_K(f,\tau))]^{\mathrm{T}}, \tag{12}$$

$$\varphi(o_k(f,\tau)) \equiv \tanh o_k^{(\mathrm{R})}(f,\tau) + i \tanh o_k^{(\mathrm{I})}(f,\tau), \tag{13}$$

where the superscripts (R) and (I) denote the real and imaginary parts, respectively. The nonlinear function given by (12) indicates that the nonlinearity is applied to the real and imaginary parts of complex-valued signals separately. This type of complex-valued nonlinear function has been introduced by Smaragdis [14] for FDICA, where it can be assumed for speech signals that the real (or imaginary) parts of the time-frequency representations of sources are mutually independent. According to Refs. [19, 23], the source separation performance of HO-ICA is almost the same as or superior to that of SO-ICA. Thus, in this chapter, HO-ICA is utilized as the basic ICA algorithm in the simulation (Sect. 3.4) and experiments (Sect. 5).

3. Analysis of ICA under non-point-source noise condition

In this section, we investigate the proficiency of ICA under a non-point-source noise condition. In relation to the performance analysis of ICA, Araki et al. have reported that ICA-based BSS has equivalence to parallel constructed ABFs [24]. However, this investigation was focused on separation with a nonsingular mixing matrix, and thus was valid for only point sources.

First, we analyze beamformers that are optimized by ICA under a non-point-source condition. In the analysis, it is clarified that beamformers optimized by ICA become specific beamformers that maximize the signal-to-noise ratio (SNR) in each output (so-called *SNR-maximize beamformers*). In particular, the beamformer for target speech estimation is optimized to be a DS beamformer, and the beamformer for noise estimation is likely to be a null beamformer (NBF) [16].

Next, a computer simulation is conducted. Its result also indicates that ICA is proficient in noise estimation under a non-point-source noise condition. Then, it is concluded that ICA is suitable for noise estimation under such a condition.

3.1 Can ICA separate any source signals?

Many previous studies on BSS provided strong evidence that conventional ICA could perform source separation, particularly in the special case of speech-speech mixing, i.e., all sound sources are point sources. However, such sound mixing is not realistic under common acoustic conditions; indeed the following scenario and problem are likely to arise (see Fig. 3):

- The target sound is the user's speech, which can be approximately regarded as a *point source*. In addition, the users themselves locate relatively *near the microphone array* (e.g., 1 m apart), and consequently the accompanying reflection and reverberation components are moderate.
- For the noise, we are often confronted with interference sound(s) which is *not a point source* but a widespread source. Also, the noise is usually far from the array and is heavily reverberant.

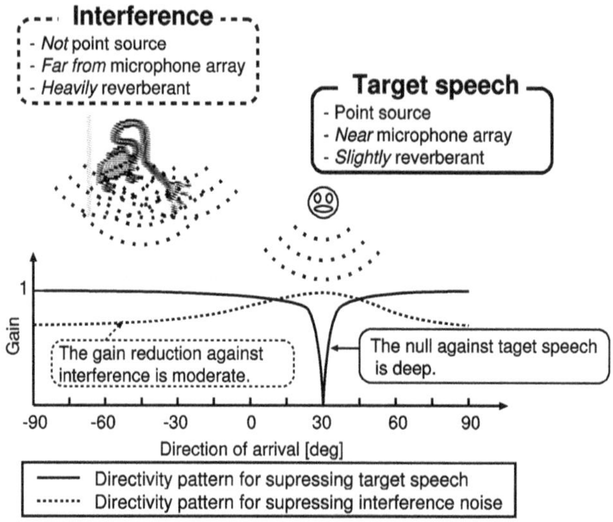

Fig. 3. Expected directivity patterns that are shaped by ICA.

In such an environment, can ICA separate the user's speech signal and a widespread noise signal? The answer is *no*. It is well expected that conventional ICA can suppress the user's speech signal to pick up the noise source, but ICA is very weak in picking up the target speech itself via the suppression of a distant widespread noise. This is due to the fact that ICA with small numbers of sensors and filter taps often provides only directional nulls against undesired source signals. Results of the detailed analysis of ICA for such a case are shown in the following subsections.

3.2 SNR-maximize beamformers optimized by ICA

In this subsection, we consider beamformers that are optimized by ICA in the following acoustic scenario: the target signal is the user's speech and the noise is not a point source. Then, the observed signal contains only one target speech signal and an additive noise. In this scenario, the observed signal is defined as

$$\boldsymbol{x}(f,\tau) = \boldsymbol{A}(f)\boldsymbol{s}(f,\tau) + \boldsymbol{n}_a(f,\tau). \tag{14}$$

Note that the additive noise $\boldsymbol{n}_a(f,\tau)$ cannot be negligible in this scenario. Then, the output of ICA contains two components, i.e., the estimated speech signal $y_s(f,\tau)$ and estimated noise signal $y_n(f,\tau)$; these are given by

$$[y_s(f,\tau), y_n(f,\tau)]^{\mathrm{T}} = \boldsymbol{W}_{\mathrm{ICA}}(f)\boldsymbol{x}(f,\tau). \tag{15}$$

Therefore, ICA optimizes two beamformers; these can be written as

$$\boldsymbol{W}_{\mathrm{ICA}}(f) = [\boldsymbol{g}_s(f), \boldsymbol{g}_n(f)]^{\mathrm{T}}, \tag{16}$$

where $g_s(f) = [g_1^{(s)}(f), \ldots, g_J^{(s)}(f)]^{\mathrm{T}}$ is the coefficient vector of the beamformer used to pick up the target speech signal, and $g_n(f) = [g_1^{(n)}(f), \ldots, g_J^{(n)}(f)]^{\mathrm{T}}$ is the coefficient vector of the beamformer used to pick up the noise. Therefore, (15) can be rewritten as

$$[y_s(f,\tau), y_n(f,\tau)]^{\mathrm{T}} = [g_s(f), g_n(f)]^{\mathrm{T}} \boldsymbol{x}(f,\tau). \tag{17}$$

In SO-ICA, the multiple second-order correlation matrices of distinct time block outputs,

$$\langle \boldsymbol{o}(f,\tau_b)\boldsymbol{o}^{\mathrm{H}}(f,\tau_b)\rangle_{\tau_b}, \tag{18}$$

are diagonalized through joint diagonalization.

On the other hand, in HO-ICA, the higher-order correlation matrix is also diagonalized. Using the Taylor expansion, we can express the factor of the nonlinear vector function of HO-ICA, $\varphi(o_k(f,\tau))$, as

$$\begin{aligned}
\varphi(o_k(f,\tau)) &= \tanh o_k^{(\mathrm{R})}(f,\tau) + i \tanh o_k^{(\mathrm{I})}(f,\tau), \\
&= \left\{ o_k^{(\mathrm{R})}(f,\tau) - \frac{\left(o_k^{(\mathrm{R})}(f,\tau)\right)^3}{3} + \cdots \right\} + i \left\{ o_k^{(\mathrm{I})}(f,\tau) - \frac{\left(o_k^{(\mathrm{I})}(f,\tau)\right)^3}{3} + \cdots \right\}, \\
&= o_k(f,\tau) - \left(\frac{\left(o_k^{(\mathrm{R})}(f,\tau)\right)^3}{3} + i \frac{\left(o_k^{(\mathrm{I})}(f,\tau)\right)^3}{3} \right) + \cdots.
\end{aligned} \tag{19}$$

Thus, the calculation of the higher-order correlation in HO-ICA, $\boldsymbol{\varphi}(\boldsymbol{o}(f,\tau))\boldsymbol{o}^{\mathrm{H}}(f,\tau)$, can be decomposed to a second-order correlation matrix and the summation of higher-order correlation matrices of each order. This is shown as

$$\langle \boldsymbol{\varphi}(\boldsymbol{o}(f,\tau))\boldsymbol{o}^{\mathrm{H}}(f,\tau)\rangle_\tau = \langle \boldsymbol{o}(f,\tau)\boldsymbol{o}^{\mathrm{H}}(f,\tau)\rangle_\tau + \Psi(f), \tag{20}$$

where $\Psi(f)$ is a set of higher-order correlation matrices. In HO-ICA, separation filters are optimized so that all orders of correlation matrices become diagonal matrices. Then, at least the second-order correlation matrix is diagonalized by HO-ICA. In both SO-ICA and HO-ICA, at least the second-order correlation matrix is diagonalized. Hence, we prove in the following that ICA optimizes beamformers as SNR-maximize beamformers focusing on only part of the second-order correlation. Then the absolute value of the normalized cross-correlation coefficient (off-diagonal entries) of the second-order correlation, C, is defined by

$$C = \frac{|\langle y_s(f,\tau) y_n^*(f,\tau)\rangle_\tau|}{\sqrt{\langle |y_s(f,\tau)|^2\rangle_\tau}\sqrt{\langle |y_n(f,\tau)|^2\rangle_\tau}}, \tag{21}$$

$$y_s(f,\tau) = \hat{s}(f,\tau) + r_s \hat{n}(f,\tau), \tag{22}$$

$$y_n(f,\tau) = \hat{n}(f,\tau) + r_n \hat{s}(f,\tau), \tag{23}$$

where $\hat{s}(f,\tau)$ is the target speech component in ICA's output, $\hat{n}(f,\tau)$ is the noise component in ICA's output, r_s is the coefficient of the residual noise component, r_n is the coefficient of the target-leakage component, and the superscript $*$ represents a complex conjugate. Therefore, the SNRs of $y_s(f,\tau)$ and $y_n(f,\tau)$ can be respectively represented by

$$\Gamma_s = \langle|\hat{s}(f,\tau)|^2\rangle_\tau / (|r_s|^2 \langle|\hat{n}(f,\tau)|^2\rangle_\tau), \tag{24}$$

$$\Gamma_n = \langle|\hat{n}(f,\tau)|^2\rangle_\tau / (|r_n|^2 \langle|\hat{s}(f,\tau)|^2\rangle_\tau), \tag{25}$$

where Γ_s is the SNR of $y_s(f,\tau)$ and Γ_n is the SNR of $y_n(f,\tau)$. Using (22), (23), (24), and (25), we can rewrite (21) as

$$C = \frac{\left|1/\sqrt{\Gamma_s}\cdot e^{j\arg r_s} + 1/\sqrt{\Gamma_n}\cdot e^{j\arg r_n^*}\right|}{\sqrt{1+1/\Gamma_s}\sqrt{1+1/\Gamma_n}} = \frac{\left|1/\sqrt{\Gamma_s} + 1/\sqrt{\Gamma_n}\cdot e^{j(\arg r_n^* - \arg r_s)}\right|}{\sqrt{1+1/\Gamma_s}\sqrt{1+1/\Gamma_n}}, \tag{26}$$

where $\arg r$ represents the argument of r. Thus, C is a function of only Γ_s and Γ_n. Therefore, the cross-correlation between $y_s(f,\tau)$ and $y_n(f,\tau)$ only depends on the SNRs of beamformers $g_s(f)$ and $g_n(f)$.

Now, we consider the minimization of C, which is identical to the second-order correlation matrix diagonalization in ICA. When $|\arg r_n^* - \arg r_s| > \pi/2$, where $-\pi < \arg r_s \leq \pi$ and $-\pi < \arg r_n^* \leq \pi$, it is possible to make C zero or minimum independently of Γ_s and Γ_n. This case is appropriate for the orthogonalization between $y_s(f,\tau)$ and $y_n(f,\tau)$, which is related to principal component analysis (PCA) unlike ICA. However, SO-ICA requires that all correlation matrices in the different time blocks are diagonalized (joint diagonalization) to maximize independence among all outputs. Also, HO-ICA requires that all order correlation matrices are diagonalized, i.e., not only $\langle \boldsymbol{o}(f,\tau)\boldsymbol{o}^{\mathrm{H}}(f,\tau)\rangle_\tau$ but also $\Psi(f)$ in (20) is diagonalized. These diagonalizations result in the prevention of the orthogonalization of $y_s(f,\tau)$ and $y_n(f,\tau)$; consequently, hereafter, we can consider only the case of $|\arg r_n^* - \arg r_s| \leq \pi/2$. Then, the partial differential of C^2 with respect to Γ_s is given by

$$\frac{\partial C^2}{\partial \Gamma_s} = \frac{(1-\Gamma_s)}{(\Gamma_s+1)^2(\Gamma_n+1)} + \frac{\Gamma_s\sqrt{\Gamma_s\Gamma_n}(1-\Gamma_s)}{(\Gamma_s+1)^2(\Gamma_n+1)} \cdot 2\mathrm{Re}\left[e^{j(\arg r_n^* - \arg r_s)}\right] < 0, \tag{27}$$

where $\Gamma_s > 1$ and $\Gamma_n > 1$. Similarly to the partial differential of C^2 with respect to Γ_n, we can also prove that $\partial C^2/\partial \Gamma_n < 0$, where $\Gamma_s > 1$ and $\Gamma_n > 1$ in the same manner. Therefore, C is a monotonically decreasing function of Γ_s and Γ_n. The above-mentioned fact indicates the following in ICA.

- The absolute value of cross-correlation only depends on the SNRs of the beamformers spanned by each row of an unmixing matrix.
- The absolute value of cross-correlation is a monotonically decreasing function of SNR.
- Therefore, the diagonalization of a second-order correlation matrix leads to SNR maximization.

Thus, it can be concluded that ICA, in a parallel manner, optimizes multiple beamformers, i.e., $g_s(f)$ and $g_n(f)$, so that the SNR of the output of each beamformer becomes maximum.

3.3 What beamformers are optimized under non-point-source noise condition?

In the previous subsection, it has been proved that ICA optimizes beamformers as SNR-maximize beamformers. In this subsection, we analyze what beamformers are optimized by ICA, particularly under a non-point-source noise condition, where we assume a two-source separation problem. The target speech can be regarded as a point source, and the noise is a non-point-source noise. First, we focus on the beamformer $g_s(f)$ that picks up the target speech signal. The SNR-maximize beamformer for $g_s(f)$ minimizes the undesired signal's power under the condition that the target signal's gain is kept constant. Thus, the desired beamformer should satisfy

$$\min_{g_s(f)} g_s^{\mathrm{T}}(f)\boldsymbol{R}(f)g_s(f) \quad \text{subject to}\; g_s^{\mathrm{T}}(f)\boldsymbol{a}(f,\theta_s) = 1, \tag{28}$$

$$\boldsymbol{a}(f,\theta_s(f)) = [\exp(i2\pi(f/M)f_s d_1 \sin\theta_s/c),\ldots,\exp(i2\pi(f/M)f_s d_J \sin\theta_s/c)]^{\mathrm{T}}, \tag{29}$$

where $\boldsymbol{a}(f,\theta_s(f))$ is the steering vector, $\theta_s(f)$ is the direction of the target speech, M is the DFT size, f_s is the sampling frequency, c is the sound velocity, and $\boldsymbol{R}(f) = \langle \boldsymbol{n}_a(f,\tau)\boldsymbol{n}_a^{\mathrm{H}}(f,\tau)\rangle_\tau$ is the correlation matrix of $\boldsymbol{n}_a(f,\tau)$. Note that $\theta_s(f)$ is a function of frequency because the DOA of the source varies in each frequency subband under a reverberant condition. Here, using the Lagrange multiplier, the solution of (28) is

$$g_s(f)^{\mathrm{T}} = \frac{\boldsymbol{a}(f,\theta_s(f))^{\mathrm{H}}\boldsymbol{R}^{-1}(f)}{\boldsymbol{a}(f,\theta_s(f))^{\mathrm{H}}\boldsymbol{R}^{-1}(f)\boldsymbol{a}(f,\theta_s(f))}. \tag{30}$$

This beamformer is called a minimum variance distortionless response (MVDR) beamformer [25]. Note that the MVDR beamformer requires the true DOA of the target speech and the noise-only time interval. However, we cannot determine the true DOA of the target source signal and the noise-only interval because ICA is an *unsupervised* adaptive technique. Thus, the MVDR beamformer is expected to be the upper limit of ICA in the presence of non-point-source noises.

Although the correlation matrix is often not diagonalized in lower-frequency subbands [25], e.g., diffuse noise, we approximate that the correlation matrix is almost diagonalized in subbands in the entire frequency. Then, regarding the power of noise signals as approximately $\delta^2(f)$, the correlation matrix results in $\boldsymbol{R}(f) = \delta^2(f)\cdot\boldsymbol{I}$. Therefore, the inverse of the correlation matrix $\boldsymbol{R}^{-1}(f) = \boldsymbol{I}/\delta^2(f)$ and (30) can be rewritten as

$$g_s(f)^{\mathrm{T}} = \frac{\boldsymbol{a}(f,\theta_s(f))^{\mathrm{H}}}{\boldsymbol{a}(f,\theta_s(f))^{\mathrm{H}}\boldsymbol{a}(f,\theta_s(f))}. \tag{31}$$

Since $\boldsymbol{a}(f,\theta_s(f))^{\mathrm{H}}\boldsymbol{a}(f,\theta_s(f)) = J$, we finally obtain

$$g_s(f) = \frac{1}{J}[\exp(-i2\pi(f/M)f_s d_1 \sin\theta_s(f)/c),\ldots,\exp(-i2\pi(f/M)f_s d_J \sin\theta_s(f)/c)]^{\mathrm{T}}. \tag{32}$$

This filter $g_s(f)$ is approximately equal to a DS beamformer [4]. Note that the filter $g_s(f)$ is not a simple DS beamformer but a *reverberation-adapted DS beamformer* because it is optimized for a distinct $\theta_s(f)$ in each frequency bin. The resultant noise power is $\delta^2(f)/J$ when the noise is spatially uncorrelated and white Gaussian. Consequently the noise-reduction performance of the DS beamformer optimized by ICA under a non-point-source noise condition is proportional to $10\log_{10} J$ [dB]; this performance is not particularly good.

Next, we consider the other beamformer $g_n(f)$, which picks up the noise source. Similar to the noise signal, the beamformer that removes the target signal arriving from $\theta_s(f)$ is the SNR-maximize beamformer. Thus, the beamformer that steers the directional null to $\theta_s(f)$ is the desired one for the noise signal. Such a beamformer is called NBF [16]. This beamformer compensates for the phase of the signal arriving from $\theta_s(f)$, and carries out subtraction. Thus, the signal arriving from $\theta_s(f)$ is removed. For instance, NBF with a two-element array is designed as

$$g_n(f) = [\exp(-i2\pi(f/M)f_s d_1 \sin\theta_s(f)/c), -\exp(-i2\pi(f/M)f_s d_2 \sin\theta_s(f)/c)]^{\mathrm{T}} \cdot \sigma(f), \quad (33)$$

where $\sigma(f)$ is the gain compensation parameter. This beamformer surely satisfies $g_n^{\mathrm{T}}(f) \cdot a(f, \theta_s(f)) = 0$. The steering vector $a(f, \theta_s(f))$ expresses the wavefront of the plane wave arriving from $\theta_s(f)$. Thus, $g_n(f)$ actually steers the directional null to $\theta_s(f)$. Note that this always occurs regardless of the number of microphones (at least two microphones). Hence, this beamformer achieves a reasonably high, ideally infinite, SNR for the noise signal. Also, note that the filter $g_n(f)$ is not a simple NBF but a *reverberation-adapted NBF* because it is optimized for a distinct $\theta_s(f)$ in each frequency bin. Overall, the performance of enhancing the target speech is very poor but that of estimating the noise source is good.

3.4 Computer simulations

We conduct computer simulations to confirm the performance of ICA under a non-point-source noise condition. Here, we used HO-ICA [14] as the ICA algorithm. We used the following 8-kHz-sampled signals as the ICA's input; the original target speech (3 s) was convoluted with impulse responses that were recorded in an actual environment, and to which three types of noise from 36 loudspeakers were added. The reverberation time (RT_{60}) is 200 ms; this corresponds to mixing filters with 1600 taps in 8 kHz sampling. The three types of noise are an independent Gaussian noise, actually recorded railway-station noise, and interference speech by 36 people. Figure 4 illustrates the reverberant room used in the simulation. We use 12 speakers (6 males and 6 females) as sources of the original target speech, and the input SNR of test data is set to 0 dB. We use a two-, three-, or four-element microphone array with an interelement spacing of 4.3 cm.

The simulation results are shown in Figs. 5 and 6. Figure 5 shows the result for the average noise reduction rate (NRR) [16] of all the target speakers. NRR is defined as the output SNR in dB minus the input SNR in dB. This measure indicates the objective performance of noise reduction. NRR is given by

$$\mathrm{NRR}\ [\mathrm{dB}] = \frac{1}{J}\sum_{j=1}^{J}(\mathrm{OSNR} - \mathrm{ISNR}_j), \quad (34)$$

where OSNR is the output SNR and ISNR_j is the input SNR of microphone j.

From this result, we can see an imbalance between the target speech estimation and the noise estimation in every noise case; the performance of the target speech estimation is significantly poor, but that of noise estimation is very high. This result is consistent with

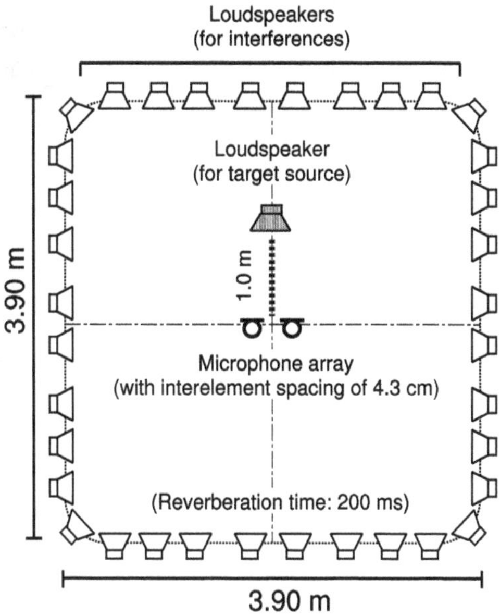

Fig. 4. Layout of reverberant room in our simulation.

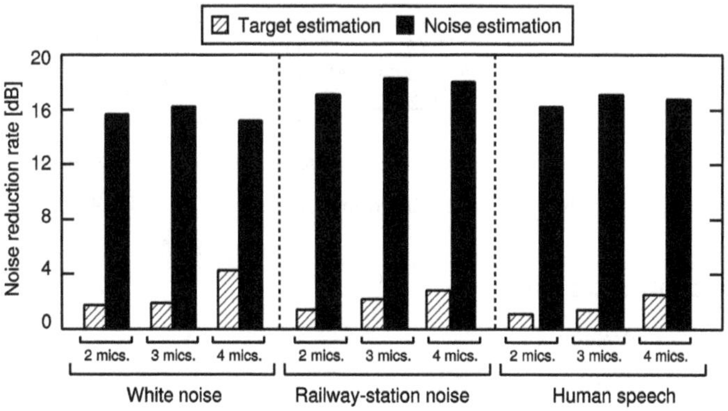

Fig. 5. Simulation-based separation results under non-point-source noise condition.

the previously stated theory. Moreover, Fig. 6 shows directivity patterns shaped by the beamformers optimized by ICA in the simulation. It is clearly indicated that beamformer $g_s(f)$, which picks up the target speech, resembles the DS beamformer, and that beamformer $g_n(f)$, which picks up the noise, becomes NBF. From these results, it is confirmed that the previously stated theory, i.e., the beamformers optimized by ICA under a non-point-source noise condition are DS and NBF, is valid.

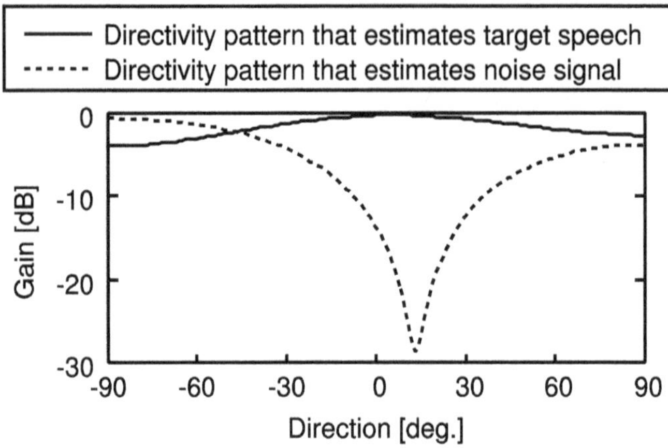

Fig. 6. Typical directivity patterns under non-point-source noise condition shaped by ICA at 2 kHz and two-element array for case of white Gaussian noise.

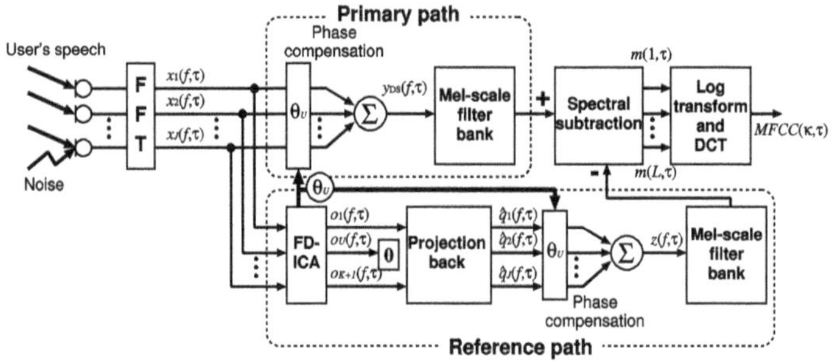

Fig. 7. Block diagram of blind spatial subtraction array.

4. Blind spectral subtraction array

4.1 Motivation and strategy

As clearly shown in Sects. 3.3 and 3.4, ICA is proficient in noise estimation rather than in target-speech estimation under a non-point-source noise condition. Thus, we cannot use ICA for direct target estimation under such a condition. However, we can still use ICA as a noise estimator. This motivates us to introduce an improved speech-enhancement strategy, i.e., BSSA [21]. BSSA consists of a DS-based primary path and a reference path including ICA-based noise estimation (see Fig. 7). The estimated noise component in ICA is efficiently subtracted from the primary path in the power-spectrum domain without phase information. This procedure can yield better target-speech enhancement than simple ICA, even with the additional benefit of estimation-error robustness in speech recognition applications. The detailed process of signal processing is shown below.

4.2 Partial speech enhancement in primary path

We again consider the generalized form of the observed signal as described in (1). The target speech signal is partly enhanced in advance by DS. This procedure can be given as

$$y_{\rm DS}(f,\tau)=\boldsymbol{w}_{\rm DS}^{\rm T}(f)\boldsymbol{x}(f,\tau)=\boldsymbol{w}_{\rm DS}^{\rm T}(f)\boldsymbol{A}(f)\boldsymbol{s}(f,\tau)+\boldsymbol{w}_{\rm DS}^{\rm T}(f)\boldsymbol{A}(f)\boldsymbol{n}(f,\tau)+\boldsymbol{w}_{\rm DS}^{\rm T}(f)\boldsymbol{n}_a(f,\tau), \quad (35)$$

$$\boldsymbol{w}_{\rm DS} = \left[w_1^{(\rm DS)}(f),\ldots,w_J^{(\rm DS)}(f)\right]^{\rm T}, \quad (36)$$

$$w_j^{(\rm DS)}(f) = \frac{1}{J}\exp\left(-i2\pi(f/M)f_{\rm s}d_j\sin\theta_U/c\right), \quad (37)$$

where $y_{\rm DS}(f,\tau)$ is the primary-path output that is a slightly enhanced target speech, $\boldsymbol{w}_{\rm DS}(f)$ is the filter coefficient vector of DS, and θ_U is the estimated DOA of the target speech given by the ICA part in Sect. 4.3. In (35), the second and third terms on the right-hand side express the remaining noise in the output of the primary path.

4.3 ICA-based noise estimation in reference path

BSSA provides ICA-based noise estimation. First, we separate the observed signal by ICA and obtain the separated signal vector $\boldsymbol{o}(f,\tau)$ as

$$\boldsymbol{o}(f,\tau) = \boldsymbol{W}_{\rm ICA}(f)\boldsymbol{x}(f,\tau), \quad (38)$$

$$\boldsymbol{o}(f,\tau) = \left[o_1(f,\tau),\ldots,o_{K+1}(f,\tau)\right]^{\rm T}, \quad (39)$$

$$\boldsymbol{W}_{\rm ICA}(f) = \begin{bmatrix} W_{11}^{(\rm ICA)}(f) & \cdots & W_{1J}^{(\rm ICA)}(f) \\ \vdots & & \vdots \\ W_{(K+1)1}^{(\rm ICA)}(f) & \cdots & W_{(K+1)J}^{(\rm ICA)}(f) \end{bmatrix}, \quad (40)$$

where the unmixing matrix $\boldsymbol{W}_{\rm ICA}(f)$ is optimized by (11). Note that the number of ICA outputs becomes $K+1$, and thus the number of sensors, J, is more than $K+1$ because we assume that the additive noise $\boldsymbol{n}_a(f,\tau)$ is not negligible. We cannot estimate the additive noise perfectly because it is deformed by the filter optimized by ICA. Moreover, other components also cannot be estimated perfectly when the additive noise $\boldsymbol{n}_a(f,\tau)$ exists. However, we can estimate at least noises (including interference sounds that can be regarded as point sources, and the additive noise) that do not involve the target speech signal, as indicated in Sect. 3. Therefore, the estimated noise signal is still beneficial.

Next, we estimate DOAs from the unmixing matrix $\boldsymbol{W}_{\rm ICA}(f)$ [16]. This procedure is represented by

$$\theta_u = \sin^{-1}\frac{\arg\left(\frac{[\boldsymbol{W}_{\rm ICA}^{-1}(f)]_{ju}}{[\boldsymbol{W}_{\rm ICA}^{-1}(f)]_{j'u}}\right)}{2\pi f_{\rm s}c^{-1}(d_j-d_{j'})}, \quad (41)$$

where θ_u is the DOA of the uth sound source. Then, we choose the Uth source signal, which is nearest the front of the microphone array, and designate the DOA of the chosen source signal as θ_U. This is because almost all users are expected to stand in front of the microphone array in a speech-oriented human-machine interface, e.g., a public guidance system. Other strategies for choosing the target speech signal can be considered as follows.

- If the approximate location of a target speaker is known in advance, we can utilize the location of the target speaker. For instance, we can know the approximate location of the target speaker at a hands-free speech recognition system in a car navigation system in advance. Then, the DOA of the target speech signal is approximately known. For such systems, we can choose the target speech signal, selecting the specific component in which the DOA estimated by ICA is nearest the known target-speech DOA.
- For an interaction robot system [26], we can utilize image information from a camera mounted on a robot. Therefore, we can estimate DOA from this information, and we can choose the target speech signal on the basis of this estimated DOA.
- If the only target signal is speech, i.e., none of the noises are speech, we can choose the target speech signal on the basis of the Gaussian mixture model (GMM), which can classify sound signals into voices and nonvoices [27].

Next, in the reference path, no target speech signal is required because we want to estimate only noise. Therefore, we eliminate the user's signal from the ICA's output signal $\boldsymbol{o}(f,\tau)$. This can be written as

$$\boldsymbol{q}(f,\tau) = [o_1(f,\tau),...,o_{U-1}(f,\tau),0,o_{U+1}(f,\tau),...,o_{K+1}(f,\tau)]^{\mathrm{T}}, \tag{42}$$

where $\boldsymbol{q}(f,\tau)$ is the "noise-only" signal vector that contains only noise components. Next, we apply the projection back (PB) [13] method to remove the ambiguity of amplitude. This procedure can be represented as

$$\hat{\boldsymbol{q}}(f,\tau) = \boldsymbol{W}_{\mathrm{ICA}}^{+}(f)\boldsymbol{q}(f,\tau), \tag{43}$$

where $\boldsymbol{M}^{+}$ denotes the Moore-Penrose pseudo-inverse matrix of $\boldsymbol{M}$. Thus, $\hat{\boldsymbol{q}}(f,\tau)$ is a good estimate of the noise signals received at the microphone positions, i.e.,

$$\hat{\boldsymbol{q}}(f,\tau) \simeq \boldsymbol{A}(f)\boldsymbol{n}(f,\tau) + \boldsymbol{W}_{\mathrm{ICA}}^{+}(f)\hat{\boldsymbol{n}}_a(f,\tau), \tag{44}$$

where $\hat{\boldsymbol{n}}_a(f,\tau)$ contains the deformed additive noise signal and separation error due to an additive noise. Finally, we construct the estimated noise signal $z(f,\tau)$ by applying DS as

$$z(f,\tau) = \boldsymbol{w}_{\mathrm{DS}}^{\mathrm{T}}(f)\hat{\boldsymbol{q}}(f,\tau) \simeq \boldsymbol{w}_{\mathrm{DS}}^{\mathrm{T}}(f)\boldsymbol{A}(f)\boldsymbol{n}(f,\tau) + \boldsymbol{w}_{\mathrm{DS}}^{\mathrm{T}}(f)\boldsymbol{W}_{\mathrm{ICA}}^{+}(f)\hat{\boldsymbol{n}}_a(f,\tau). \tag{45}$$

This equation means that $z(f,\tau)$ is a good candidate for noise terms of the primary path output $y_{\mathrm{DS}}(f,\tau)$ (see the 2nd and 3rd terms on the right-hand side of (35)). Of course this noise estimation is not perfect, but we can still enhance the target speech signal via oversubtraction in the power-spectrum domain, as described in Sect. 4.4. Note that $z(f,\tau)$ is a function of the frame index τ, unlike the constant noise prototype in the traditional spectral subtraction method [2]. Therefore, the proposed BSSA can deal with *nonstationary* noise.

4.4 Noise reduction processing in BSSA

In BSSA, noise reduction is carried out by subtracting the estimated noise power spectrum (45) from the partly enhanced target speech signal power spectrum (35). This procedure is given as

$$y_{\mathrm{BSSA}}(f,\tau) = \begin{cases} \{|y_{\mathrm{DS}}(f,\tau)|^2 - \beta \cdot |z(f,\tau)|^2\}^{\frac{1}{2}} \\ \quad (\text{ if } |y_{\mathrm{DS}}(f,\tau)|^2 - \beta \cdot |z(f,\tau)|^2 \geq 0\ ), \\ \gamma \cdot |y_{\mathrm{DS}}(f,\tau)| \qquad (\text{otherwise}), \end{cases} \tag{46}$$

where $y_{\mathrm{BSSA}}(f,\tau)$ is the final output of BSSA, β is the oversubtraction parameter, and γ is the flooring parameter. Their appropriate setting, e.g., $\beta > 1$ and $\gamma \ll 1$, results in efficient noise reduction. For example, a larger oversubtraction parameter ($\beta \gg 1$) leads to a larger SNR improvement. However, the target signal would be distorted. On the other hand, a smaller oversubtraction parameter ($\beta \ll 1$) gives a less-distorted target signal. However, the SNR improvement is decreased. In the end, a trade-off between SNR improvement and the distortion of the output signal exists with respect to the parameter β; $1 < \beta < 2$ is usually used.

The system switches between two equations depending on the conditions in (46). If the calculated noise components using ICA in (45) are underestimated, i.e., $|y_{\mathrm{DS}}(f,\tau)|^2 > \beta|z(f,\tau)|^2$, the resultant output $y_{\mathrm{BSSA}}(f,\tau)$ corresponds to power-spectrum-domain subtraction among the primary and reference paths with an oversubtraction rate of β. On the other hand, if the noise components are overestimated in ICA, i.e., $|y_{\mathrm{DS}}(f,\tau)|^2 < \beta|z(f,\tau)|^2$, the resultant output $y_{\mathrm{BSSA}}(f,\tau)$ is floored with a small positive value to avoid a negative-valued unrealistic spectrum. These *oversubtraction* and *flooring* procedures enable error-robust speech enhancement in BSSA rather than a simple linear subtraction. Although the nonlinear processing in (46) often generates an artificial distortion, so-called *musical noise*, it is still applicable in the speech recognition system because the speech decoder is not very sensitive to such a distortion. BSSA involves mel-scale filter bank analysis and directly outputs the mel-frequency cepstrum coefficient (MFCC) [28] for speech recognition. Therefore, BSSA requires no transformation into the time-domain waveform for speech recognition.

In BSSA, DS and SS are processed in addition to ICA. In HO-ICA or SO-ICA, to calculate the correlation matrix, at least hundreds of product-sum operations are required in each frequency subband. On the other hand, in DS, at most J product-sum operations are required in each frequency subband. A mere 4 or 5 products are required for SS. Therefore, the complexity of BSSA does not increase by as much as 10% compared with ICA.

4.5 Variation and extension in noise reduction processing

As mentioned in the previous subsection, the noise reduction processing of BSSA is mainly based on SS, and therefore it often suffers from the problem of musical noise generation due to its nonlinear signal processing. This becomes a big problem in any audio applications aimed for human hearing, e.g., hearing-aids, teleconference systems, etc.

To improve the sound quality of BSSA, many kinds of variations have been proposed and implemented in the post-processing part in (46). Generalized SS and parametric Wiener filtering algorithms [29] have been introduced to successfully mitigate musical

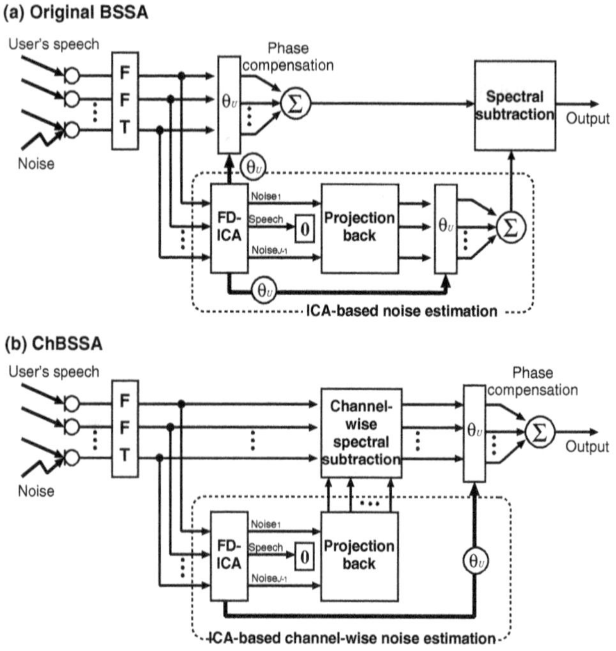

Fig. 8. Configurations of (a) original BSSA and (b) chBSSA.

noise generation [30]. Furthermore, the minimum mean-square error (MMSE) short-time spectral amplitude (STSA) estimator [31] can be used for achieving low-distortion speech enhancement in BSSA [32]. In addition, this MMSE-STSA estimator with ICA-based noise estimation has been modified to deal with binaural signal enhancement, where the spatial cue of the target speech signal can be maintained in the output of BSSA [33].

In recent studies, an interesting extension in the signal processing structure has been addressed [34, 35]. Two types of the BSSA structures are shown in Fig. 8. One is the original BSSA structure that performs SS after DS (see Fig. 8(a)), and another is that SS is channelwisely performed before DS (chBSSA; see Fig. 8(b)). It has been theoretically clarified that chBSSA is superior to BSSA in the mitigation of the musical noise generation via higher-order statistics analysis.

5. Experiment and evaluation

5.1 Experiment in reverberant room

In this experiment, we present a comparison of typical blind noise reduction methods, namely, the conventional ICA [14] and the traditional SS [2] cascaded with ICA (ICA+SS). We utilize the HO-ICA algorithm as conventional ICA [14]. Hereafter, 'ICA' simply indicates HO-ICA. For ICA+SS, we first obtain the estimated noise from the speech pause interval in the target speech estimation by ICA. The noise reduction achieved by SS is

$$y_{\mathrm{ICA+SS}}(f,\tau) = \begin{cases} \{|o_U(f,\tau)|^2 - \beta|\hat{n}_{\mathrm{remain}}(f)|^2\}^{\frac{1}{2}} & (\text{where } |o_U(f,\tau)|^2 - \beta|\hat{n}_{\mathrm{remain}}(f,\tau)|^2 \geq 0), \\ \gamma|o_U(f,\tau)| & (\text{otherwise}), \end{cases} \tag{47}$$

where $\hat{n}_{\mathrm{remain}}(f)$ is the noise signal from the speech pause in the target speech estimated by ICA. Moreover, a DOA-based permutation solver[16] is used in conventional ICA and in the ICA part in BSSA.

We used 16-kHz-sampled signals as test data; the original speech (6 s) was convoluted with impulse responses recorded in an actual environment, to which cleaner noise or a male's interfering speech recorded in an actual environment was added. Figure 9 shows the layout of the reverberant room used in the experiment. The reverberation time of the room is 200 ms; this corresponds to mixing filters of 3200 taps in 16 kHz sampling. The cleaner noise is not a simple point source signal but consists of several *nonstationary* noises emitted from a motor, an air duct, and a nozzle. Also, the male's interfering speech is not a simple point source but is slightly moving. In addition, these interference noises involve background noise. The SNR of the background noise (power ratio of target speech to background noise) is about 28 dB. We use 46 speakers (200 sentences) as the source of the target speech. The input SNR is set to 10 dB at the array. We use a four-element microphone array with an interelement spacing of 2 cm. The DFT size is 512. The oversubtraction parameter β is 1.4 and the flooring coefficient γ is 0.2. Such parameters were experimentally determined. The speech recognition task and conditions are shown in Table 1.

Regarding the evaluation index, we calculate NRR described in (34), cepstral distortion (CD), and speech recognition, which is the final goal of BSSA, in which the separated sound quality is fully considered. CD [36] is a measure of the degree of distortion via the cepstrum domain. It indicates the distortion among two signals, which is defined as

$$\mathrm{CD}\ [\mathrm{dB}] \equiv \frac{1}{T}\sum_{\tau=1}^{T} D_b \sqrt{\sum_{\rho=1}^{B} 2(C_{\mathrm{out}}(\rho;\tau) - C_{\mathrm{ref}}(\rho;\tau))^2}, \tag{48}$$

$$D_b = \frac{20}{\log 10}, \tag{49}$$

where T is the frame length, $C_{\mathrm{out}}(\rho;\tau)$ is the ρth cepstrum coefficient of the output signal in the frame τ, $C_{\mathrm{ref}}(\rho;\tau)$ is the ρth cepstrum coefficient of the speech signal convoluted with the impulse response, and D_b is a constant that transforms the measure into dB. Moreover, B is the number of dimensions of the cepstrum used in the evaluation. Moreover, we use the word accuracy (WA) score as a speech recognition performance. This index is defined as

$$\mathrm{WA}\ [\%] \equiv \frac{W_{\mathrm{WA}} - S_{\mathrm{WA}} - D_{\mathrm{WA}} - I_{\mathrm{WA}}}{W_{\mathrm{WA}}} \times 100, \tag{50}$$

where W_{WA} is the number of words, S_{WA} is the number of substitution errors, D_{WA} is the number of dropout errors, and I_{WA} is the number of insertion errors.

First, actual separation results obtained by ICA for the case of cleaner noise and interference speech are shown in Fig. 10. We can confirm the imbalanced performance between target estimation and noise estimation, similar to the simulation-based results (see Sect. 3.4).

Database	JNAS [37], 306 speakers (150 sentences/speaker)
Task	20 k newspaper dictation
Acoustic model	phonetic tied mixture (PTM) [37], clean model
Number of training speakers for acoustic model	260 speakers (150 sentences/speaker)
Decoder	JULIUS [37] ver 3.5.1

Table 1. Conditions for Speech Recognition

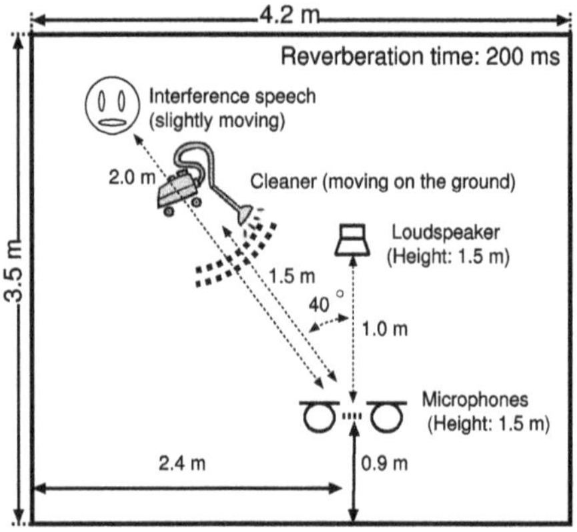

Fig. 9. Layout of reverberant room used in our experiment.

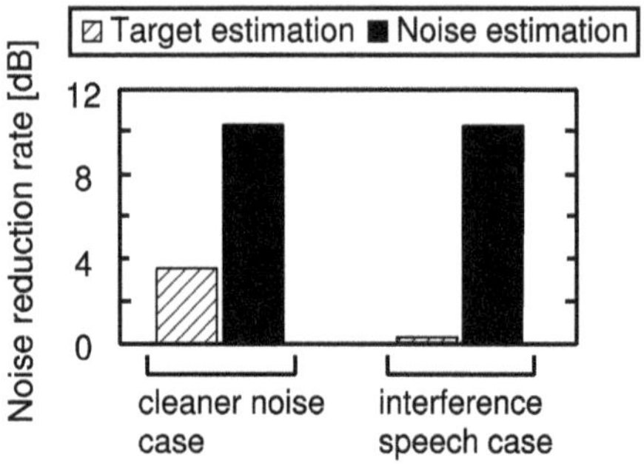

Fig. 10. NRR-based separation performance of conventional ICA in environment shown in Fig. 9.

Next, we discuss the NRR-based experimental results shown in Figs. 11(a) and 12(a). From the results, we can confirm that the NRRs of BSSA are more than 3 dB greater than those of conventional ICA and ICA+SS. However, we can see that the distortion of BSSA is slightly higher from Figs. 11(b) and 12(b). This is due to the fact that the noise reduction of BSSA

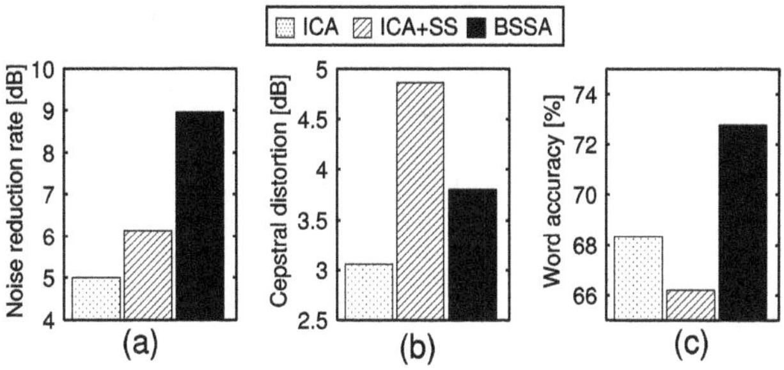

Fig. 11. Results of (a) noise reduction rate, (b) cepstral distortion, and (c) speech recognition test for each method (cleaner noise case).

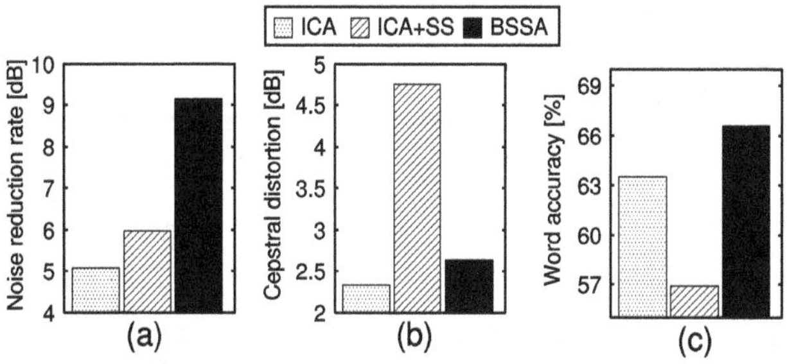

Fig. 12. Results of (a) noise reduction rate, (b) cepstral distortion, and (c) speech recognition test using each method (interference speech case).

is performed on the basis of spectral subtraction. However, the increase in the degree of distortion is expected to be negligible.

Finally, we show the speech recognition result in Figs. 11(c) and 12(c). It is evident that BSSA is superior to conventional ICA and ICA+SS.

5.2 Experiment in real world

An experiment in an actual railway-station environment is discussed here. Figure 13 shows the layout of the railway-station environment used in this experiment, where the reverberation time is about 1000 ms; this corresponds to mixing filters of 16000 taps in 16 kHz sampling. We used 16-kHz-sampled signals as test data; the original speech (6 s) was convoluted with impulse responses recorded in the same railway-station environment, to which a real-recorded noise was added. We use 46 speakers (200 sentences) as the original source of the target speech. The noise in the environment is nonstationary and is almost a non-point-source; it consists of various kinds of interference noise, namely, background noise and the sounds of trains, ticket-vending machines, automatic ticket gates, footsteps, cars, and wind. Figure 14 shows two typical noises, i.e., noises 1 and 2, which are recorded in distinct time periods and used in this experiment. A four-element array with an interelement spacing of 2 cm is used.

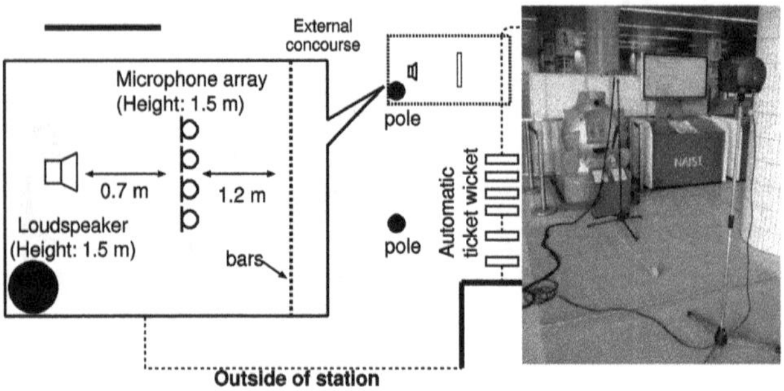

Fig. 13. Layout of railway-station environment used in our experiment.

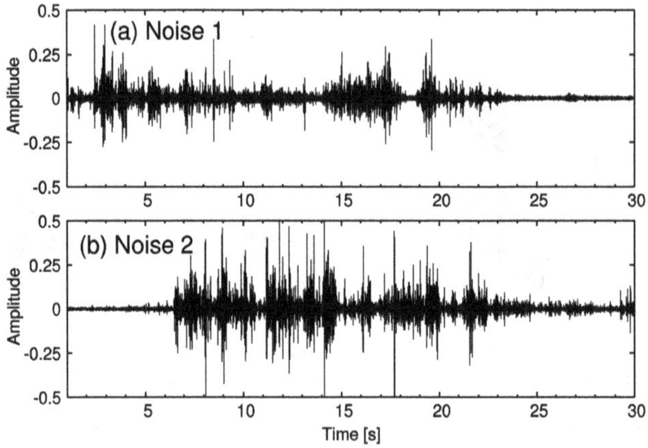

Fig. 14. Two typical noises in railway-station environment.

Figure 15 shows the real separation results obtained by ICA in the railway-station environment. We can ascertain the imbalanced performance between target estimation and noise estimation, similar to the simulation-based results (see Sect. 3.4).

In the next experiment, we compare conventional ICA, ICA+SS, and BSSA in terms of NRR, cepstral distortion, and speech recognition performance. Figure 16(a) shows the results of the average NRR for whole sentences. From these results, we can see that the NRR of BSSA that utilizes ICA as a noise estimator is superior to those of conventional methods. However, we find that the cepstral distortion in BSSA is greater than compared with that in ICA from Fig. 16(b).

Finally, we show the results of speech recognition, where the extracted sound quality is fully considered, in Fig. 16(c). The speech recognition task and conditions are the same as those in Sect. 5.1, as shown in Table 1. From this result, it can be concluded that the target-enhancement performance of BSSA, i.e., the method that uses ICA as a noise estimator, is evidently superior to the method that uses ICA directly as well as ICA+SS.

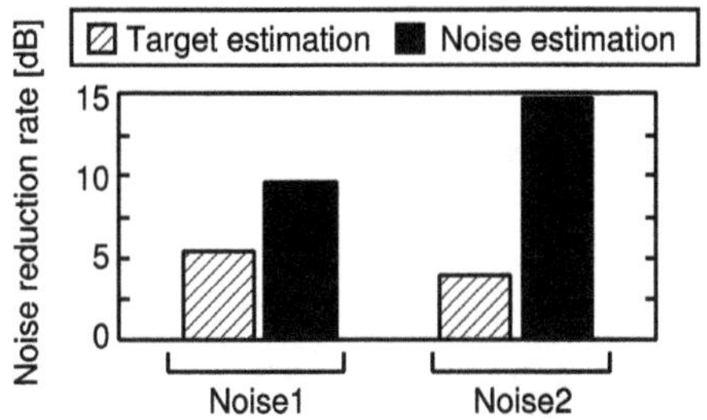

Fig. 15. NRR-based noise reduction performance of conventional ICA in railway-station environment.

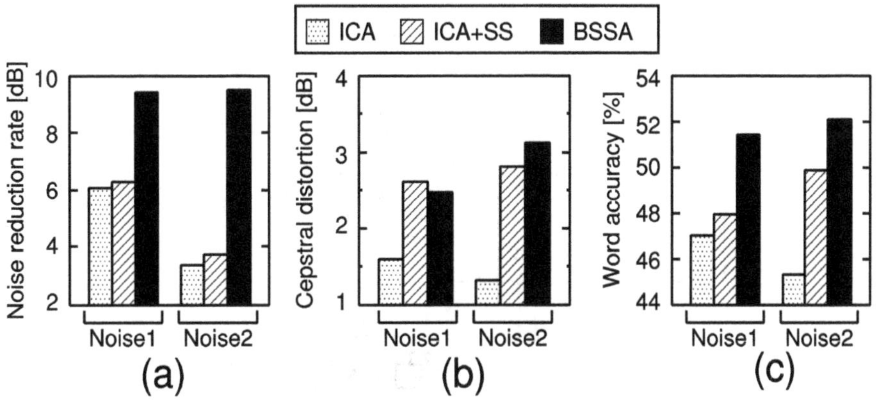

Fig. 16. Experimental results of (a) noise reduction rate, (b) cepstral distortion, and (c) speech recognition test in railway-station environment.

6. Real-time implementation of BSS

Several recent studies [19, 38, 39] have dealt with the issue of real-time implementation of ICA. The methods used, however, require high-speed personal computers, and BSS implementation on a small LSI still receives much attention in industrial applications. As a recent example of the implementation of real-time BSS, a real-time BSSA algorithm and its development are described in the following.

In BSSA's signal processing, the DS, SS, and separation filtering parts are possible to work in real-time. However, it is toilsome to optimize (update) the separation filter in real-time because the optimization of the unmixing matrix by ICA consumes huge amount of computations. Therefore, we should introduce a strategy in which the separation filter optimized by using the past time period data is applied to the current data. Figure 17 illustrates the configuration of the real-time implementation of BSSA. Signal processing in this implementation is performed as follows.

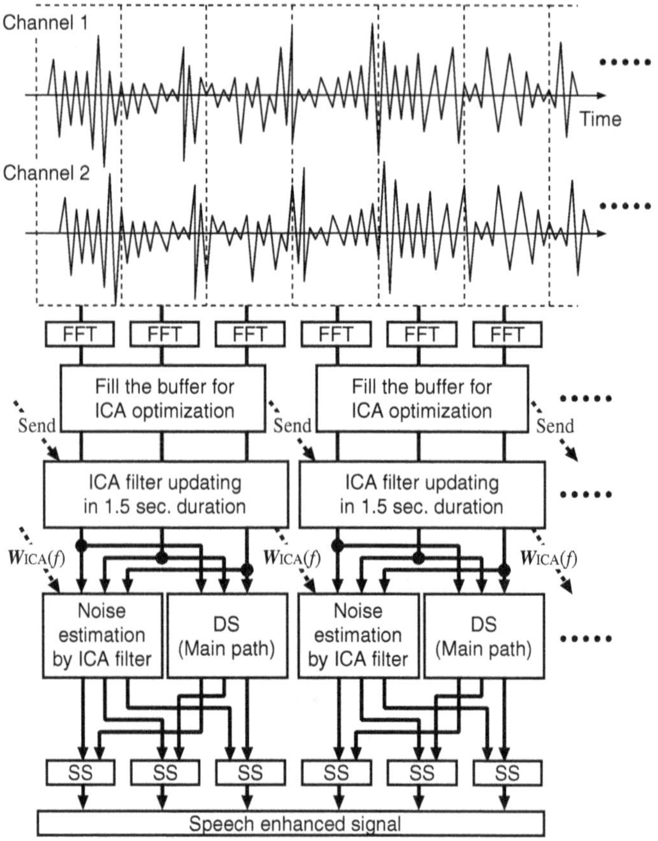

Fig. 17. Signal flow in real-time implementation of BSSA.

Step 1: Inputted signals are converted into time-frequency domain series by using a frame-by-frame fast Fourier transform (FFT).

Step 2: ICA is conducted using the past 1.5-s-duration data for estimating the separation filter while the current 1.5 s. The optimized separation filter is applied to the next (*not current*) 1.5 s samples. This staggered relation is due to the fact that the filter update in ICA requires substantial computational complexities and cannot provide an optimal separation filter for the current 1.5 s data.

Step 3: Inputted data is processed in two paths. In the primary path, the target speech is partly enhanced by DS. In the reference path, ICA-based noise estimation is conducted. Again, note that the separation filter for ICA is optimized by using the past time period data.

Step 4: Finally, we obtain the target-speech-enhanced signal by subtracting the power spectrum of the estimated noise signal in the reference path from the power spectrum of the primary path's output.

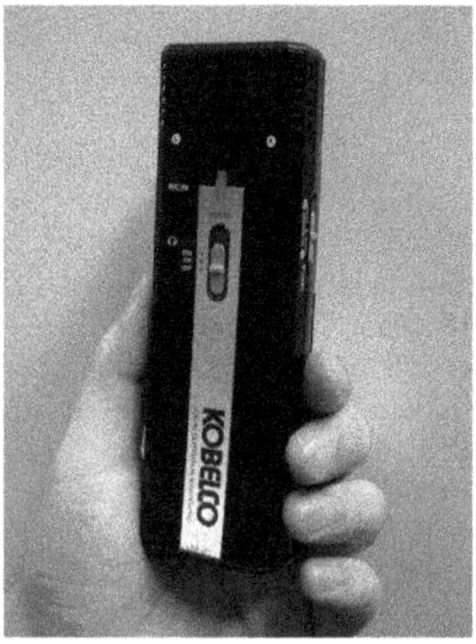

Fig. 18. BSS microphone (SSM-001 by KOBELCO Ltd., Japan) based on BSSA algorithm [40].

Although the update of the separation filter in the ICA part is not real-time processing, but involves a total latency of 3.0 s, the entire system still seems to run in real-time because DS, SS, and separation filtering can be carried out in the current segment with no delay. In the system, the performance degradation due to the latency problem in ICA is mitigated by oversubtraction in spectral subtraction.

Figure 18 shows an example of the hardware implementation of BSSA, which was developed by KOBELCO Ltd., Japan [40]. They have fabricated a pocket-size real-time BSS microphone, where the BSSA algorithm can work on a general-purpose DSP (TEXAS INSTRUMENTS TMS320C6713; 200 MHz clock, 100 kB program size, 1 MB working memory). This microphone was made commercially available in 2007 and has been adopted for the purpose of surveillance by the Japanese National Police Agency.

7. Conclusion

This chapter addressed the BSS problem for speech applications under real acoustic environments, particularly focusing on BSSA that utilizes ICA as a noise estimator. Under a non-point-source noise condition, it was pointed out that beamformers optimized by ICA are a DS beamformer for extracting the target speech signal that can be regarded as a point source and NBF for picking up the noise signal. Thus, ICA is proficient in noise estimation under a non-point-source noise condition. Therefore, it is valid to use ICA as a noise estimator. In experiments involving computer-simulation-based and real-recording-based data, the SNR improvement and speech recognition results of BSSA are superior to those of conventional methods. These results indicate that the ICA-based noise estimation is beneficial for speech enhancement in adverse environments. Also, the hardware implementation of BSS was discussed with a typical example of a real-time BSSA algorithm.

8. References

[1] B. H. Juang and F. K. Soong, "Hands-free telecommunications," *Proc. International Conference on Hands-Free Speech Communication*, pp. 5–10, 2001.

[2] S. Boll, "Suppression of acoustic noise in speech using spectral subtraction," *IEEE Transactions on Acoustics, Speech, and Signal Processing*, vol. ASSP-27, no. 2, pp. 113–120, 1979.

[3] G. W. Elko, "Microphone array systems for hands-free telecommunication," *Speech Communication*, vol. 20, pp. 229–240, 1996.

[4] J. L. Flanagan, J. D. Johnston, R. Zahn, and G. W. Elko, "Computer-steered microphone arrays for sound transduction in large rooms," *Journal of the Acoustical Society of America*, vol. 78, no. 5, pp. 1508–1518, 1985.

[5] M. Omologo, M. Matassoni, P. Svaizer, and D. Giuliani, "Microphone array based speech recognition with different talker-array positions," *Proc. ICASSP'97*, pp. 227–230, 1997.

[6] H. F. Silverman and W. R. Patterson, "Visualizing the performance of large-aperture microphone arrays," *Proc. ICASSP'99*, pp. 962–972, 1999.

[7] O. Frost, "An algorithm for linearly constrained adaptive array processing," *Proceedings of the IEEE*, vol. 60, pp. 926–935, 1972.

[8] L. J. Griffiths and C. W. Jim, "An alternative approach to linearly constrained adaptive beamforming," *IEEE Transactions on Antennas and Propagation*, vol. 30, no. 1, pp. 27–34, 1982.

[9] Y. Kaneda and J. Ohga, "Adaptive microphone-array system for noise reduction," *IEEE Trans. Acoust. Speech, Signal Process.*, pp. 2109–2112, 1986.

[10] P. Comon, "Independent component analysis, a new concept?" *Signal Processing*, vol. 36, pp. 287–314, 1994.

[11] J. F. Cardoso, "Eigenstructure of the 4th-order cumulant tensor with application to the blind source separation problem," *Proc. ICASSP'89*, pp. 2109–2112, 1989.

[12] C. Jutten and J. Herault, "Blind separation of sources part i: An adaptive algorithm based on neuromimetic architecture," *Signal Processing*, vol. 24, pp. 1–10, 1991.

[13] S. Ikeda and N. Murata, "A method of ICA in the frequency domain," *Proc. International Workshop on Independent Component Analysis and Blind Signal Separation*, pp. 365–371, 1999.

[14] P. Smaragdis, "Blind separation of convolved mixtures in the frequency domain," *Neurocomputing*, vol. 22, no. 1-3, pp. 21–34, 1998.

[15] L. Parra and C. Spence, "Convolutive blind separation of non-stationary sources," *IEEE Transactions on Speech and Audio Processing*, vol. 8, pp. 320–327, 2000.

[16] H. Saruwatari, S. Kurita, K. Takeda, F. Itakura, and T. Nishikawa, "Blind source separation combining independent component analysis and beamforming," *EURASIP Journal on Applied Signal Processing*, vol. 2003, pp. 1135–1146, 2003.

[17] D.-T. Pham, C. Serviere, and H. Boumaraf, "Blind separation of convolutive audio mixtures using nonstationarity," *International Symposium on Independent Component Analysis and Blind Signal Separation (ICA2003)*, pp. 975–980, 2003.

[18] H. Saruwatari, T. Kawamura, T. Nishikawa, A. Lee, and K. Shikano, "Blind source separation based on a fast-convergence algorithm combining ICA and beamforming," *IEEE Transactions on Speech and Audio Processing*, vol. 14, no. 2, pp. 666–678, 2006.

[19] Y. Mori, H. Saruwatari, T. Takatani, S. Ukai, K. Shikano, T. Hiekata, Y. Ikeda, H. Hashimoto, and T. Morita, "Blind separation of acoustic signals combining SIMO-model-based independent component analysis and binary masking," *EURASIP Journal on Applied Signal Processing*, vol.2006, Article ID 34970, 17 pages, 2006.

[20] B. Sallberg, N. Grbic, and I. Claesson, "Online maximization of subband kurtosis for blind adaptive beamforming in realtime speech extraction," *Proc. IEEE Workshop DSP 2007*, pp. 603–606, 2007.

[21] Y. Takahashi, T. Takatani, K. Osako, H. Saruwatari, and K. Shikano, "Blind spatial subtraction array for speech enhancement in noisy environment," *IEEE Transactions on Audio, Speech and Language Processing*, vol.17, no.4, pp.650–664, 2009.

[22] T.-W. Lee, *Independent Component Analysis*. Norwell, MA: Kluwer Academic, 1998.

[23] S. Ukai, T. Takatani, T. Nishikawa, and H. Saruwatari, "Blind source separation combining SIMO-model-based ICA and adaptive beamforming," *Proc. ICASSP2005*, vol. III, pp. 85–88, 2005.

[24] S. Araki, S. Makino, Y. Hinamoto, R. Mukai, T. Nishikawa, and H. Saruwatari, "Equivalence between frequency domain blind source separation and frequency domain adaptive beamforming for convolutive mixtures," *EURASIP Journal on Applied Signal Processing*, vol. 2003, no. 11, pp. 1157–1166, 2003.

[25] M. Brandstein and D. Ward, Eds., *Microphone Arrays: Signal Processing Techniques and Applications*. Springer-Verlag, 2001.

[26] H. Saruwatari, N. Hirata, T. Hatta, R. Wakisaka, K. Shikano, and T. Takatani, "Semi-blind speech extraction for robot using visual information and noise statistics," *Proc. of the 11th IEEE International Symposium on Signal Processing and Information Technology (ISSPIT2011)*, pp.238–243, 2011.

[27] A. Lee, K. Nakamura, R. Nishimura, H. Saruwatari, and K. Shikano, "Noise robust real world spoken dialogue system using GMM based rejection of unintended inputs," *Proc. 8th International Conference on Spoken Language Processing (ICSLP2004)*, vol.I, pp.173–176, 2004.

[28] S. B. Davis and P. Mermelstein, "Comparison of parametric representations for monosyllabic word recognition in continuously spoken sentences," *IEEE Trans. Acoustics, Speech, Signal Proc.*, vol. ASSP-28, no. 4, pp. 357–366, 1982.

[29] T. Inoue, H. Saruwatari, Y. Takahashi, K. Shikano, and K. Kondo, "Theoretical analysis of musical noise in generalized spectral subtraction based on higher-order statistics," *IEEE Transactions on Audio, Speech and Language Processing*, vol.19, no.6, pp.1770–1779, 2011.

[30] R. Miyazaki, H. Saruwatari, R. Wakisaka, K. Shikano, and T. Takatani, "Theoretical analysis of parametric blind spatial subtraction array and its application to speech recognition performance prediction," *Proc. of Joint Workshop on Hands-free Speech Communication and Microphone Arrays 2011 (HSCMA2011)*, pp.19–24, 2011.

[31] Y. Ephraim, and D. Malah, "Speech enhancement using a minimum mean-square error short-time spectral amplitude estimator," *IEEE Transactions on Acoustics, Speech, Signal Processing*, vol.32, no.6, pp.1109–1121, 1984.

[32] R. Okamoto, Y. Takahashi, H. Saruwatari, and K. Shikano, "MMSE STSA estimator with nonstationary noise estimation based on ICA for high-quality speech enhancement," *Proc. of IEEE International Conference on Acoustics, Speech, and Signal Processing (ICASSP2010)*, pp.4778–4781, 2010.

[33] H. Saruwatari, M. Go, R. Okamoto, and K. Shikano, "Binaural hearing aid using sound-localization-preserved MMSE STSA estimator with ICA-based noise estimation," *Proc. of International Workshop on Acoustic Echo and Noise Control (IWAENC2010)*, 2010.

[34] Y. Takahashi, H. Saruwatari, K. Shikano, and K. Kondo, "Musicalnoise analysis in methods of integrating microphone array and spectral subtraction based on higher-order statistics," *EURASIP Journal on Advances in Signal Processing*, vol.2010, Article ID 431347, 25 pages, 2010.

[35] R. Miyazaki, H. Saruwatari, and K. Shikano, "Theoretical analysis of amounts of musical noise and speech distortion in structure-generalized parametric spatial subtraction

array," *IEICE Transactions on Fundamentals of Electronics, Communications and Computer Sciences*, vol.95-A, no.2, pp.586–590, 2012.

[36] L. Rabiner and B. Juang, *Fundamentals of Speech Recognition*. Upper Saddle River, NJ: Prentice Hall PTR, 1993.

[37] A. Lee, T. Kawahara, and K. Shikano, "Julius – an open source real-time large vocabulary recognition engine," *European Conference on Speech Communication and Technology*, pp. 1691–1694, 2001.

[38] R. Mukai, H. Sawada, S. Araki, and S. Makino, "Blind source separation for moving speech signals using blockwise ICA and residual crosstalk subtraction," *IEICE Transactions on Fundamentals of Electronics, Communications and Computer Sciences*, vol.E87-A, no.8, pp.1941–1948, 2004.

[39] H. Buchner, R. Aichner, and W. Kellermann, "A generalization of blind source separation algorithms for convolutive mixtures based on second-order statistics," *IEEE Transactions on Speech and Audio Processing*, vol.13, no.1, pp.120–134, 2005

[40] T. Hiekata, Y. Ikeda, T. Yamashita, T. Morita, R. Zhang, Y. Mori, H. Saruwatari, and K. Shikano, "Development and evaluation of pocket-size real-time blind source separation microphone," *Acoustical Science and Technology*, vol.30, no.4, pp.297–304, 2009.

6

Unsupervised and Neural Hybrid Techniques for Audio Signal Classification

Andrés Ortiz, Lorenzo J. Tardón, Ana M. Barbancho and Isabel Barbancho
Dept. Ingeniería de Comunicaciones, ETSI Telecomunicación-University of Malaga, Campus Universitario de Teatinos s/n, Malaga Spain

1. Introduction

Audio signal analysis and classification have arisen as an important research topic that has been developed by many authors in different areas over the time. A main development context has been speech recognition Holmes & Huckvale (1994); Juang & Rabiner (2005); Kimura (1999). This specific topic is, in fact, an important source of references to applications of artificial intelligence techniques Juang & Rabiner (2005); Prasad & Prasanna (2008). Many classical problems encountered in this research field have been addressed from this perspective by many authors Farahani & Ahadi (2005); Minematsu et al. (2006).

However, in this same context, a different point of view can be adopted to deal with the analysis and classification of music signals. The tasks in this framework are varied, including the detection of pitch, tempo or rhythm Thornburg et al. (2007), but also other tasks like the identification of musical instruments Müller et al. (2011) or musical genre recognition Tzanetakis & Cook (2002) can be considered. Also the classification of audio samples as music or speech has been thoroughly considered in the literature Panagiotakis & Tziritas (2005), Tardón et al. (2010).

In this specific context of analysis of musical signals, we will expose some ideas regarding signal classification and their application to a real task.

2. Models and applications

Audio signal classification involves the extraction of a number of descriptive features from the sound and the proper utilization of them as input for a classifier. Artificial Intelligence (AI) techniques provide a way to deal with signal processing and pattern classification tasks from a different point of view of classical techniques.

AI techniques play an important role in the signal processing and classification context as they have been widely used by a number of authors in the literature for very different tasks Haykin (1999); Kohonen et al. (1996); Ortiz, Górriz, Ramírez & Salas-Gonzalez (2011a); Ortiz, Gorriz, Ramirez & Salas-Gonzalez (2011b); Ortiz, Ortega, Diaz & Prieto (2011); Riveiro et al. (2008). Also, we must be aware of the fact that the current trend in artificial intelligence systems points to the utilization of both classical and AI-based methods in conjunction to improve the overall system performance. This leads to build hybrid artificial intelligence systems. These systems can include neural-based techniques, evolutive computation, or statistical classifiers as well as other statistical techniques such as multivariate or stochastic methods.

Through this chapter, we will present the problem of classification of musical signals. The problem will be defined to be handled in a supervised or unsupervised way. It is be important to point out that a preprocessing stage to determine the features to be used for the classification task must be done although the classifiers can be able to properly deal with the different discrimination capability of the different features. Afterwards, a classifier will assign a different label related to the features employed to any different sound class in a process performed in a supervised or unsupervised way.

Taking into account that unsupervised classification strategies are able to organize training samples into suitable groups according to certain classes without using any a priori information (the samples to be classified are not labelled), we will describe how to apply these ideas in this context. We will present features related to the analysis framework and samples of the performance.

Unsupervised techniques include the Self-Organizing Maps (SOM) Kohonen (2001), a vector quantization method with a competitive learning process. SOMs provide a generalization of the input space through the prototypes computed during the training phase. In this model, each prototype represents a set of input vectors on the basis of a given similarity measurement. Very often, the similarity measure selected is the Euclidean distance. Moreover, it will be specially interesting to observe that SOMs group the prototypes maintaining the more similar ones close to each other in the output space while the less similar prototypes ones are kept apart. In this way, SOMs provide valuable topological information that can be exploited to make them more flexible than other vector quantization methods. This means that a description of a SOM model for a certain application must be complemented by additional details. Specifically, we are referring to two different aspects: the first one consists on the modelling of the activation process of the neurons of the SOM which can be done by means of statistical techniques such as Gaussian Mixture Models (GMM) to account for a probabilistic behaviour of the map. The second aspect to consider is related to the techniques that allow the extraction of valuable topological information from the SOM; this process can be accomplished by classical clustering techniques such as k-means or other techniques Therrien (1989) specially developed for clustering the SOM.

On the other hand, supervised classification Murtagh (1991) techniques use a priori information of at least some of the samples. It means that the training samples have labels according to their corresponding class. Classical methods include different clustering techniques or statistical classifiers. We will introduce some labels and related features for classification and describe the utilization of them to attain certain classification objectives.

3. Audio signal features for classification: Clarinet music example

In this section we describe five features extracted from audio signal. These features will compose the input data used in the unsupervised classifiers shown in the following sections. The examples shown aims to detect four different playing techniques from clarinet music. The features extracted from the audio signal attempts to characterize the signal in both, time and frequency domains in a simple way, being effective enough to perform classification experiments.

Classification examples have been performed using three different clarinets from real recordings taken from the Musical Instrument Sound Data Base RWC-MDB-1-2001-W08 Goto (2004). Clarinet 1 is a french clarinet made by Buffet, Clarinet 2 is a french clarinet made by Selmer and Clarinet 3 is a japanese clarinet many by Yamaha. For each clarinet, we have 120 note samples that contain the whole clarinet note range played with different dynamics for each playing technique (Normal, Staccato, Vibrato and Trill). This gives a total of 1440 note samples.

3.1 Time domain characterization

In this Section, we describe the features used to characterize the audio signal in time domain. The duration and the shape of the envelope of the clarinet audio signals contain information about the technique used to play the notes.

Thus, as in Barbancho et al. (2009); Jenses (1999), the attack time (T_a) is considered from the first sample that reaches a 10% of the maximum of the amplitude of the waveform of the note until it reaches the 90% of that amplitude. The release time (T_r) is considered from the last sample that reaches 70% of the maximum of the amplitude of the waveform to the last one over the 10% of the amplitude. On the other hand, the time between the attack time and the release time is called the sustain time T_S. T_a, T_r and T_s depend on the playing technique.

The signal envelope is obtained by filtering the signal with a 5th order Butterworth filter with cut-off frequency of 66 Hz. After the application of the low-pass filter, the signal is normalized so that the amplitude is 1. On the other hand, the samples with amplitude under 2.5% of the maximum are removed.

Additionally, we include another time domain feature T_f based on the presence of signal fading (i.e.: if the signal envelope is fading, $T_f = 1$, otherwise $T_s = 0$).

3.2 Frequency domain characterization

In order to characterize the clarinet in the frequency domain, we use the *Fast Fourier Transform* (FFT). In the frequency domain, the frequency axis is converted into MIDI numbers according to the following equation:

$$MIDI = 69 + 12\log_2(f/440) \tag{1}$$

Taking into account that the fundamental frequency of the notes of the clarinet ranges from 146.83 Hz to 1975 Hz, the MIDI range of interest is 50 to 94.

However, it is necessary to remove redundant information from the FFT in order to simplify the signal spectrum. Thus, for a certain MIDI number n_{MIDI} the spectrum between $n_{MIDI} - 0.5$ and $n_{MIDI} + 0.5$ is considered and the maximum value of the spectrum in that interval is assigned to n_{MIDI}. Thus, the MIDI number spectrum (or MIDI simplified spectrum) of each note will have 45 samples. From this simplified spectrum, the pitch and the spectrum width around the pitch will be calculated. At this point, we have three time domain features (T_a, T_r, T_s and T_f) and two frequency domain features, F_p and F_w. F_p is the pitch of the played note, and F_w is the spectral width of the fundamental frequency defined and calculated as the number of significative samples around the fundamental frequency. We consider as significant samples those over 30% of the value of the fundamental frequency. Thus, the feature space is composed by six-dimensional vectors in the form (T_a, T_r, T_s, T_f, F_p, F_w). These six features may be discriminant enough to characterize the playing technique.

3.3 Playing technique characterization

The time and frequency domain features previously described can be used for characterizing different playing techniques from clarinet music samples. Thus, we will recognize four playing techniques as *normal (NO)*, *staccato (ST)*, *vibrato (VI)* and *trill (TL)*.

Normal playing technique presents a short attack time (T_a) (see Fig. 1). Figure 1, presents the audio waveform, the envelope and the simplified spectrum of an *A*4 played normal. This note corresponds to the middle range of the notes played with a violin. There not special characteristics if notes with higher or lower pitches are played. The shortness of the attack time, and the lengths of the sustain time (T_s) and the release time (T_r) (the time it takes the sound to die when the air pressure, that maintains the clarinetist, finishes) can be observed.

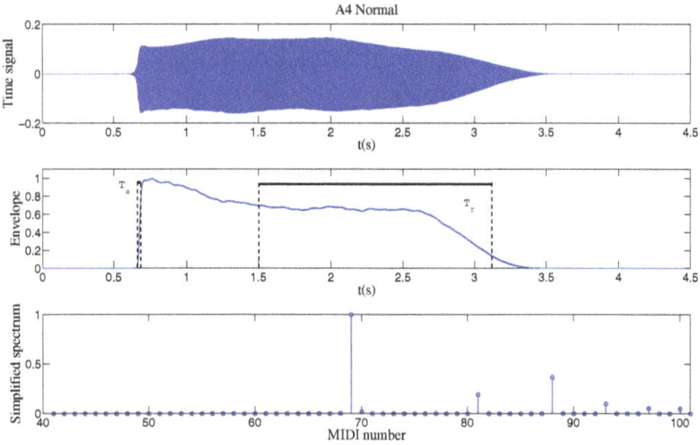

Fig. 1. Waveform, envelope and simplified spectrum of an A4 played normal.

In *staccato*, the duration of the note is reduced to a half, that is, the main characteristic of a note played in staccato will be its short duration. In this technique the clarinettist, produces an constant air pressure, but his tongue will be in the reed to avoid any vibration. When the clarinettist removes the tongue from the reed, a fast attack is get and just after put again his tongue in the reed, so the sustain time and the release time are reduced. So, in this technique, the attack, sustain and release times are going to be reduced. Figure 2, presents the audio waveform, the envelope and the simplified spectrum of an *G*4 played in staccato.

Vibrato stands for a slight fluctuation of the amplitude and the pitch of the notes. This technique can be an undesirable effect when it is produced unwittingly due to the nervousness of the clarinettist. Vibrato is a difficult playing technique and it requires certain level of expertise. In this technique, the clarinettist produces a periodic variation of the air pressure by means of contractions of the diaphragm or lip pressure over the reed. Consequently, when this technique is used, the envelope shows periodic oscillations. Figure 3, shows the envelopes of *E*3 and *D*6 played in normal mode and in vibrato and Figure 4, shows the simplified spectrum of *E*3 and *D*6 played in normal mode and in vibrato.

Trill is a quavering or vibratory sound, specially a rapid alternation of sung or played notes. This technique consists on changing between one note and the following one very quickly by moving fingers. So, it should be expected that the envelope and the spectrum present notable

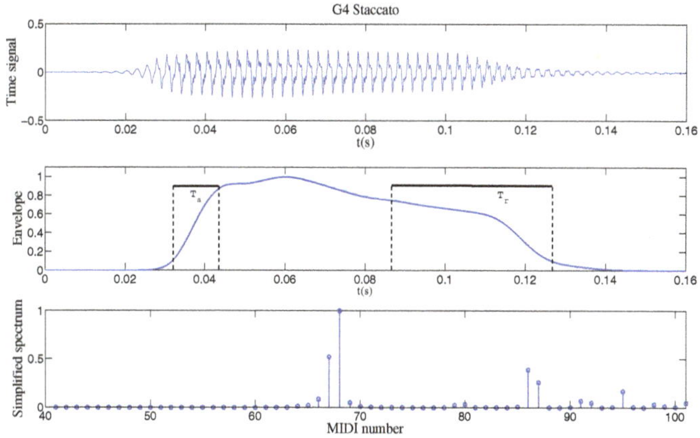

Fig. 2. Waveform, envelope and simplified spectrum of an *G4* played staccato.

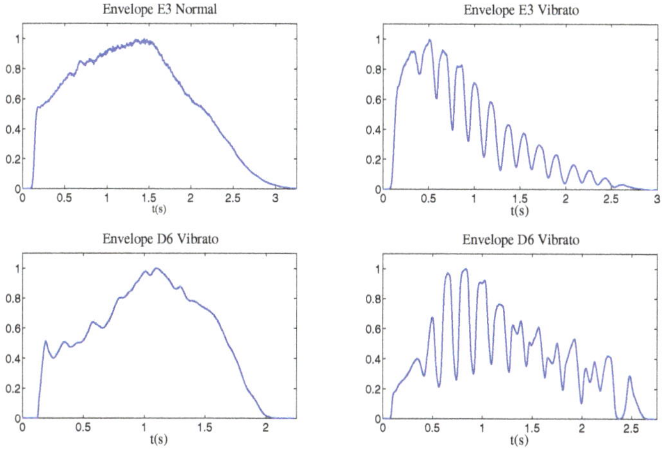

Fig. 3. Envelopes of *E3* and *D6* played in normal mode and in vibrato.

differences with respect to the ones found when the other playing techniques are employed. Figure 5, presents the time signal, the envelope and the simplified spectrum of an *G5* played with trill.

4. Classification with Self-Organizing Maps

The Self-Organizing Map (SOM) is one of the most commonly used artificial neural network models for unsupervised learning. This model, proposed by Kohonen Kohonen (2001), is a biologically inspired algorithm based on the search for the most economic representation of data and its relationships, as in the animal brain.

Sensory experience consists in capturing features from the surrounding world, and these features usually are multidimensional. For instance, the human visual system captures

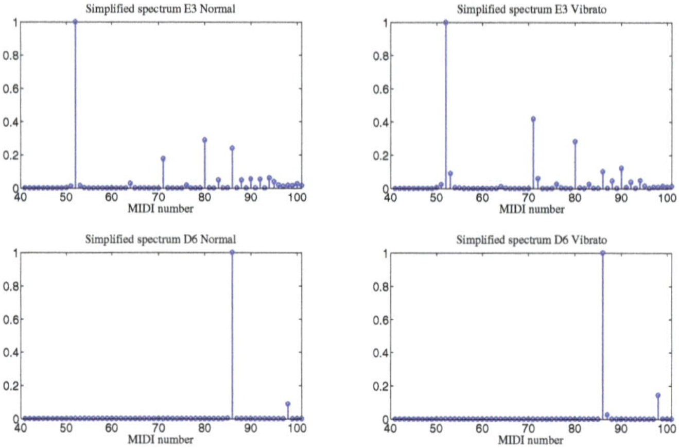

Fig. 4. Simplified spectrum of *E3* and *D6* played in normal mode and in vibrato.

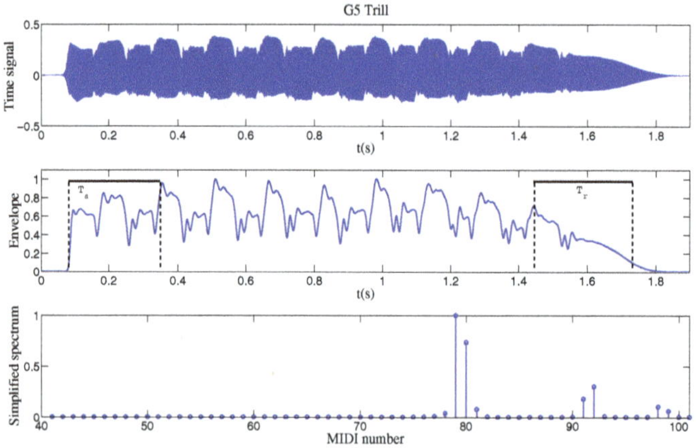

Fig. 5. Waveform, envelope and simplified spectrum of an *G5* played with trill.

features from different objects such as colour, texture, size or shape, which will provide enough information to ensure further recognition of that object. The learning process in the human brain projects the extracted features onto the neural structures of the brain to creating different maps corresponding to different types of features. Moreover, the learning process stores the prototypes of the features in the maps created. These maps are continuously modified as the learning process progresses. Prototypes could be considered as the smallest feature set needed to be able to represent the sensory information acquired. In other words, prototypes represent generalizations of the learnt features, being possible to distinguish between different objects, and associate similar ones. Moreover, sensory pieces of information coming from different organs are topologically ordered in the brain cortex (topology preservation in brain mapping).

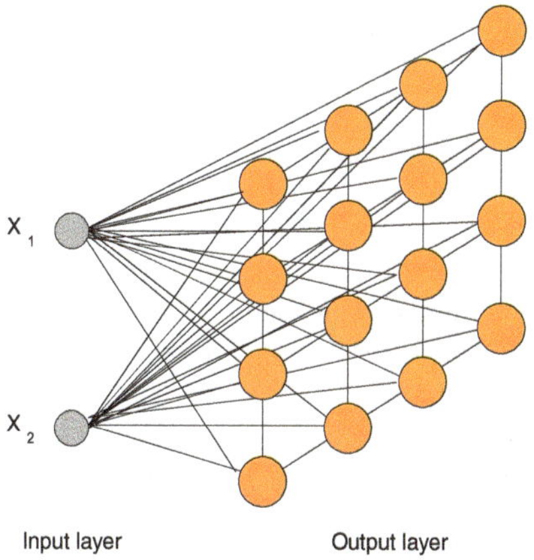

Fig. 6. Self-Organizing Map architecture.

Thus, as in the brain learning process, the main purpose of SOMs is to group the similar data instances close into a two or three dimensional lattice (output map), keeping apart the different ones. Moreover, as the difference between data instances increases, the distance in the output map also increases.

SOMs consist of a number or neurons, also called units, which are arranged following a previously determined 2D or 3D lattice, and each unit stores a multidimensional prototype. Thus, topology is preserved as in the human brain, which is a unique feature of the SOM. Figure 6 shows the SOM architecture in which the output layer is arranged in a 2D lattice.

Units on the SOM are self organized during the training phase. During this stage, the distance between any input vector and the weights associated to the units on the output map are calculated. Usually, the Euclidean distance is used as shown in Equation 2:

$$U_{\omega(t)} = \underset{i}{\operatorname{argmin}} \parallel x(t) - \omega_i(t) \parallel \tag{2}$$

where $x(t) \in X$, is the input vector at time t and $\omega_i(t)$ is the prototype vector associated to the unit i.

The unit closest to the input vector $U_\omega(t)$ is referred to as winning unit and the associated prototype is updated. To complete the adaptive learning process of the SOM, the prototypes of the units in the neighbourhood of the winning unit are also updated according to equation 3 Then, the unit nearest the input vector is referred to as winning unit and the associated weight is updated. Moreover, the weights of the units in the neighbourhood of the winning unit are also updated according to Equation 3:

$$\omega_i(t+1) = \omega_i(t) + \alpha(t) h_{U_i}(t) (x(t) - \omega_i(t) \tag{3}$$

where $\alpha(t)$ is the exponential decay learning factor and $h_{U_i(t)}$ is the neighborhood function associated to the unit i. Both, the learning factor and the neighborhood function decay with time, thus the prototypes adaptation becomes slower as the neighborhood of the unit i contains less number of units. This is a competitive process in which the winning neuron each iteration is called Best Matching Unit (BMU).

The neighbour function defines the shape of the neighbourhood and it is usually a Gaussian function which shrinks at each iteration, as defined by the next equations:

$$h_{U_i}(t) = e^{-\frac{\|r_U - r_i\|^2}{2\sigma^2(t)}} \tag{4}$$

$$\sigma(t) = \sigma_0 e^{\frac{-t}{\tau_1}} \tag{5}$$

In Equation 4, r_i represents the position on the output space (2D or 3D) and $\| r_U - r_i \|$ is the distance between the winning unit and the i-neuron on the output space. On the other hand, $\sigma(t)$ controls the reduction of the Gaussian neighborhood at each iteration. $\sigma(t)$ usually takes the form of exponential decay function as in Equation 5.

Similarly, the learning factor in Equation 3, also diminishes in time. However, α may decay in a linear or exponential fashion. Therefore, during the training phase, the prototypes associated to each unit are computed at each iteration. At the same time, the position of the units in the output space changes according to the similarity among the actual prototypes.

As previously commented, topology preservation is a unique feature of the SOM related to the goodness of the clustering process performed during training. This way, the calculation of the quality of the output map, results essential to evaluate the overall process.

In order to perform a quantitative evaluation of the goodness of the SOM, two measures can be accomplished. The first one is the quantization error, which is a measure of the resolution of the map. This quantization error can be calculated by computing the average distance between all the BMUs and the input data vectors as shown in Equation 6:

$$qe_i = \sum_{x_j \in c:i} \| \omega_i - x_j \| \tag{6}$$

The second one measures the topographic error, i.e.: how the SOM preserves the topology. This error can be computed using Equation 7:

$$t_e = \frac{1}{N} \sum_{i=1}^{N} u(\overrightarrow{x_i}) \tag{7}$$

In this equation, N is the total number of input vectors. $u(\overrightarrow{x_i})$ is 1 if the first and the second BMU for the input vector $\overrightarrow{x_i}$ are adjacent units and 0 otherwise Arsuaga & Díaz (2005); Kohonen (2001). Then, the lower q_e and t_e, the better the SOM is adapted to the input patterns.

5. SOM clustering

In order to deal with fully unsupervised classification using SOMs, it is necessary to compute the clusters brought up during the training phase.

This clustering process consist on grouping the prototypes in different classes according to a similarity criterion. This similarity criterion depends on the clustering technique used. Thus, several clustering approaches which use the Euclidean distance as similarity criterion have been developed Murtagh (1995); Rossi & Villa (2009); Vesanto et al. (2000); Wu & Chow (2004). In this chapter, we deal with SOM clustering using two different approaches.

5.1 SOM clustering with K-means

k-means algorithm MacQueen (1967) is a well-known and widely used clustering method due to its performance and simplicity. *k-means* aims to create k partitions from n observations in such a way that each observation will belong to the cluster with the nearest mean to that observation. In other words, the algorithm computes the centroids of each class at each iteration, and computes the euclidean distance to each observation. Then, the nearest observations to each centroid are considered as belonging to the centroid class.

Thus, given a $n - dimensional$ data collection $(x_1...x_n)$, *k-means* creates $k, k < n$ classes, while minimizing the *mean squared error* of the euclidean distance between each data point and the corresponding cluster centroid, as shown in equation 8:

$$\underset{C}{\operatorname{argmin}} \sum_{i=1}^{k} \sum_{x_j \in m_i} \| x_j - s_i \|^2 \tag{8}$$

where m_i represents the mean (centroid) of all the points belonging to the class i.

Although k-means constitutes a usual way to cluster the SOMs in an unsupervised manner, it presents two main drawbacks:

- The number of clusters to be found has to be determined in advance. In other words, it is necessary to know beforehand the value of k.
- k-means only uses the Euclidean distance between prototypes, and does not take into account the relationship between the output and the data topology to compute the clusters. This underutilizes the knowledge available at the output layer of the SOM given by data topology and data distribution Tasdemir & Merenyi (2009); Tasdemir et al. (2011).

Hence, to deal with fully unsupervised classification using SOMs, supervised clustering techniques have to be used at the output layer. In this sense, the CONN clustering method Tasdemir et al. (2011), which is based on SOM visualization of a weighted Delaunay graph, not only performs unsupervised clustering but also takes into account data distribution and topology.

5.2 CONN clustering

CONN clustering computes clusters on the SOM output layer using a new similarity metric for the prototypes Tasdemir & Merenyi (2009). This similarity measure is based on the receptive field of each unit, instead of using the Euclidean distance, unlike other clustering algorithms. Moreover, this technique does not need to know the number of clusters to be found, as in k-means. CONN clustering computes the number of clusters during the clustering process.

As described in Section 4, each SOM unit has an associated prototype ω_i. Each of these units is the centroid of its voronoi polyhedron containing the receptive field of unit i, as described in equation 9:

$$RF_i = \{x_k \in X : \| x_k - \omega_i \| \leq \| x_k - \omega_j \forall j \in S\} \tag{9}$$

where X is the data manifold and S is the set of units on the SOM layer.

Then, it is possible to define the connectivity strength between two prototypes ω_i and ω_j as data vectors for which ω_i and ω_j are the BMU and second BMU Tasdemir & Merenyi (2009);

Tasdemir et al. (2011):

$$RF_{ij} = \{x_k \in RF_i / \parallel x_k - \omega_j \parallel \leq \parallel x_k - \omega_l \parallel, \forall l \neq i\} \tag{10}$$

Then, the connectivity strength matrix $CONN(i,j)$ is created as $CONN(i,j) = |RF_{ij} - RFji|$ where each element indicates the connectivity between the prototypes ω_i and ω_j. In this way, if $CONN(i,j) = 0$, it indicates that ω_i and ω_j are not connected.

Thus, it is possible to determine not only the similar prototypes which are included in a cluster but also the number of clusters with a fully unsupervised clustering technique. In addition, this clustering technique exploits the relationship between the data topology and the SOM layer in a hierarchical agglomerative clustering method.

Figure 7a shows the labels assigned to each SOM unit when audio signal features are extracted from clarinet music, and Figure 7b shows the clustering of the output layer using k-means, computed using the *SOM toolbox* Vesanto et al. (2000). As commented above, it is necessary to know the number of clusters (k) beforehand to run the k-means algorithm. Thus, it is possible to compute the number of clusters that attain the best performance using the k-means algorithm, using a measure of the validity of the clustering.

There are several metrics to evaluate the validity of the clustering process, such as the Davies-Boulding index Davies & Bouldin (1979), the Generalized Dunn index (GDI) Bezdek & Pal (1998), PBM index Hassar & Bensaid (1999) or the silhouette width criterion Kaufman & Rosseauw (1990). These validity indexes provide lower or higher values as the clustering improves depending on the specific algorithm. Particularly, DBI provides lower values for better clustering results. This index is defined by the next equation:

$$DBI = \frac{1}{K} \sum_{k=1}^{K} \max_{i \neq j} \left(\frac{S_K Q_i + S_K Q_j}{S(Q_i, Q_j)} \right) \tag{11}$$

where K is the number of clusters, S_K is the average distance of all objects from the cluster to their cluster centre and $S(Q_i, Q_j)$ the distance between clusters centroids.

Then after several runs of the k-means algorithm the DBI has been computed and represented in Figure 7c. According to this Figure, $k = 4$ leads to the best clustering scheme since it attains the minimum DBI.

On the other hand, Figure 8 shows the clustering found using the CONN clustering algorithm described above.

6. SOM modelling

In Section 4, the SOM learning model was described. In that model, the unit corresponding to the prototype which is closest to the data instance is activated (BMU). This imposes with a binary response of each unit, since each unit is activated or deactivated but intermediate states are not considered.

A variant of the SOM consist on measuring the response of the map units instead of calculating the BMU as the unit which is closest to the input data. This is related to a probabilistic view of the output layer. In order to provide the SOM with this probabilistic behaviour, a Gaussian Mixture Model (GMM) is built over the output layer Kohonen (2001). Thus, the

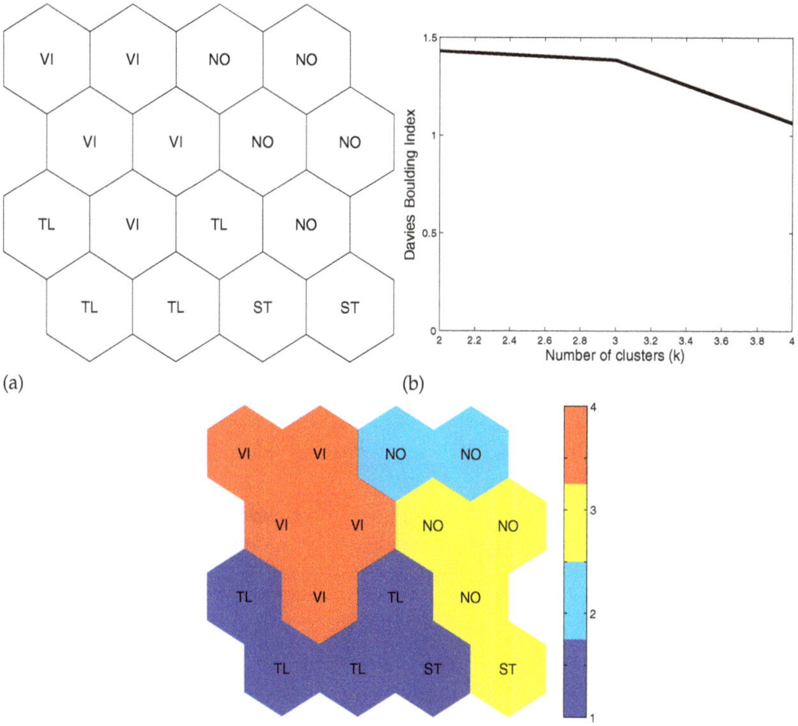

Fig. 7. (a) Unit labels, (b) Davies-Boulding index for different values of k (c) Clustering result for minimum DBI.

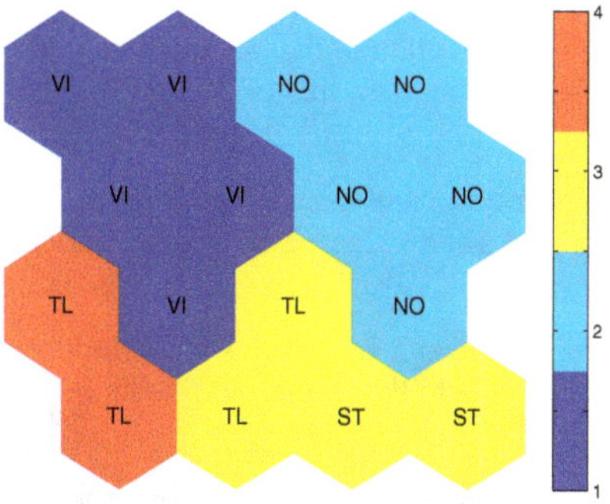

Fig. 8. SOM clustering using the CONN Tasdemir et al. (2011) algorithm.

BMU is determined computing not only the minimum distance from an input vector but also taking into account the likelihood of a unit of being the BMU. This way,the responses of the units surrounding the BMU can be taken into account.

Thus, the prior probability of each map unit i is computed in a similar way as in Alhoniemi et al. (1999), as shown in equation 12:

$$p(i) = \frac{\#\widetilde{X_i}}{\#\widetilde{X}} \tag{12}$$

where $\#\widetilde{X}$ is the total number of input vectors and $\#\widetilde{X_i}$ is the number of vectors whose closest prototype is ω_i. More specifically, $\#\widetilde{X_i}$ is the number of sample vectors found by equation 13:

$$\widetilde{X_i} = \{x \in V \ / \ \|x - m_i\| \leq \|x - m_k\| \ k = 1, ...N\} \tag{13}$$

Thus, $\#\widetilde{X_i}$ can be defined as the set of data samples whose first BMU is the unit i (Voronoi set of unit i).

The GMM is built according to the equation 14:

$$P(x_1, ..., x_n) = \sum_N p_i P_i(x_1, ..., x_n) \tag{14}$$

where the weights p_i for each Gaussian component correspond to the prior probabilities computed in equation 12. In 14, each individual Gaussian component P_i corresponds to the $n - dimensional$ weights associated to each unit (prototype vectors) Alhoniemi et al. (1999); Riveiro et al. (2008). The mean of each individual Gaussian component (kernel centre) is the weight vector of the corresponding unit itself, while the covariance matrix for the i-component is given by the dispersion of the data samples around the prototype i.

Once the GMM model has been built, the response of the unit i can be computed as the posterior probability by using the Bayes theorem.

$$p(\omega_k|x) = \frac{p(x|\omega_k)P(\omega_k)}{p(x)} \tag{15}$$

In equation 15, $p(\omega_k|x)$ represents the probability that a sample vector x belongs to class ω_k. $p(x|\omega_k)$ is the probability density function of the prototype ω_k computed from the GMM and $p(x)$ is a normalization constant. This way, this posterior probability can be used to classify new samples.

Figure 9 shows the mixing proportions of each component in the GMM that correspond to each unit on the SOM. Thus, peaks and valleys can be used to identify th units to be activated with largest probability.

In this way, SOM modelling provides a framework to include a probabilistic behaviour to the SOM, making it possible to modify the activation likelihood of each unit by means of the mixing proportions.

On the other hand, 6-dimensional SOM prototypes can be projected into a 2 or 3 dimensional space using PCA and storing the 2 or 3 principal components with the largest eigenvalues, respectively. Then, also the reduced prototypes can be modelled using a GMM. Figure 10a shows the clusters computed by the projected prototypes considering a 2-dimensional GMM. Similarly, Figure 10b shows the clusters computed using a 3-dimensional GMM.

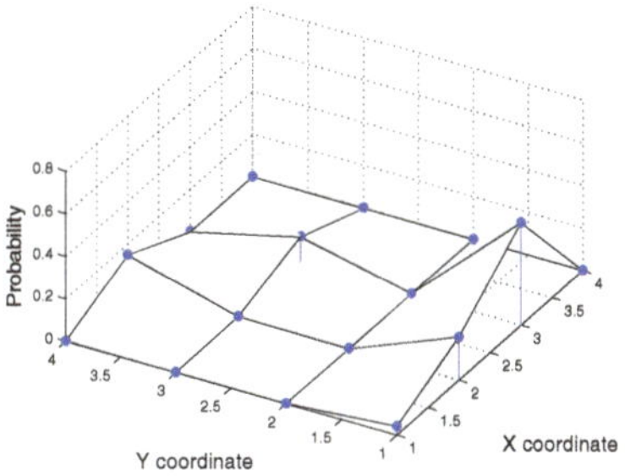

Fig. 9. Activation probability of each SOM unit.

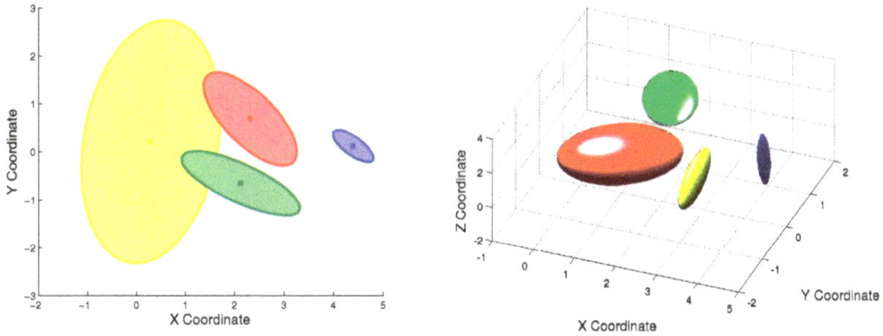

(a) (b)

Fig. 10. GMM modelling of the projected prototyped into (a) 2D space and (b) 3D space.

6.1 SOM labeling

Once the map is trained, a label can be assigned to each unit in order to identify the receptive field of this unit. This labelling process can be addressed in two main ways:

1. Labelling with majority voting scheme Vesanto & Alhoniemi (2000); Vesanto et al. (2000). Each unit is labelled with the most frequent label on its 1-neighbourhood. In this case, the size of the neighbourhood for labelling is fixed to radius 1. This process is performed once after the SOM is trained and the assigned labels are used in further classification tasks.
2. Dynamic labelling Cruz et al. (2012); Ortiz, Górriz, Ramírez & Salas-Gonzalez (2011a); Ortiz, Gorriz, Ramirez & Salas-Gonzalez (2011b); Ortiz, Ortega, Diaz & Prieto (2011). As commented in Section 6, the response of the SOM units computed from the posterior probabilities are used to label those units which remain unlabelled during the training process. This way, when a new sample arrives, the BMU with the maximum *a posteriori* probability is selected. However, this BMU could be unlabelled if the map is big enough. In that case, the label of this unit is computed taking into account the response of the units in the neighbourhood of the BMU. Hence, the label assigned to the BMU will be the label of the unit in the neighbourhood $\mathcal{L}_p$ which provides the strongest response at the unlabelled BMU, as shown in equation 16:

$$\mathcal{L}_p = \mathcal{L}\{argmax_i\{p(\omega_i|x)\ \forall i\ \in \mathcal{B}_N\}\} \tag{16}$$

This leads to a dynamic method which labels the units according to the response strength of the neighbourhood units.

Figure 11a shows *a posteriori* probabilities associated to each unit in the 1-neighbourhood of the BMU, which is currently unlabelled. As indicated in the figure legend, darker color correspond to larger *a posteriori* activation probability. Then, the label of the unit in the neighbourhood with the largest activation probability is assigned to the BMU, as depicted in Figure 11b.

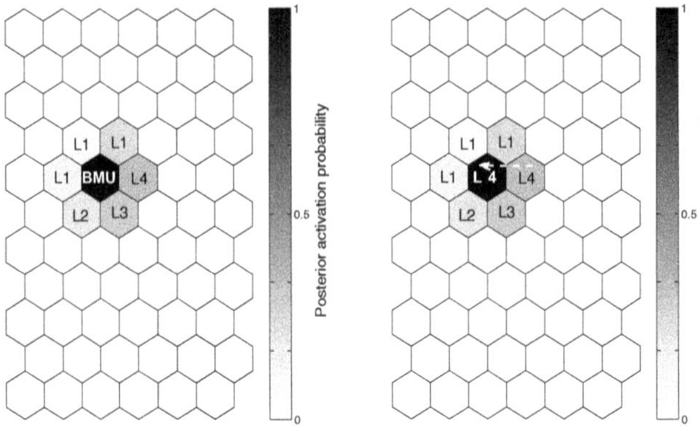

(a) (b)

Fig. 11. Probabilistic labelling procedure Cruz et al. (2012).

7. SOM modelling and clustering using ICA

SOM modelling has been addressed by means of Gaussian Mixture Models. This assumes Gaussian distributions on SOM prototypes. An alternative for SOM modelling consist in using Independent Component Analysis (ICA) Mixture Models Ghahramani & Beal (2000); Lee et al. (2000); Tipping & Bishop (1999). This deals with an unsupervised classification of SOM prototypes by modelling them as a mixture of several independent classes. Each of these classes are described by linear combinations of independent non-Gaussian densities Lee et al. (2000). Thus, in the same way the previously model uses multivariate Gaussian to model the SOM prototypes, ICA mixture models can be used to estimate the probability of each data instance to be generated by a non-Gaussian structure. Let $X = \{x_1, ..., x_n\}$ be the data set generated by a mixture model. The likelihood of the data can be expressed in terms of the component densities in a similar way that in Equation 14, but in this case, each component has a non-gaussian density derived from the ICA model. Then, the data belonging to each class are described as

$$x_t = A_k s_k \tag{17}$$

where A_k is the mixing matrix in ICA and s_k is the source vector.

Equation 17 assumes that the individual sources are independent, and each class was generated from 17 using a different mixing matrix A_k. This way, it is possible to classify the input data and to compute the probability of each class for each data point. In other word, this allows calculating the probability of a data instance to be generated from an independent component. Nevertheless, as each SOM unit is considered as a kernel center, gaussian components are usually more suitable to provide a probabilistic measurement of the SOM units activation.

8. Growing Hierarchical Self-Organizing Map

The main drawback of SOMs is the size of the map, which has to be selected beforehand. In addition, the performance of the classification process depends on the size of the SOM. Additionally, the performance of SOMs with highly dimensional input data highly depends on the specific features and the calculation of the clusters borders may be not optimally defined. Taking into account these issues, the Growing Hierarchical Self-organizing Map (GHSOM) Dittenbach et al. (2000); Rauber et al. (2002) arises as a convenient variant of SOMs. GHSOMs dynamically grow to overcomes the limitations of the SOMs and to discover inherent hierarchies on the data.

GHSOM is a hierarchical and non-fixed structure developed to overcome the main limitations of classical SOM Rauber et al. (2002). GHSOM structure (shown in Figure 12 consists of multiple layers in which each layer is composed by several independent SOM maps. Hence, during the training process, the number of the SOM maps on each layer and the size of each of these SOMs is determined. This constitutes an adaptive growing process in horizontal and vertical ways. The growing process is controlled by two parameters that control the depth of the hierarchy and the breadth of each map. Therefore, these two parameters are the ones which have to be determined beforehand.

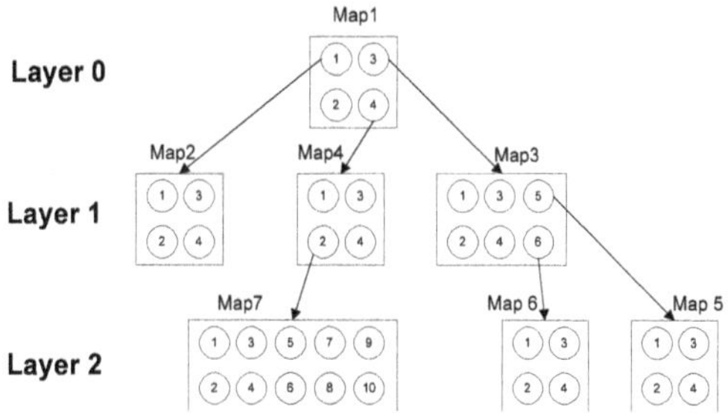

Fig. 12. GHSOM growing example.

In order to determine the limit of the growth of the GHSOM, the quantization error of each unit is calculated according to the next equation:

$$qe_i = \sum_{x_j \in C_i} \parallel \omega_i - x_j \parallel \tag{18}$$

where C_i is the set of input vectors mapped into the unit i, x_j is the j-th input vector belonging to C_i and is the weight associated to the unit i.

Initially, all the input vectors are considered to belong to C_0. This means that the whole available input data are used to compute the initial quantization error, qe_0. Then, the quantization errors qe_i for each neuron are calculated. Thus, if $qe_i < \tau_2 \times qe_0$, then the neuron i is expanded in a new map on the next level of the hierarchy. Each new map is trained as an independent SOM and the BMU calculation is performed as shown in Equation 3 by using the Euclidean distance metric. Once the new map is trained, the quantization error of each neuron on this map is computed as

$$q_i = \sum_{x_j \in C_i} \| \omega_i - x_j \| \tag{19}$$

(where q_i is the quantization error of the unit u on the upper layer). Then, the mean quantization error MQE_m of the new map is computed. If $MQE_m \leq \tau_1 q_i$ the map stops growing.

All the growing process is depicted in Figure 12. If τ_1 and τ_2 are selected in such a way that the GHSOM is sightly oversized, some of the units on the GHSOM maps may remain unlabelled after training.

Classification using GHSOM can be accomplished in two ways:

1. Using the clusters computed by GHSOM. These clusters correspond to the receptive field of each map on the lower level of the hierarchy. In this case, since labelling information has not been used, it is possible to distinguish different data instances because the BMU of these instances will be in the corresponding map. However, the identification of a specific playing technique is not possible. This is shown in Figure 13.
2. Using the labelled data for GHSOM training. In this case, although the clustering process is still competitive, the units will contain a label that identifies the playing technique represented by its receptive field.

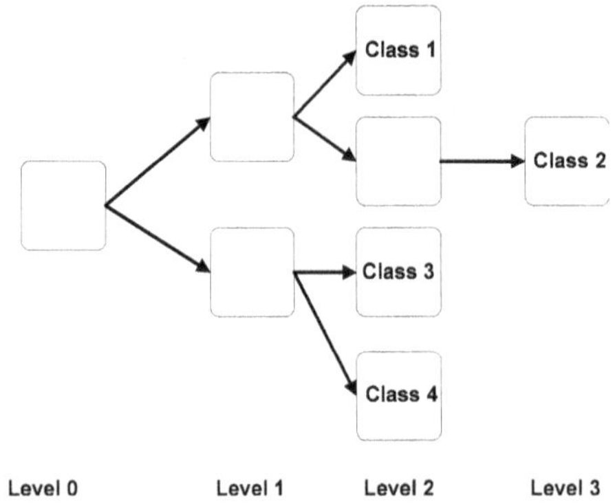

Fig. 13. GHSOM clustering.

Also, BMU calculation can be performed in GHSOMs in a similar way as in classical SOMs by simply following the hierarchy level until the deepest map is found.

8.1 BMU calculation on GHSOM

Once the GHSOM is trained, the BMU is computed for every data sample.

In the case of SOM, the BMU is calculated in the same way it was found during the training phase. Since several SOM layers have been created during the GHSOM training, we have to follow the whole SOM hierarchy in order to determine the winning unit and the map it belongs to.

To this end, an iterative algorithm, depicted in Figure 14, has been developed. In this figure, an example of BMU calculation on a three-level GHSOM is represented. Assume that the distances between an input pattern and the weight vectors of the level 0 map are calculated, then compute the minimum of these distances. As a result, the winning neuron on map 1 is found. Since other maps could be grown from this winning neuron, we have to check if the wining neuron is a parent unit. This test can be carried out making use of the parent vectors resulting from the GHSOM training process. If a new map arose from the wining neuron, the BMU on this map is calculated. This process is repeated until a BMU with no growing map is found. Thus, the BMU in the GHSOM is associated to a map in a level of the hierarchy.

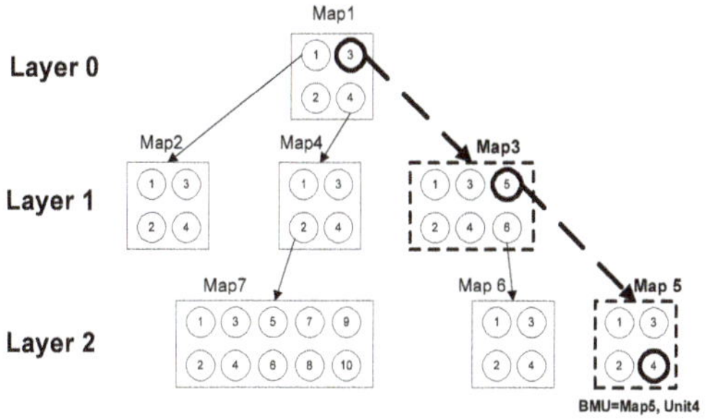

Fig. 14. GHSOM BMU calculation example.

At his point, a probability-based relabelling method can be applied using a 2D Gaussian kernel centered at each BMU Ortiz, Górriz, Ramírez & Salas-Gonzalez (2011a); Ortiz, Gorriz, Ramirez & Salas-Gonzalez (2011b); Ortiz, Ortega, Diaz & Prieto (2011). Thus, a mayority-voting scheme with the units inside the Gaussian kernel is used to relabel the unlabelled units, assigning the calculated label to the data samples as shown in Figure 15.

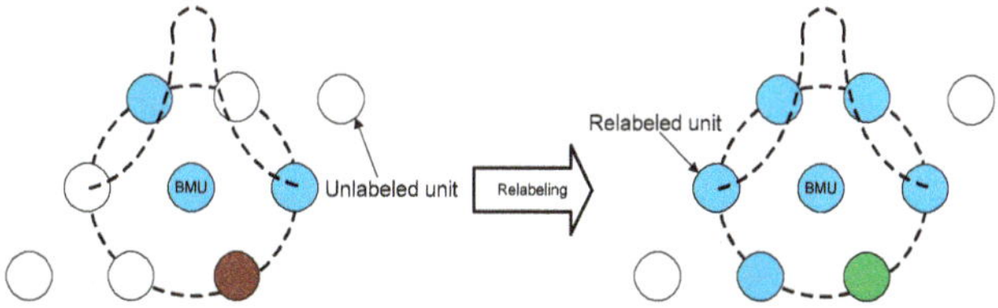

Fig. 15. GHSOM map relabeling method.

In Equation 20, the Gaussian kernel used to estimate the label for unlabelled units is shown:

$$L(x, y, \sigma) = \frac{1}{2\pi\sigma^2} e^{-\frac{x^2+y^2}{2\sigma^2}} \tag{20}$$

In this equation, σ determines the width of the Gaussian kernel. In other words, it defines the neighbourhood considered for the relabelling process. On the other hand, (x, y) is the position of the BMU in the SOM grid.

9. Concluding summary

In this chapter, we have presented the utilization of artificial intelligence techniques in the context of musical signal analysis.

We have described some features related to a specific classification problem and, then, we have focused on the utilization of neural models for the development of specific tasks in musical audio classification.

The classification task analysed in the context of the application of self-organizing maps and some of its variants which make use of the competitive learning paradigm to implement unsupervised classification techniques applied to audio signals.

As it has been shown in this chapter, these performance of these models can be improved by means of hybridizing with other clustering and dimension reduction techniques like Principal Component Analysis methodologies.

Also, the application of Gaussian Mixture Models provides a probabilistic behaviour of the maps instead of a binary activation of the neurons.

10. Acknowledgments

This work was supported by the Ministerio de Economía y Competitividad of the Spanish Government under Project No. TIN2010-21089-C03-02 and Project No. IPT-2011-0885-430000.

11. References

Alhoniemi, E., Himberg, J. & Vesanto, J. (1999). Probabilistic measures for responses of self-organizing map units alhoniemi, *Proceedings of the International ICSC Congress on Computational Intelligence Methods and Applications (CIMA'99)*.

Arsuaga, E. & Díaz, F. (2005). Topology preservation in som, *International Journal of Mathematical and Computer Sciences* 1(1): 19–22.

Barbancho, I., de la Bandera, C., Barbancho, A. M. & Tardon, L. J. (2009). Transcription and expressiveness detection system for violin music, *Proc. IEEE Int. Conf. Acoustics, Speech and Signal Processing ICASSP 2009*, pp. 189–192.

Bezdek, J. C. & Pal, N. R. (1998). Some new indexes of cluster validity, 28(3): 301–315.

Cruz, R., Ortiz, A., Barbancho, A. & Barbancho, I. (2012). Unsupervised classification of audio signals, *7th International Conference on Hybrid Artificial Intelligence (HAIS2012)*.

Davies, D. L. & Bouldin, D. W. (1979). A cluster separation measure, (2): 224–227.

Dittenbach, M., Merkl, D. & Rauber, A. (2000). The growing hierarchical self-organizing map, *Proc. IEEE-INNS-ENNS Int Neural Networks IJCNN 2000 Joint Conf*, Vol. 6, pp. 15–19.

Farahani, G. & Ahadi, S. M. (2005). Robust features for noisy speech recognition based on filtering and spectral peaks in autocorrelation domain, *Proc. of the European Signal Processing Conference*, Antalya (Turkey).

Ghahramani, Z. & Beal, M. (2000). Variational inference for bayesian mixtures of factor analysers, *Advances in Neural Information Processing Systems* 12: 449–455.

Goto, M. (2004). Development of the rwc music database, *Proceedings of the 18th International Congress on Acoustics*.

Hassar, H. & Bensaid, A. (1999). Validation of fuzzy and crisp c-partitions, *Proc. NAFIPS Fuzzy Information Processing Society 18th Int. Conf. of the North American*, pp. 342–346.

Haykin, S. (1999). *Neural Networks: a comprehensive foundation*, 2nd edn, Prentice-Hall.

Holmes, W. J. & Huckvale, M. (1994). Why have HMMs been so successful for automatic speech recognition and how might they be improved?, *Speech, Hearing and Language, UCL Work in Progress*, Vol. 8, pp. 207–219.

Jenses, J. (1999). Envelope model of isolated musical sounds, *Proceedings of the 2nd COST G-6 workshop on digital audio effects, Trondheim (Norway).*

Juang, B. H. & Rabiner, L. R. (2005). Automatic speech recognition – a brief history of the technology, *in* K. Brown (ed.), *Encyclopedia of Language and Linguistics*, Elsevier.

Kaufman, L. & Rosseauw, P. (1990). *Finding groups in data*, Wiley, New York.

Kimura, S. (1999). Advances in speech recognition technologies, *Fujitsu Sci. Tech. J.* 35(2): 202–211.

Kohonen, T. (2001). *Self-Organizing Maps*, Springer.

Kohonen, T., Oja, E., Simula, O., Visa, A. & Kangas, J. (1996). Engineering applications of the self-organizing map, *Proc. of the IEEE.*

Lee, T., Lewicki, M. & Sejnowski, T. (2000). Ica mixture models for unsupervised classification of non-gaussian classes and automatic context switching in blind signal separation, *IEEE Transactions on Patt* 22(10): 1078–1089.

MacQueen, J. B. (1967). Some methods for classification and analysis of multivariate observations, *Proceedings of 5th Berkeley Symposium on Mathematical Statistics and Probability. University of California Press. pp. 281–297.*

Minematsu, N., Nishimura, T., Murakami, T. & Hirose, K. (2006). Speech recognition only with suprasegmental features - hearing speech as music, *Proc. of the International Conference on Speech Prosody*, Dresden (Germany).

Müller, M., Ellis, D. P. W., Klapuri, A. & Richard, G. (2011). Signal processing for music analysis, *IEEE Journal on Selected Topics in Signal Processing* 5(6): 1088–1110.

Murtagh, F. (1991). Multilayer perceptrons for classification and regression, *Neurocomputing* 2(5–6): 183–197.

Murtagh, F. (1995). Interpreting the kohonen self-organizing map using contiguity-constrained clustering, *Pattern Recognition Letters* 16(4): 399–408.

Ortiz, A., Górriz, J., Ramírez, J. & Salas-Gonzalez, D. (2011a). Mr brain image segmentation by growing hierarchical som and probability clustering, *Electronics Letters* 47(10): 585–586.

Ortiz, A., Gorriz, J., Ramirez, J. & Salas-Gonzalez, D. (2011b). Mri brain image segmentation with supervised som and probability-based clustering method, *Lecture Notes in Computer Science, LNCS-6686*, pp. 49–58.

Ortiz, A., Ortega, J., Diaz, A. & Prieto, A. (2011). Network intrusion prevention by using hierarchical self-organizing maps and probability-based labeling, *Proceedings of the 11th international conference on Artificial neural networks conference on Advances in computational intelligence.*

Panagiotakis, C. & Tziritas, G. (2005). A speech/music discriminator based on rms and zero-crossings, *IEEE Transaction on Multimedia* pp. 155–166.

Prasad, B. & Prasanna, S. R. M. (eds) (2008). *Speech, Audio, Image and Biomedical Signal Processing using Neural Networks*, Vol. 83 of *Studies in Computational Intelligence*, Springer.

Rauber, A., Merkl, D. & Dittenbach, M. (2002). The growing hierarchical self-organizing map: exploratory analysis of high-dimensional data, 13(6): 1331–1341.

Riveiro, M., Johansson, F., Falkman, G. & Ziemke, T. (2008). Supporting maritime situation awareness using self organizing maps and gaussian mixture models, *Proceedings of the 2008 conference on Tenth Scandinavian Conference on Artificial Intelligence: SCAI 2008.*

Rossi, F. & Villa, N. (2009). Topologically ordered graph clustering via deterministic annealing, *Proceesdings of the 17th European Symposium on Artificial Neural Networks.*

Tardón, L. J., Sammartino, S. & Barbancho, I. (2010). Design of an efficient music-speech discriminator, *J. Acoustical Society of America* 127(1): 271–279.

Tasdemir, K. & Merenyi, E. (2009). Exploiting data topology in visualization and clustering of self-organizing maps, 20(4): 549–562.

Tasdemir, K., Milenov, P. & Tapsall, B. (2011). Topology-based hierarchical clustering of self-organizing maps, 22(3): 474–485.

Therrien, C. (1989). *Decision estimation and classification: an introduction to pattern recognition and related topics*, John Wiley & Sons, Inc.

Thornburg, H., Leistikow, R. & Berger, J. (2007). Melody extraction and musical onset detection via probabilistic models of framewise stft peak data, *IEEE Transactions on Audio, Speech, and Language Processing* 15(4): 1257–1272.

Tipping, M. & Bishop, C. (1999). Mixtures of probabilistic preservationincipal component analyzers, *Neural Computation* 11(2): 443–482.

Tzanetakis, G. & Cook, P. R. (2002). Musical genre classification of audio signals, *IEEE Transactions on Speech and Audio Processing* 10(5): 293–302.

Vesanto, J. & Alhoniemi, E. (2000). Clustering of the self-organizing map, 11(3): 586–600.

Vesanto, J., Himberg, J., Alhoniemi, E. & Parhankangas, J. (2000). Som toolbox, Helsinki University of Technology.
URL: *http://www.cis.hut.fi/somtoolbox/*

Wu, S. & Chow, T. (2004). Clustering of the self organizing map using a clustering validity index based on inter-cluster and intra-cluster density, *Pattern Recognition Letters* 37(2): 175–188.

7

Convolutive ICA for Audio Signals

Masoud Geravanchizadeh and Masoumeh Hesam
Faculty of Electrical and Computer Engineering, University of Tabriz, Tabriz, Iran

1. Introduction

The goal of Blind Source Separation (BSS) is to estimate latent sources from their mixed observations without any knowledge of the mixing process. Under the assumption of statistical independence of hidden sources, the task in BSS is to obtain Independent Components (IC) from the mixed signals. Such algorithms are called ICA-based BSS algorithms [1, 2]. ICA-based BSS has been well studied in the fields of statistics and information theory for different applications, including wireless communication and biomedicine. However, as speech and audio signal mixtures in a real reverberant environment are generally convolutive mixtures, they involve a structurally much more challenging task than instantaneous mixtures, which are prevalent in many other applications [3, 4]. Such a mixing situation is generally modeled with impulse responses from sound sources to microphones. In a practical room situation, such impulse responses can have thousands of taps even with an 8 kHz sampling rate, and this makes the convolutive problem difficult to solve. Blind speech separation is applicable to the realization of noise-robust speech recognition, high quality hands-free telecommunication systems and hearing aids.

Various efforts have been devoted to the separation of convolutive mixtures. They can be classified into two major approaches: time-domain BSS [5, 6] and frequency-domain BSS [7]. With time-domain BSS, a cost function is defined for time-domain signals, and optimized with convolutive separation filters. However, the optimization with convolutive separation filters is not as simple as BSS for instantaneous mixtures, and generally computationally expensive. With frequency-domain BSS, time-domain mixed signals observed at microphones are converted into frequency-domain time-series signals by a short-time Fourier transform (STFT). However, choosing the length of STFT has relationship with the length of room impulse response [8]. The merit of these approaches is that the ICA algorithm becomes simple and can be performed separately at each frequency by any complex-valued instantaneous ICA algorithm [9-11]. However, the drawbacks of frequency-domain ICA are the permutation and scaling ambiguities of an ICA solution. In the frequency-domain ICA, different permutations at different frequencies lead to re-mixing of signals in the final output. Also, different scaling at different frequencies leads to distortion of the frequency spectrum of the output signal. For the scaling problem, in one method, the output is filtered by the inverse of the separation filter [12]. For the permutation problem, spatial information, such as the direction-of-arrivals (DOA) of sources, can be estimated and used [13, 14]. Another method utilizes the coherency of the mixing matrices in several adjacent frequencies [15]. For non-stationary sources such as speech, many methods exploit the dependency of separated signals across frequencies to solve the permutation problem [16, 17]. We propose a method for the permutation problem, by maximizing the correlation of power ratio measure of each bin frequency with the average of previous bin frequencies [18].

This chapter deals with the frequency-domain BSS for convolutive mixtures of speech signals. We begin by formulating the BSS problem for convolutive mixtures in Section 2. Section 3 provides an overview of the frequency-domain BSS. Section 4 discusses Principal Component Analysis (PCA) as a pre-processing step. Fast ICA algorithm for complex-valued signals is discussed in Section 5. We then present several important techniques along with our proposed method for solving the permutation problem in Section 6. Section 7 introduces a common method for the scaling problem. Section 8 considers ways of choosing the STFT length for a better performance of the separation problem. In Section 9, we compare our proposed method in the permutation problem with some other conventional methods by conducting several experiments. Finally, Section 10 concludes this chapter.

2. Mixing process and convolutive BSS

Convolutive mixing arises in acoustic scenarios due to time delays resulting from sound propagation over space and the multipath generated by reflections of sound from different objects, particularly in rooms and other enclosed settings. If we denote by $s_j(t)$ the signal emitted by the j-th source $(1 \le j \le N)$, $x_i(t)$ the signal recorded by the i-th microphone $(1 \le i \le M)$, and $h_{ij}(t)$ the impulse response from source j to sensor i, we have:

$$x_i(t) = \sum_{j=1}^{N} \sum_{\tau} h_{ij}(\tau) s_j(t-\tau) . \tag{1}$$

We can write this equation into a more elegant form as:

$$\mathbf{x}(t) = \sum_{\tau} \mathbf{h}(\tau) \mathbf{s}(t-\tau), \tag{2}$$

where $\mathbf{h}(t)$ is an unknown $M \times N$ mixing matrix. Now, the goal of a convolutive BSS is to obtain separated signals $y_1(t),...,y_N(t)$, each of which corresponds to each of the source signals. The task should be performed only with M observed mixtures, and without information on the sources and the impulse responses:

$$y_j(t) = \sum_{i=1}^{M} \sum_{\tau} b_{ji}(\tau) x_i(t-\tau), \tag{3}$$

where $b_{ji}(t)$ represents the impulse response of the multichannel separation system. Convolutive BSS as applied to speech signal mixtures involves relatively-long multichannel FIR filters to achieve separation with even moderate amounts of room reverberation. While time-domain algorithms can be developed to perform this task, they can be difficult to code primarily due to the multichannel convolution operations involved [5, 6]. One way to simplify the conceptualization of the convolutive BSS algorithms is to transform the task into the frequency domain, as convolution in time becomes multiplication in frequency. Ideally, each frequency component of the mixture signal contains an instantaneous mixture of the corresponding frequency components of the underlying source signals. One of the advantages of the frequency-domain BSS is that we can employ any ICA algorithm for instantaneous mixtures, such as the information maximization (Infomax) approach [19] combined with the natural gradient [20], Fast ICA [21], JADE [22], or an algorithm based on non-stationarity of signals [23].

3. Frequency-domain convolutive BSS

This section presents an overview of the frequency-domain BSS approach that we consider in this chapter. First, each of the time-domain microphone observations $x_j(t)$ is converted into frequency-domain time-series signals $X_j(k,f)$ by a short-time Fourier transform (STFT) with a K-sample frame and its S-sample shift:

$$X_j(k,f) = \sum_t x_j(t)\,\mathbf{win}\left(t - k\frac{S}{f_s}\right) e^{-i2\pi ft}, \tag{4}$$

for all discrete frequencies $f \in \left\{0, \frac{1}{K} f_S, \ldots, \frac{K-1}{K} f_S\right\}$, and for frame index k. The analysis window $\mathbf{win}(t)$ is defined as being nonzero only in the K-sample interval $\left[-\frac{K}{2}\frac{1}{f_s}, (\frac{K}{2}-1)\frac{1}{f_s}\right]$ and tapers smoothly to zero at each end of the interval, such as a Hanning window $\mathbf{win}(t) = \frac{1}{2}\left(1 + \cos\frac{2\pi f_s}{K}t\right)$.

If the frame size K is long enough to cover the main part of the impulse responses h_{ij}, the convolutive model (2) can be approximated as an instantaneous model at each frequency [8, 24]:

$$\mathbf{X}(k,f) = \mathbf{H}(f)\mathbf{S}(k,f), \tag{5}$$

where $\mathbf{H}(f)$ is an $M \times N$ mixing matrix in frequency domain, and $\mathbf{X}(k,f)$ and $\mathbf{S}(k,f)$ are vectors of observations and sources in frequency domain, respectively. Notice that, the convolutive mixture problem is reduced to a complex but instantaneous mixture problem and separation is performed at each frequency bin by:

$$\mathbf{Y}(k,f) = \mathbf{B}(f)\mathbf{X}(k,f), \tag{6}$$

where $\mathbf{B}(f)$ is an $N \times M$ separation matrix. As a basic setup, we assume that the number of sources N is no more than the number of the microphones M, i.e., $N \le M$. However, in a case with $N > M$ that is referred to as underdetermined BSS, separating all the sources is a rather difficult problem [25].

We can limit the set of frequencies to perform the separations by $\left\{0, \frac{1}{K} f_s, \ldots, \frac{1}{2} f_s\right\}$ due to the relationship of complex conjugate:

$$X_j(k,\frac{m}{K}f_s)=X_j^*\left(k,\frac{K-m}{K}f_s\right),\quad (m=1,\ldots,K/2-1). \tag{7}$$

We employ the complex-valued instantaneous ICA to calculate the separation matrix $\mathbf{B}(f)$. Section 5 describes the detailed procedure for the complex-valued ICA used in our implementation and experiments. However, the ICA solution at each frequency bin has permutation and scaling ambiguity. In order to construct proper separated signals in time domain, frequency-domain separated signals originating from the same source should be grouped together. This is the permutation problem. Also, different scaling at different frequencies leads to distortion of the frequency spectrum of the output signal. This is the scaling problem. There are some methods to solve the permutation and scaling problems [12-18]. After solving the permutation and the scaling problem, the time-domain output signals $y_i(t)$ are calculated with an inverse STFT (ISTFT) of the separated signals $Y_i(k,f)$. The flow of the frequency-domain BSS is shown in Figure 1.

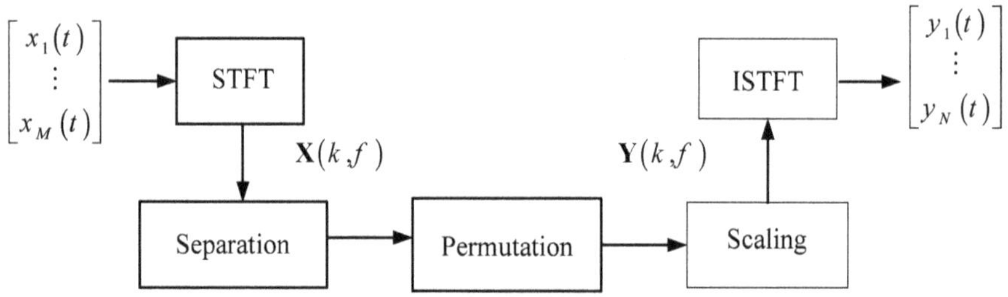

Fig. 1. System structure for the frequency-domain BSS

4. Pre-processing with principal component analysis

It is known that choosing the number of microphones more than the number of sources improves the separation performance. This is termed as the overdetermined case, in which the dimension of the observed signals is greater than the number of sources. Many methods have been proposed to solve the overdetermined problem. In a typical method, the subspace procedure is used as a pre-processing step for ICA in the framework of BSS [15, 26, 27]. The subspace method can be understood as a special case of principal component analysis (PCA) with $M \geq N$, where M and N denote the number of observed signals and source signals, respectively. This technique reduces room reflections and ambient noise [15]. Also, as pre-processing, PCA improves the convergence speed of ICA. Figure 2 shows the use of PCA as pre-processing to reduce the dimension of microphone signals.

In the PCA process, the input microphone signals are assumed to be modeled as:

$$\mathbf{X}(k,f)=\mathbf{A}(f)\mathbf{S}(k,f)+\mathbf{n}(k,f), \tag{8}$$

where the (m,n)-th element of $\mathbf{A}(f)$ is the transfer function from the n-th source to the m-th microphone as:

$$A_{m,n}(f)=T_{m,n}(f)e^{-i2\pi f\tau_{m,n}}. \tag{9}$$

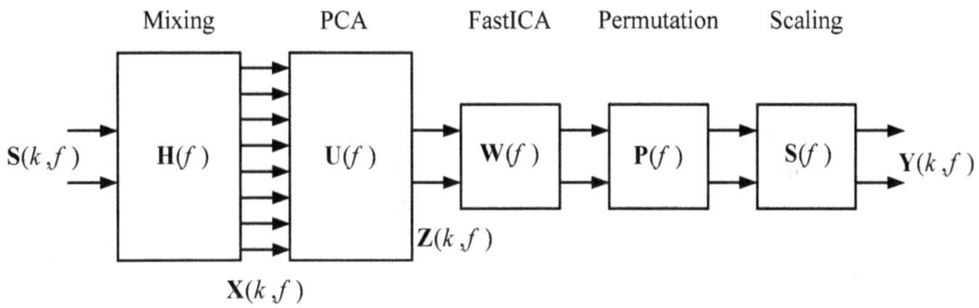

Fig. 2. The use of PCA as a pre-processing step in the frequency-domain BSS

Here, the symbol $T_{m,n}(f)$ is the magnitude of the transfer function. The symbol $\tau_{m,n}$ denotes the propagation time from the n-th source to the m-th microphone. The first term in Eq. (8), $\mathbf{A}(f)\mathbf{S}(k,f)$, expresses the directional components in $\mathbf{X}(k,f)$ and the second term, $\mathbf{n}(k,f)$, is a mixture of less-directional components which includes room reflections and ambient noise.

The spatial correlation matrix $\mathbf{R}(f)$ of $\mathbf{X}(k,f)$ is defined as:

$$\mathbf{R}(f)=E[\mathbf{X}(k,f)\mathbf{X}^H(k,f)]. \tag{10}$$

The eigenvalues of $\mathbf{R}(f)$ are denoted as $\lambda_1(f),...,\lambda_M(f)$ with $\lambda_1(f)\geq,...,\geq\lambda_M(f)$ and the corresponding eigenvectors are denoted as $\mathbf{e}_1(f),...,\mathbf{e}_M(f)$. Assuming that $\mathbf{s}(t)$ and $\mathbf{n}(t)$ are uncorrelated, the energy of the N directional signals $\mathbf{s}(t)$ is concentrated on the N dominant eigenvalues and the energy of $\mathbf{n}(t)$ is equally spread over all eigenvalues. In this case, it is generally satisfied that:

$$\lambda_1(f),...,\lambda_N(f)>>\lambda_{N+1}(f),...,\lambda_M(f). \tag{11}$$

The vectors $\mathbf{e}_1(f),...,\mathbf{e}_N(f)$ and $\mathbf{e}_{N+1}(f),...,\mathbf{e}_M(f)$ are the basis of the signal and noise subspaces, respectively.

In the PCA method, the input signal is processed as:

$$\mathbf{Z}(k,f)=\mathbf{U}(f)\mathbf{X}(k,f), \tag{12}$$

that reduces the energy of $\mathbf{n}(t)$ in the noise subspace, and the PCA filter is defined as:

$$\mathbf{U}(f)=\mathbf{\Lambda}^{-\frac{1}{2}}(f)\,\mathbf{E}^H(f), \tag{13}$$

where

$$\mathbf{\Lambda}(f)=diag(\lambda_1(f),...,\lambda_N(f)),\ \mathbf{E}(f)=[e_1(f),...,e_N(f)]. \tag{14}$$

The PCA filtering of $\mathbf{X}(k,f)$ reduces the dimension of input signal to the number of sources N which is equivalent to a spatially whitening operation, i.e., $E\{\mathbf{Z}(k,f)\mathbf{Z}^H(k,f)\} = \mathbf{I}$ where $\mathbf{I}$ is the $N \times N$ identity matrix.

5. Complex-valued fast fixed-point ICA

The ICA algorithm used in this chapter is fast fixed-point ICA (fast ICA). The fast ICA algorithm for the separation of linearly mixed independent source signals was presented in [21]. This algorithm is a computationally efficient and robust fixed-point type algorithm for independent component analysis and blind source separation. However, the algorithm in [21] is not applicable to frequency-domain ICA as these are complex-valued. In [9], the fixed-point ICA algorithm of [21] has been extended to involve complex-valued signals. The fast fixed-point ICA algorithm is based on the assumption that when the non-Gaussian signals get mixed, it becomes more Gaussian and thus its non-Gaussianization can yield independent components. The process of non-Gaussianization consists of two-steps, namely, pre-whitening or sphering and rotation of the observation vector. Sphering is half of the ICA task and gives spatially decorrelated signals. The process of sphering (pre-whitening) is accomplished by the PCA stage as described in the previous section. The task remaining after whitening involves rotating the whitened signal vector $\mathbf{Z}(k,f)$ such that $\mathbf{Y}(k,f) = \mathbf{W}(f)\,\mathbf{Z}(k,f)$ returns independent components. For measuring the non-Gaussianity, we can use the negentropy-based cost function:

$$J = E\left\{G\left(|\mathbf{w}^H\mathbf{Z}|^2\right)\right\}, \tag{15}$$

where $G(t) = \log(0.01 + t)$ [9].

The elements of the matrix $\mathbf{W} = (\mathbf{w}_1, \ldots, \mathbf{w}_N)$ are obtained in an iterative procedure. The fixed-point iterative algorithm for each column vector $\mathbf{w}$ is as follows (the frequency index f and frame index k are dropped hereafter for clarity):

$$\mathbf{w} = E\left\{\mathbf{y}(\mathbf{w}^H\mathbf{Z})^* g\left(\left|\mathbf{w}^H\mathbf{Z}\right|^2\right)\right\} - E\left\{g\left(\left|\mathbf{w}^H\mathbf{Z}\right|^2\right) + \left|\mathbf{w}^H\mathbf{Z}\right|^2 g'\left(\left|\mathbf{w}^H\mathbf{Z}\right|^2\right)\right\}\mathbf{w}, \tag{16}$$

where $g(.)$ and $g'(.)$ are first- and second-order derivatives of G:

$$g(t) = \frac{1}{(0.01 + t^2)}, \quad g'(t) = \frac{0.5}{(0.01 + t^2)^2}. \tag{17}$$

After each iteration, it is also essential to decorrelate $\mathbf{W}$ to prevent its convergence to the previously converged point. The decorrelation process to obtain $\mathbf{W}$ for the next iteration is obtained as [9]:

$$\mathbf{W} = \mathbf{W}(\mathbf{W}^H\mathbf{W})^{-1/2}. \tag{18}$$

Then, the separation matrix is obtained by the product of $\mathbf{U}(f)$ and $\mathbf{W}(f)$:

$$\mathbf{B}(f) = \mathbf{W}(f)\mathbf{U}(f). \tag{19}$$

6. Solving the permutation problem

In order to get separated signals correctly, the order of separation vectors (position of rows) in $\mathbf{B}(f)$ must be the same at each frequency bin. This is called permutation problem. In this section, we review various methods which have already been proposed to solve permutation problem.

6.1 Solving permutation by Direction of Arrival (DOA) estimation

Some methods for permutation problem use the information of source locations, such as direction of arrival (DOA). In the totally blind setup, DOA cannot be known so it is estimated from the directivity pattern of the separation matrix. In this method, the effect of room reverberation is neglected, and the elements of the mixing matrix in Eq. (9) can be written as the following expression:

$$A_{m,n}(f) = T_{m,n}(f)e^{-i2\pi f \tau_{m,n}}, \qquad (\tau_{m,n} \equiv \frac{1}{c} d_m \sin\theta_n), \tag{20}$$

where $\tau_{m,n}$ is the arriving lag with respect to the n-th source signal from the direction of θ_n, observed at the m-th microphone located at d_m, and c is the velocity of sound. Microphone array and sound sources are shown in Figure 3.

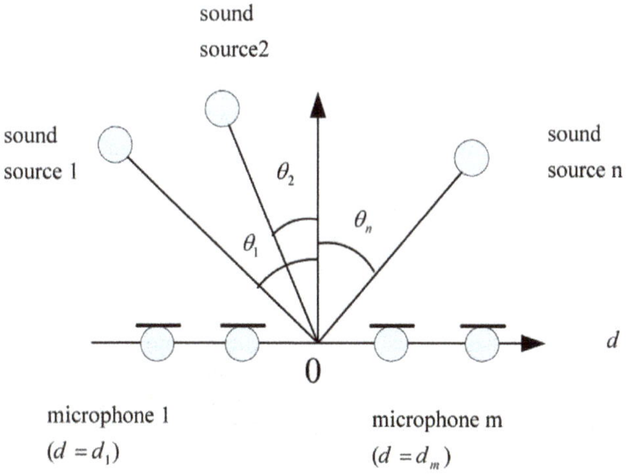

Fig. 3. Configuration of a microphone array and sound sources

From the standpoint of array signal processing, directivity patterns (DP) are produced in the array system. Accordingly, directivity patterns with respect to $B_{nm}(f)$ are obtained at every frequency bin to extract the DOA of the n-th source signal. The directivity pattern $F_n(f,\theta)$ is given by [13]:

$$F_n(f,\theta) = \sum_{m=1}^{M} B_{nm}(f).\exp[i2\pi f d_m \sin\theta / c]\ . \tag{21}$$

The DP of the separation matrix contains nulls in each source direction. Figure 4 shows an example of directivity patterns at frequency bins f_1 and f_2 plotted for two sources. As it is observed, the positions of the nulls vary at each frequency bin for the same source direction. Hence, in order to solve the permutation problem and sort out the different sources, the separation matrix at each frequency bin is arranged in accordance with the directions of nulls.

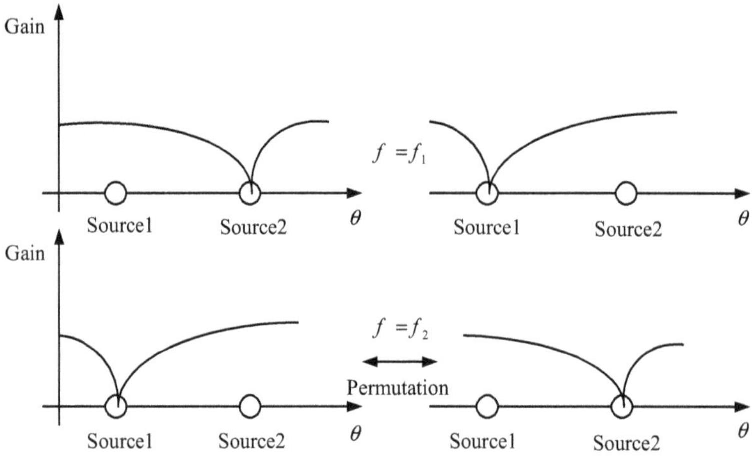

Fig. 4. Examples of directivity patterns

This method is not always effective in the overdetermined case, because the directions giving the nulls of the directivity patterns of the separation matrix $\mathbf{B}(f)$ do not always correspond to the source directions. Figure 5 shows the directivity pattern for the case ($M = 2,\ N = 2$), and the overdetermined case ($M = 8,\ N = 2$).

6.1.1 Closed-form formula for estimating DOAs

The DOA estimation method by the directivity pattern has three problems, a high computational cost, the difficulty of using it for mixtures of more than two sources, and for overdetermined case in which the number of microphones is more than the number of sources. Instead of plotting directivity patterns and searching for the minimum as a null direction, some propose a closed-form formula for estimating DOAs [16]. In principle, this method can be applied to any number of source signals as well as to the overdetermined case. It can be shown that the DOAs for sources are estimated by the following relation [16]:

$$\theta_k = \arccos \frac{\arg\left(\dfrac{\left[\mathbf{B}^{-1}\right]_{jk}}{\left[\mathbf{B}^{-1}\right]_{j'k}} \right)}{2\pi f c^{-1}\left(d_j - d_{j'}\right)}, \tag{22}$$

where, d_j and $d_{j'}$ are the positions of sensors x_j and $x_{j'}$.

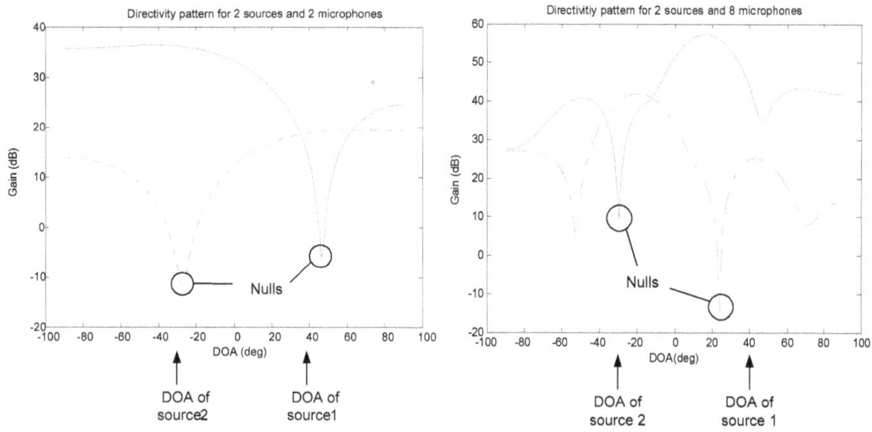

Fig. 5. The directivity patterns for the case ($M = 2, N = 2$), and the overdetermined case ($M = 8, N = 2$)

If the absolute value of the input variable of arccos(.) is larger than 1, θ_k becomes complex and no direction is obtained. In this case, formula (22) can be tested with another pair j and j'.

If $N < M$, the Moore-Penrose pseudoinverse $\mathbf{B}^{+}$ is used instead of $\mathbf{B}^{-1}$. Based on these DOA estimations, the permutation matrix is determined. In this process, no reverberation is assumed for the mixing signals. Therefore, for the reverberant case the method based on DOA estimation is not efficient.

6.2 Permutation by interfrequency coherency of mixing matrix

Another method to solve the permutation problem utilizes the coherency of the mixing matrices in several adjacent frequencies [15]. For the mixing matrix $\mathbf{A}(f)$ in the Eq. (8), the n-th column vector (location vector of the n-th source) at frequency f has coherency with that at the adjacent frequency $f_0 = f - \Delta f$. Therefore, the location vector $\mathbf{a}_n(f)$ is $\mathbf{a}_n(f_0)$ which is rotated by the angle θ_n as depicted in Figure 6(a). Accordingly, θ_n is expected to be the smallest for the correct permutation as shown in Figure 6. Based on this assumption, permutation is solved so that the sum of the angles $\{\theta_1, \theta_2, ..., \theta_N\}$ between the location vectors in the adjacent frequencies is minimized. An estimate of the mixing matrix $\hat{\mathbf{A}}(f) = [\hat{\mathbf{a}}_1(f), ..., \hat{\mathbf{a}}_N(f)]$ can be obtained as the pseudoinverse of the separation matrix as:

$$\hat{\mathbf{A}}(f) = \mathbf{B}^{+}(f). \tag{23}$$

For this purpose, we define a cost function as [15]:

$$F(\mathbf{P}) = \frac{1}{N}\sum_{n=1}^{N}\cos\theta_n, \qquad \cos\theta_n = \frac{\hat{\mathbf{a}}_n^H(f)\hat{\mathbf{a}}_n(f_0)}{\|\hat{\mathbf{a}}_n(f)\| \cdot \|\hat{\mathbf{a}}_n(f_0)\|}. \tag{24}$$

This cost function is calculated for all arranges of columns of mixing matrix $\hat{\mathbf{A}}(f) = [\hat{\mathbf{a}}_1(f),...,\hat{\mathbf{a}}_N(f)]$ and the permutation matrix $\mathbf{P}$ is obtained by maximizing it.

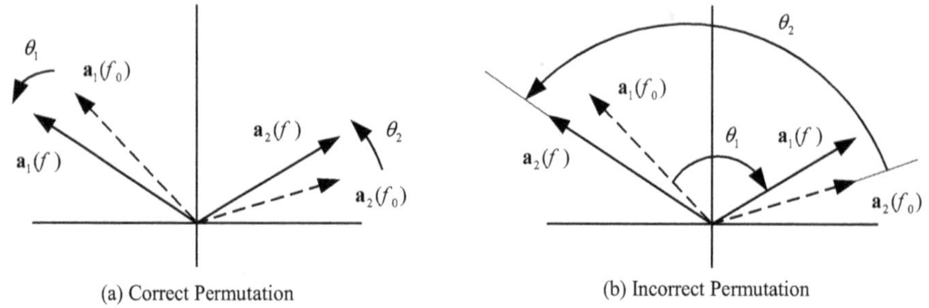

Fig. 6. The column vectors of the mixing matrix in two adjacent frequencies, with correct and incorrect permutations

To increase the accuracy of this method, the cost function is calculated for a range of frequencies instead of the two adjacent frequencies and a confidence measure is used to determine which permutation is correct [15].

The mixing matrix is defined as the transfer function of direct path from each source to each microphone where the coherency of mixing matrices is used in several adjacent frequencies to obtain the permutation matrix.

This method assumes that the spectrum of microphone signals consists of the directional components and reflection components of sources and employs the subspace method to reduce the reflection components. However, if the reflection components are not reduced by the subspace method, the mixing matrix consists of indirect path components, and the method will not be efficient.

6.3 A new method to solve the permutation problem based on power ratio measure

Another group of permutation methods use the information on the separated signals which are based on the interfrequency correlation of separated signals. Conventionally, the correlation coefficient of separated signal envelopes has been employed to measure the dependency of bin-wise separated signals. Envelopes have high correlations at neighboring frequencies if separated signals correspond to the same source signal. Thus, calculating such correlations helps us to align permutations. A simple approach to the permutation alignment is to maximize the sum of the correlations between neighboring frequencies [16]. The method in [12] assumes high correlations of envelopes even between frequencies that are not close neighbors and so it does not limit the frequency range in which correlations are calculated.

However, this assumption is not satisfied for all pairs of frequencies. Therefore, the use of envelopes for maximizing correlations in this way is not a good choice. Recently, the power ratio between the *i*-th separated signal and the total power sum of all separated signals has been proposed as another type of measure [17]. In this approach, the dependence of bin-wise separated signals can be measured more clearly by calculating correlation coefficients with power ratio values rather than with envelopes. This is shown by comparing Figures 7 and 8.

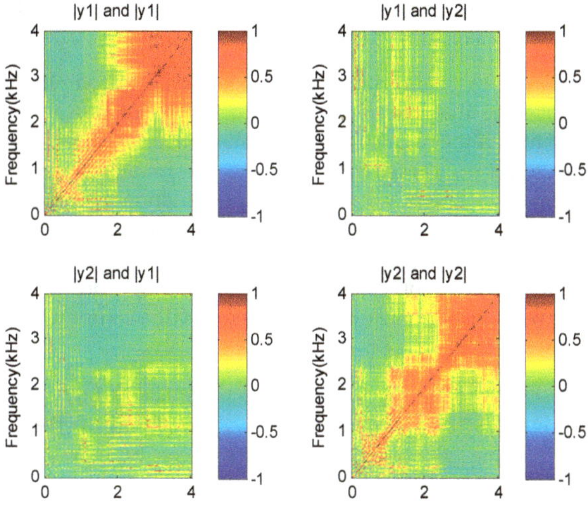

Fig. 7. Correlation coefficients between the separated signal envelopes

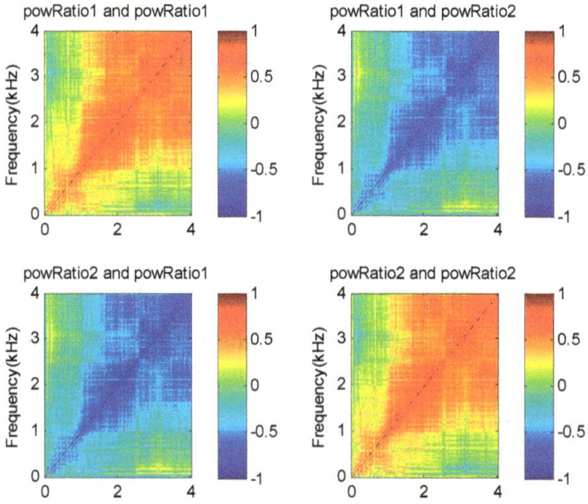

Fig. 8. Correlation coefficients between the power ratios of separated signals

This method uses two optimization techniques for permutation alignment; a rough global optimization and a fine local optimization. In rough global optimization, a centroid is calculated for each source as the average value of power ratio with the current permutation. The permutations are optimized by an iterative maximization between the power ratio measures and the current centroid. In fine local optimization, the permutations are obtained by maximizing the correlation coefficients over a set of frequencies consisting of adjacent frequencies and harmonic frequencies. Here, the experiments show that the fine local optimization alone does not provide good results in permutation alignment. But using both global and local optimization achieves almost optimal results. This method, however, is somewhat complicated for calculating the permutations.

In our proposed method, we take a rather simple technique to compute the permutation matrices. Here, we assume that the correlation coefficients of power ratios of bin-wise separated signal to be high if they come from the same source for each two frequencies even if they are not close together. Therefore, we extend the frequency range for calculating correlation to all previous frequencies, where the permutation was solved for them. We decide on the permutation by maximizing the correlation of power ratio measure of each bin frequency with the average of power ratio measures of previous bin frequencies, iteratively with increasing frequency. Therefore, this criterion is not based on local information and does not have the drawback of propagation of mistakes by the computation of permutation at each frequency.

If the separation works well, the bin-wise separated signals $\mathbf{Y}_1(k,f),\dots,\mathbf{Y}_N(k,f)$ are the estimations of the original source signals $\mathbf{S}_1(k,f),\dots,\mathbf{S}_N(k,f)$ up to the permutation and scaling ambiguity. Thus, the observation vector $\mathbf{X}(k,f)$ can be represented by the linear combination of the separated signals as:

$$\mathbf{X}(k,f) = \mathbf{A}(f)\,\mathbf{Y}(k,f) = \sum_{i=1}^{N} \mathbf{a}_i(f) Y_i(k,f), \tag{25}$$

where the mixing matrix $\mathbf{A}(f) = [\mathbf{a}_1(f),\dots,\mathbf{a}_N(f)]$ is the pseudoinverse of the separation matrix $\mathbf{B}(f)$:

$$\mathbf{A}(f) = \mathbf{B}(f)^{+}. \tag{26}$$

Now, we use the power ratio measure as given by [17]:

$$powRatio_i(k,f) = \frac{\|\mathbf{a}_i(f) Y_i(k,f)\|^2}{\sum_{n=1}^{N} \|\mathbf{a}_n(f) Y_n(k,f)\|^2}. \tag{27}$$

In the following, $v_i^{f_l}(k) = powRatio_i(k,f_l)$ denotes the power ratio measure obtained at frequency $f_l = (l/K) f_s$ $(l = 0,\dots,K/2)$, where f_s is the sampling rate.

The details of the proposed method are as follows:

1. Obtain $v_i^{f_0}(k)$, $(i = 1,\dots,N)$, set $l = 1$.
2. Obtain $v_i^{f_l}(k)$ and $c_i^{f_l}(k) = \sum_{g \in T} v_i^{g}(k)$ $(i = 1,\dots,N)$, where $T = \{f_0,\dots,f_{l-1}\}$.
3. Obtain all permutation matrices $\mathbf{P}_e$ $(e = 1,2,\dots,N)$. Permutation matrix is an $N \times N$ matrix where in each row and each column there is one nonzero element of unit value. For example, for a case of 2 sources, the permutation matrices are:

$$\mathbf{P}_1 = \begin{bmatrix} 1 & 0 \\ 0 & 1 \end{bmatrix}, \ \mathbf{P}_2 = \begin{bmatrix} 0 & 1 \\ 1 & 0 \end{bmatrix}. \tag{28}$$

4. Obtain $\mathbf{u}^{f_l} = \mathbf{P}_e \mathbf{v}^{f_l}$ for all permutation matrices.

5. Determine the permutation matrix that maximizes the correlation of power ratio measure of current frequency bin with the average of power ratio measures of previous bin frequencies:

$$\mathbf{P} = \underset{\mathbf{P}_e}{\arg\max} \sum_{i=1}^{N} \rho\left(u_i^{f_l}, c_i^{f_l}\right). \tag{29}$$

6. Then, process the separated signal $\mathbf{Y}(k, f_l)$ with the permutation matrix at the bin frequency f_l:

$$\mathbf{Y}(k, f_l) \leftarrow \mathbf{P}(f_l)\, \mathbf{Y}(k, f_l). \tag{30}$$

7. Set $l = l + 1$, and return to step 2), if $l < K/2$.

The steps of the proposed method are shown in the block diagram of Figure 9.

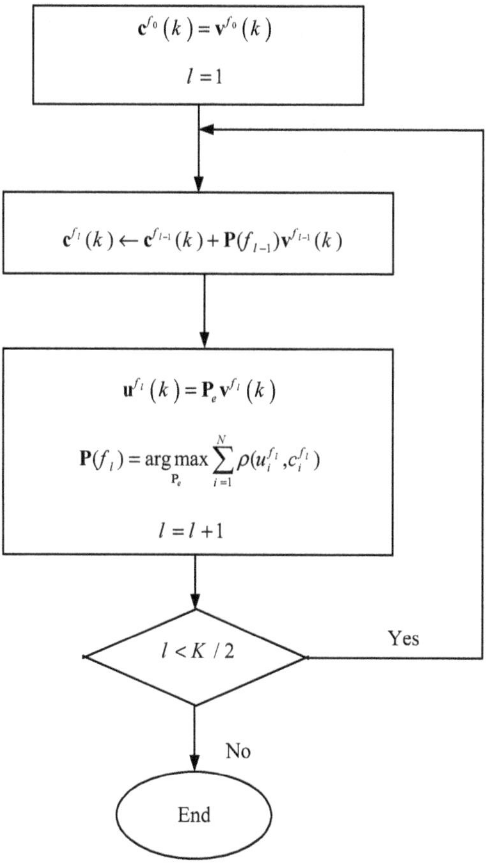

Fig. 9. The block diagram that describes our proposed method for solving the permutation problem

7. Scaling problem

The scaling problem can be solved by filtering individual outputs of the separation filter by the inverse of $\mathbf{B}(f)$ separately [12]. In the overdetermined case, (i.e., $M > N$), the pseudoinverse of $\mathbf{B}(f)$, denoted as $\mathbf{B}(f)^{+}$, is used instead of the inverse of $\mathbf{B}(f)$. This is due to the fact that in this case, because of employing the subspace method $\mathbf{B}(f)$ is not square. The scaling matrix can be expressed as:

$$\mathbf{S}(f) = diag[B^{+}_{m,1},\ldots,B^{+}_{m,N}], \tag{31}$$

where $B^{+}_{m,n}$ denotes the (m,n) - th element of $\mathbf{B}(f)^{+}$.

8. Suitable length of STFT for better separation

It is commonly believed that the length of STFT (i.e., frame size), K, must be longer than P to estimate the unmixing matrix for a P-point room impulse response. The reasons for this belief are: 1) A linear convolution can be approximated by a circular convolution if $K > 2P$, and 2) If we want to estimate the inverse system of a system with impulse response of P-taps long, we need an inverse system that is Q-taps long, where $Q > P$. If we assume that the frame size is equal to the length of unmixing filter, then we should have $K > P$. Moreover, when the filter length becomes longer, the number of separation matrices to be estimated increases while the number of samples for learning at each frequency bin decreases. This violates the assumption of independence in the time series at each bin frequency, and the performance of the ICA algorithm becomes poor [8]. Therefore, there is an optimum frame size determined by a trade-off between maintaining the assumption of independence and the length of STFT that should be longer than the room impulse response length in the frequency-domain BSS. Section 9 shows this understanding by some experiments.

9. Experimental results

The experiments are conducted to examine the effectiveness of the proposed permutation method [18]. We use two experimental setups. Setup A is considered to be a basic one, in which there are two sources and two microphones. In setup B, we have two sources and eight microphones, and discuss the effect of a background interference noise on our proposed method. Table 1 summarizes the configurations common to both setups. As the original speech, we use the wave files from 'TIMIT speech database' [28] to test the performance of different BSS algorithms. The lengths of the speech signals are 4 seconds. We have the voice of three male and three female speakers in our experiments and the investigations are carried out for nine different combinations of speakers. The image method

room dimension	L = 3.12 m, W = 5.73 m, H = 2.70 m
direction of arrivals	30° and -40°
window function	Hamming
sample rate	16000 Hz

Table 1. Common Experimental Configuration

has been used to generate multi-channel Room Impulse Responses [29]. Microphone signals are generated by adding the convolutions of source signals with their corresponding room impulse responses. Figure 10 shows the layout of the experimental room for setup *B*. For the setup *A*, we use only two microphones m1 and m2 shown in the figure.

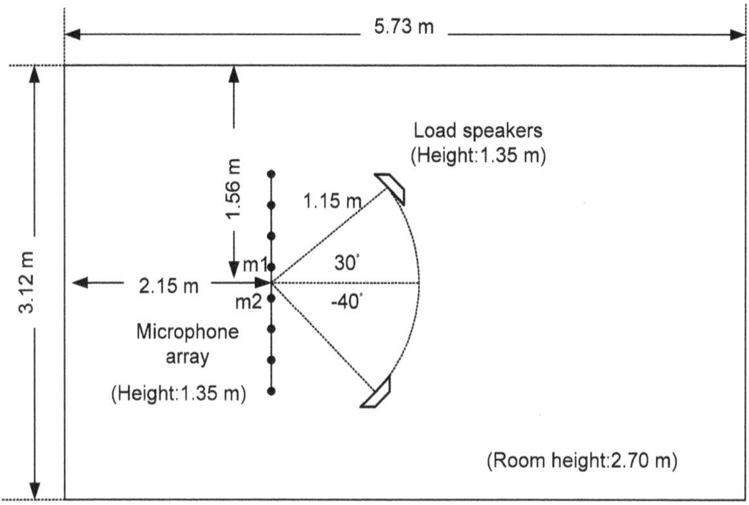

Fig. 10. Experimental Setup *B*

9.1 Evaluation criterion

For the computation of the evaluation criterion, we start by the decomposition of $y_i(t)$ (i.e., the estimation of $s_i(t)$):

$$y_i = s_{\text{target}} + e_{\text{interf}} + e_{\text{noise}}, \tag{32}$$

where s_{target} is a version of $s_i(t)$ modified by mixing and separating system, and e_{interf} and e_{noise} are respectively the interference and noise terms. Figure 11 shows the source, the microphone, and the separated signals.

We use Signal-to-Interference Rate (SIR) as performance criterion by computing energy ratios between the target signal and the interference signal expressed in decibels [30]:

$$\text{SIR}_i = 10\log_{10}\frac{\left\|s_{\text{target}}\right\|^2}{\left\|e_{\text{interf}}\right\|^2}. \tag{33}$$

To calculate s_{target}, we set the signals of all sources and noises to zero except $s_i(t)$ and measure the output signal. In the same way, to calculate e_{interf}, we set $s_i(t)$ and all noise signals to zero and obtain the output signal.

Setup A: The case of 2-Sources and 2-Microphones

In this experiment, we use only two microphones m1 and m2 in Figure 10. In this case, the reverberation time of the room is set to 130 ms. The frame length and frame shift in the STFT analysis are set to 2048 and 256 samples, respectively. Three different methods for the permutation problem are applied on 9 pairs of speech signals. The results of our simulations are shown in Figure 12. In the MaxSir approach, we select the best permutation by maximizing SIR at each frequency bin for solving perfectly the permutation ambiguity [16]. This gives a rough estimate of the upper bound of the performance.

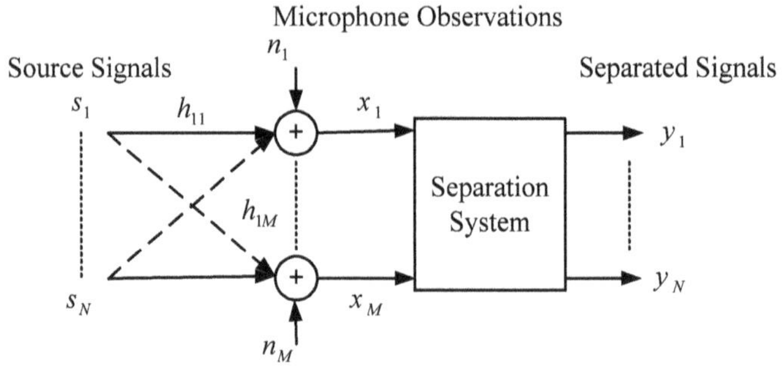

Fig. 11. Block diagram of the separating system

As seen from Figure 12, the results with Murata's method [12] are sometimes very poor, but our proposed method [18] offers almost the same results as that of MaxSir. Figure 13 shows SIRs at each frequency for the 8th pair of speech signals, obtained by the proposed method and Murata's method. The change of signs of SIRs in this figure shows the regions of permutation misalignments. Here, we see the permutation misalignments below 500 Hz obtained by Murata's method, whereas the proposed method has almost perfect permutation alignment. This shows that it is not always true to assume that frequencies not in close proximity have a high correlation of envelopes.

Setup B: The case of 2-Sources and 8-Microphones

In this experiment, we compare the separation performance of our proposed method with those of three other methods, namely, the Interfrequency Coherency method (IFC) [15], the DOA approach with a closed-form formula [17], and MaxSir for the case of 2-Sources and 8-Microphones [17]. To avoid aliasing in the DOA method, we select the distance between the microphones to be 2 cm. All these experiments are performed for three reverberation times $T_R = 100$ ms, 130 ms, and 200 ms. Before assessing different separation techniques, we first obtain the optimum frame length of STFT at each reverberation time. Then, we evaluate the proposed method in noisy and noise-free cases.

Optimum length of STFT for better separation

To show what frame length of STFT is suitable for better performance of BSS, we perform separation experiments at three reverberation times of $T_R = 100$ ms, 130 ms, and 200 ms, and by different lengths of STFT. Since the sampling rate is 16 kHz, these reverberation times correspond to $P = 1600$, 2080, and 3200 taps, respectively.

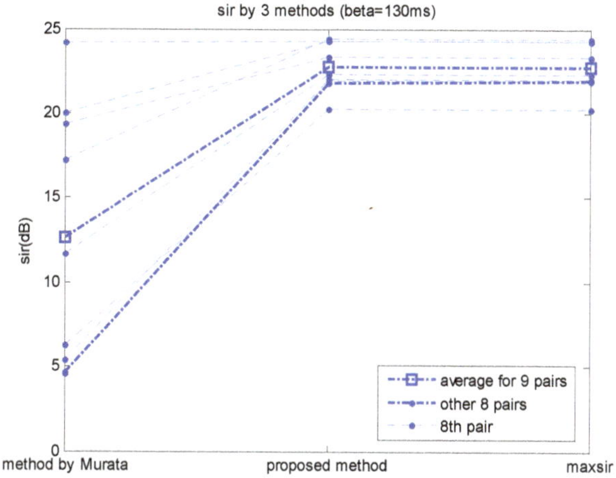

Fig. 12. The separation results of 9 pairs of speech signals for three different methods of permutation problem: Murata's method, proposed method, and MaxSir

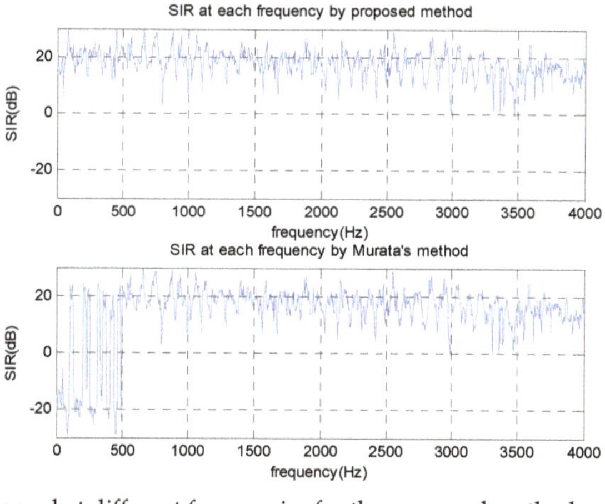

Fig. 13. SIRs measured at different frequencies for the proposed method and the Murata's method

Figure 14 shows the room impulse responses h_{11} for $T_R = 100\,\text{ms}, 130\,\text{ms}, \text{and } 200\,\text{ms}$. We vary the length of STFT by $K = 512$, 1024, 2048, 4096, and 8192 with corresponding frame shifts of $S = 64$, 128, 256, 512, and 1024, respectively. The best permutation is selected by maximizing SIR at each frequency bin. In this way, the results are ideal under the condition that the permutation problem is solved perfectly. The experimental results of SIR for different lengths of STFT are shown in Figure 15. These values are averaged over all nine combinations of speakers to obtain average values of SIR_1 and SIR_2. As it is observed from this figure, in the case of $T_R = 100\,\text{ms}$ we obtain the best performance with $K = 1024$. For

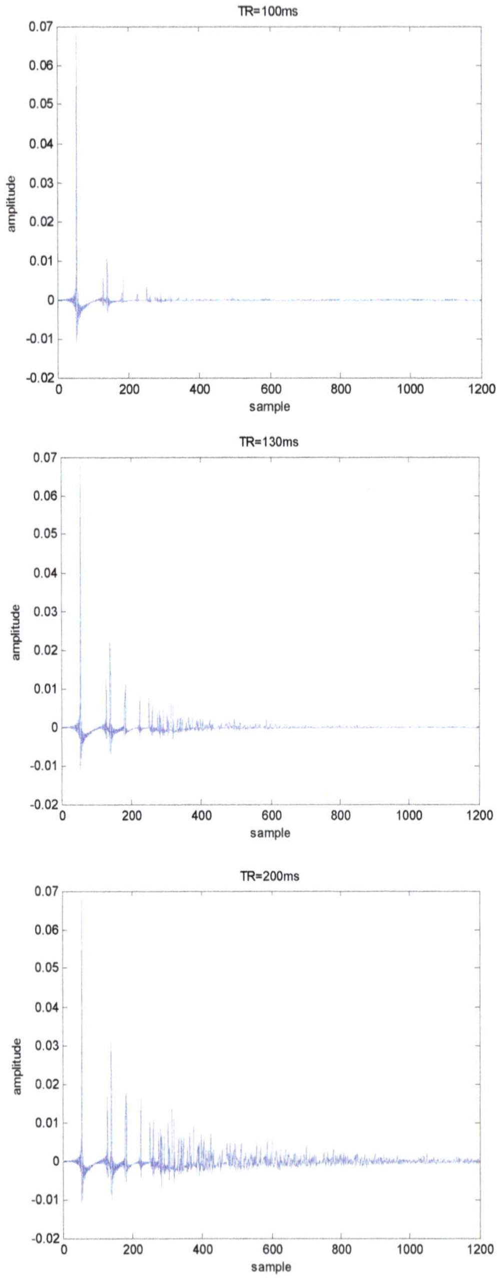

Fig. 14. The room impulse responses h_{11} for $T_R = 100\,\text{ms}, 130\,\text{ms}, \text{and } 200\,\text{ms}$

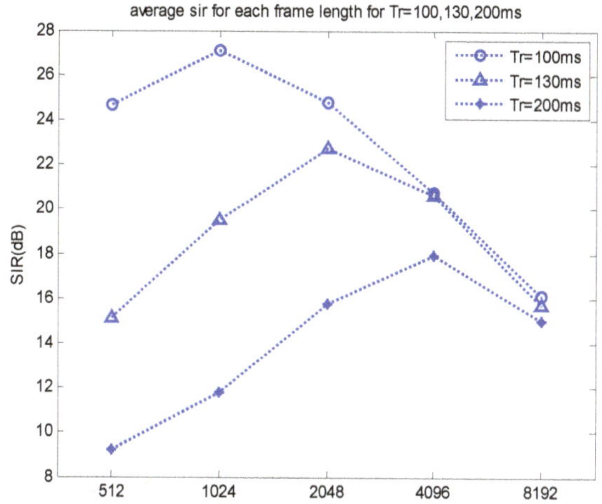

Fig. 15. The experimental results of SIR for different lengths of STFT

the reverberant conditions with $T_R = 130$, and 200 ms, the best performance is realized with $K = 2048$, and 4096, respectively. Figure 16 shows the average of neg-entropy (Eq. 15) as a measurement of independence. We see that by longer lengths of STFT the independence is smaller, and the performance of the fixed-point ICA is poorer [8].

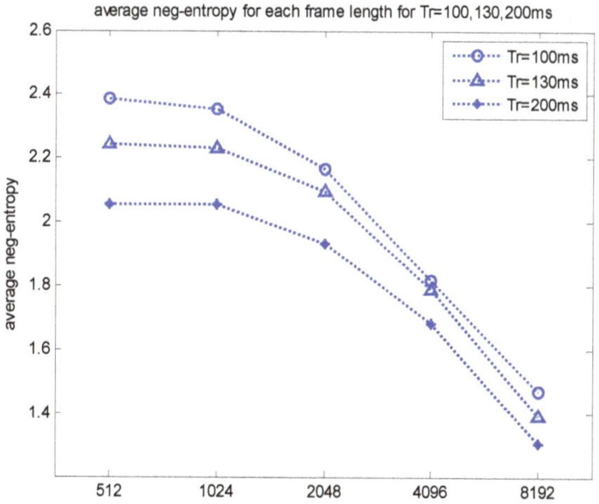

Fig. 16. The average of neg-entropy as a measurement of independence

Evaluation results without background noise

In this section, we compare our proposed method with three methods, namely, IFC, DOA, and MaxSir in the case of 2-Sources and 8-Microphones without the background noise. We select the optimum length of STFT obtained in the previous experiment for each of the three reverberation times. Figure 17 shows the separation results for nine pairs of speech signals

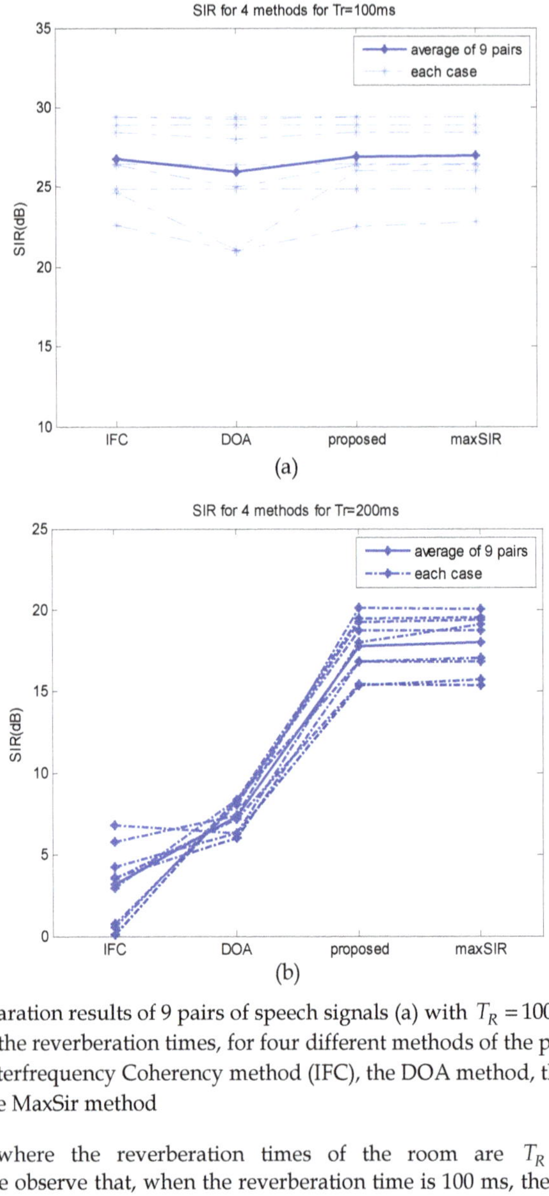

Fig. 17. The separation results of 9 pairs of speech signals (a) with $T_R = 100$ ms and (b) with $T_R = 200$ ms as the reverberation times, for four different methods of the permutation problem: the Interfrequency Coherency method (IFC), the DOA method, the proposed method, and the MaxSir method

in the cases where the reverberation times of the room are $T_R = 100$, and 200 ms, respectively. We observe that, when the reverberation time is 100 ms, the separation results for each of the three methods, i.e., IFC, DOA, and proposed methods, are close to the perfect solution obtained by MaxSir. For the reverberant case of $T_R = 200$ ms, the separation performances of IFC and DOA are not good, but the results of SIR for the proposed method are close to the MaxSir approach.

In the IFC method, to use the coherency of the mixing matrices in adjacent frequencies, the mixing matrix should have the form of the transfer function of direct path from each source to each microphone. However, this condition can hold, if the subspace filter reduces the energy of the reflection terms. The performance of the subspace method depends on both the array configuration and the sound environment. In our experiments, the subspace method could not reduce the reflection components, and the performance of the IFC method is poor for the reverberant case. However, in the case of $T_R = 100\,\text{ms}$ the energy of the reflection components is low and the IFC method has good performance. The SIRs at each frequency for four methods in the case of $T_R = 200\,\text{ms}$ are shown in Figure 18. We see a large

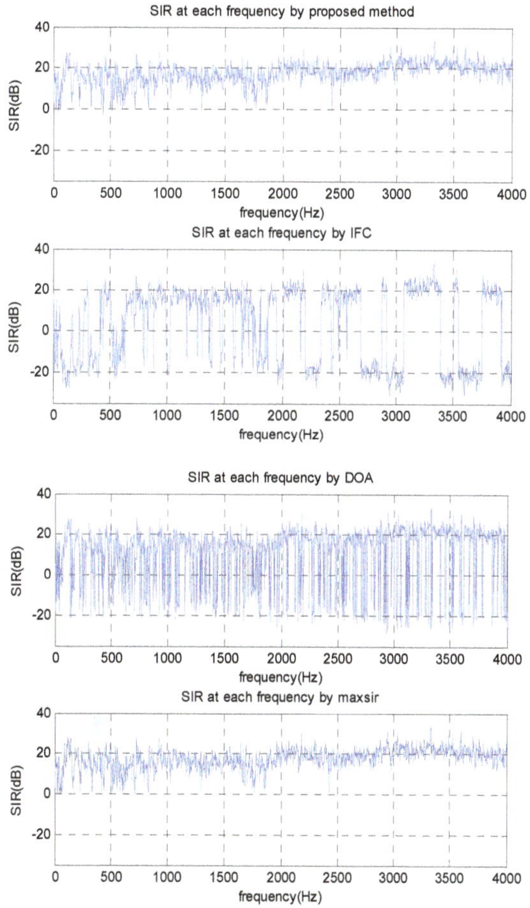

Fig. 18. SIRs measured at each bin frequency for 4 methods: the proposed method, the Interfrequency Coherency method (IFC), the DOA method, and the MaxSir method for the case of $T_R = 200\,\text{ms}$

number of frequencies with permutation misalignments for the IFC and DOA methods. As observed from the simulation results, the proposed approach outperforms the IFC and the DOA methods, where we achieve the best performance in the sense of SIR-improvement.

Evaluation results with background noise

In this part of experiments, we add the restaurant noise from the Noisex-92 database [31] with input SNRs of 5 dB, and 20 dB to the microphone signals. Here, again the optimum window length for the STFT analysis is chosen for each three reverberation times. Figures 19 and 20 show the average SIRs obtained for the proposed, IFC, DOA, and MaxSir methods for the reverberation times of $T_R = 100\text{ ms}, 130\text{ ms}, \text{and } 200\text{ ms}$, respectively with input SNRs of 5 dB and 20 dB. It is observed that under the experimental conditions of input SNR = 20 dB and reverberation time of 100 ms, all of the methods, i.e., the proposed, IFC, and DOA give the same separation results. However, as the reverberation time increases, the performance of IFC and DOA decreases. At the reverberation time of 200 ms, the average SIR of the proposed method is slightly reduced. Also, as it is expected, the comparison of Figures 19 and 20 shows that in lower values of input SNRs, the performance of source separation methods decreases. This shows that the ICA-based methods have in general poor separation results in noisy conditions.

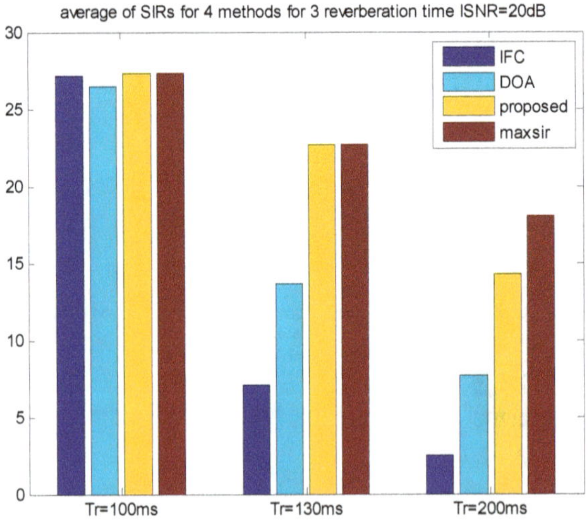

Fig. 19. Average of SIRs for the proposed, IFC, DOA, and MaxSir methods for three reverberation times of $T_R = 100\text{ ms}, 130\text{ ms}, \text{and } 200\text{ ms}$, respectively, obtained at the input SNR of 20 dB

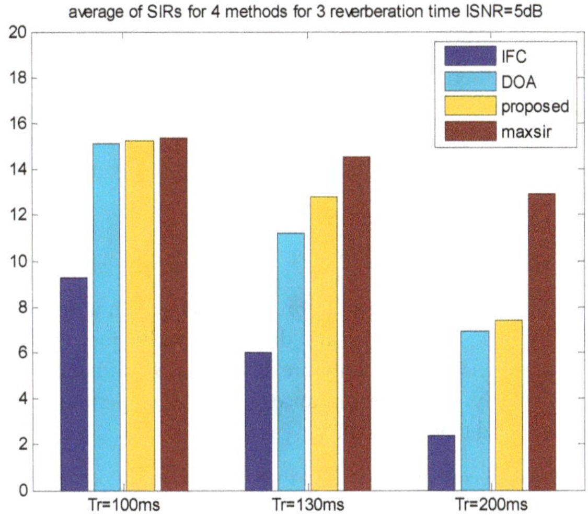

Fig. 20. Average of SIRs for the proposed, IFC, DOA, and MaxSir methods for three reverberation times of $T_R = 100\text{ ms}, 130\text{ ms}, \text{and } 200\text{ ms}$, respectively, obtained at the input SNR of 5 dB

10. Conclusion

This chapter presents a comprehensive description of frequency-domain approaches to the blind separation of convolutive mixtures. In frequency-domain approach, the short-time Fourier transform (STFT) is used to convert the convolutive mixtures in time domain to instantaneous mixtures at each frequency. In this way, we can use each of the complex-valued ICA at each frequency bin. We use the fast ICA algorithm for complex-valued signals. The key feature of this algorithm is that it converges faster than other algorithms, like natural gradient-based algorithms, with almost the same separation quality. We employ PCA as pre-processing for the purpose of decreasing the noise effect and dimension reduction. Also, we see that the length of STFT affects the performance of frequency-domain BSS. If the length of STFT becomes longer, the number of coefficients to be estimated increases while the number of samples for learning at each frequency bin decreases. This causes that the assumption of independence in the time series at each bin frequency to collapse, and the performance of the ICA algorithm to become poor. As a result, we select for the frame size an optimum value which is obtained by a trade-off between maintaining the assumption of independence and the length of STFT in the frequency-domain BSS.

We focus on the permutation alignment methods and introduce some conventional methods along with our proposed method to solve this problem. In the proposed method, we maximize the correlation of power ratio measure of each bin frequency with the average of power ratio measures of previous bin frequencies, iteratively with increasing frequency. In the case of 2-sources and 2-microphones, by conducting source separation experiments, we compare the performance of our proposed method with Murata's method which is based on envelope correlation. The results of this comparison show that it is not always true to assume that frequencies not in close proximity have a high correlation of envelopes. In another overdetermined case of experiment, the proposed method is compared with the DOA, IFC and MaxSir methods. Here, we see that in the reverberant room with high SNR values, the proposed method outperforms other methods. Finally, even though the performance of our proposed method degrades under reverberant conditions with high background noise (low SNRs), the experiments show that the separation results of the proposed method are still satisfactory.

11. Future directions

In this chapter, we have used PCA as a pre-processing technique for the purpose of decreasing the effect of background noise and dimension reduction. This approach assumes that the noise and the signal components are uncorrelated and the noise component is spatially white. Practically, the performance of PCA depends on both the array configuration and the sound environment.

From the results of the experiments, it is clear that two factors affect the performance of BSS methods; background noise and room reverberation. These factors are those that significantly influence the enhancement of audio signals. Therefore, as a future work, we should consider other pre-processing techniques in ICA-based BSS that besides performing dimension reduction also help to decrease the effect of colored noise as well as room reverberation.

12. References

[1] Lee T. W (1998) Independent Component Analysis - Theory and Applications. Norwell, MA: Kluwer.

[2] Comon P (1994) Independent Component Analysis, A New Concept? Signal Processing vol. 36 no. 3: 287-314.

[3] Benesty J, Makino S, Chen J (2005) Speech Enhancement. Springer-Verlag, Berlin, Heidelberg.

[4] Makino S, Lee T. W, Sawada H (2007) Blind Speech Separation. Springer.

[5] Douglas S. C, Sun X (2003) Convolutive Blind Separation of Speech Mixtures Using the Natural Gradient. Speech Communication, vol. 39: 65-78.

[6] Aichner R, Buchner H, Yan F, Kellermann W (2006) A Real-Time Blind Source Separation Scheme and its Application to Reverberant and Noisy Acoustic Environments. Signal Processing, vol. 86, no. 6: 1260-1277.

[7] Smaragdis P (1998) Blind Separation of Convolved Mixtures in the Frequency Domain. Neurocomputing, vol. 22: 21-34.

[8] Araki S, Mukai R, Makino S, Nishikawa T, Saruwatari H (2003) The Fundamental Limitation of Frequency-Domain Blind Source Separation for Convolutive Mixtures of Speech. IEEE Trans. on Speech and Audio Processing, vol. 11, no. 2.

[9] Bingham E, et al. (2000) A Fast Fixed-Point Algorithm for Independent Component Analysis of Complex-Valued Signal. Int. Journal of Neural Systems, vol. 10 (1) 1:8.

[10] Sawada H, Mukai R, Araki S, Makino S (2003) Polar Coordinate-Based Nonlinear Function for Frequency-Domain Blind Source Separation. IEICE Trans. Fundamentals, vol. E86-A, no. 3.

[11] Prasad R, Saruwatari H, Shikano K (2007) An ICA Algorithm for Separation of Convolutive Mixture of Speech Signals. International Journal of Information Technology, vol. 2, no. 4.
[12] Murata N, Ikeda S, Ziehe A (2001) An Approach to Blind Source Separation Based on Temporal Structure of Speech Signals. Neurocomput., vol. 41: 1-24.
[13] Kurita S, Saruwatari H, Kajita S, Takeda K, Itakkura F (2000) Evaluation of Blind Signal Separation Method Using Directivity Pattern Under Reverberant Conditions. ICASSP2000: 3140-3143.
[14] Saruwatari H, Kurita S, Takeda K, Itakura F, Nishikawa T, Shikano K (2003) Blind Source Separation Combining Independent Component Analysis and Beamforming. EURASIP2003: 1135-1146.
[15] Asano F, Ikeda S, Ogawa M, Asoh H, Kitawaki N (2003) Combined Approach of Array Processing and Independent Component Analysis for Blind Separation of Acoustic Signals. IEEE Trans. on Speech and Audio Processing, vol. 11, no. 3: 204-215.
[16] Sawada H, Mukai R, Araki S, Makino S (2004) A Robust and Precise Method for Solving the Permutation Problem of Frequency-Domain Blind Source Separation. IEEE Trans. on Speech and Audio Processing, vol. 12: 530-538.
[17] Sawada H, Araki S, Makino S (2007) Measuring Dependence of Bin-Wise Separated Signals for Permutation Alignment in Frequency-domain BSS. in Proc. ISCAS2007: 3247-3250.
[18] Hesam M, Geravanchizadeh M (2010) A New Solution for the Permutation Problem in the Frequency-Domain BSS Using Power-Ratio Correlation. IEEE Int. Symp. on telecommunications (IST 2010).
[19] Bell A. J, Sejnowski T. J (1995) An Information-Maximization Approach to Blind Separation and Blind Deconvolution. Neural Computation, vol. 7, no. 6: 1129-1159.
[20] Amari S (1998) Natural Gradient Works Efficiently in Learning. Neural Computation, vol. 10: 251-76.
[21] Hyvärinen A (1999) Fast and Robust Fixed-Point Algorithms for Independent Component Analysis. IEEE Trans. on Neural Networks, vol. 10: 626-634.
[22] Cardoso J.-F (1993) Blind Beamforming for Non-Gaussian Signals. IEE Proceedings-F, vol.140: 362-370.
[23] Matsuoka K, Ohya M, Kawamoto M (1995) A Neural Net for Blind Separation of Nonstationary Signals. Neural Networks, vol. 8: 411-419.
[24] Oppenhaim A. V, Schafer R. W, Buck J. R (1999) Discrete-Time Signal Processing. Prentice Hall, 1999.
[25] Araki S, Makino S, Blin A, Mukai R, Sawada H (2004) Underdetermined Blind Separation for Speech in Real Environments With Sparseness and ICA. in Proc. ICASSP 2004, vol. III: 881-884.
[26] Joho M, Mathis H, Lambert R. H (2000) Overdetermined Blind Source Separation: Using More Sensors Than Source Signals in a Noisy Mixture. Proceedings of ICA 2000: 81-86.
[27] Asano F, Motomura Y, Asoh H, Matsui T (2000) Effect of PCA Filter in Blind Source Separation. Proc. of Int. conf. on Independent Component Analysis (ICA2000).
[28] Allen J. B, Berkley D. A (1979) Image Method for Efficiently Simulating Small Room Acoustics. J. Acoust. Soc. Amer., vol. 65, no. 4: 943-950.
[29] http://www.ldc.upenn.edu/

[30] Vincent E, Gribonval R, Févotte C (2006) Performance Measurement in Blind Audio Source Separation. IEEE Trans. On Audio, Speech, and Language Processing, vol. 14, no. 4.

[31] http://www.speech.cs.cmu.edu/comp.speech/Section1/Data/noisex.html.

Section 3

ICA: Biomedical Applications

8

Nonlinear Independent Component Analysis for EEG-Based Brain-Computer Interface Systems

Farid Oveisi[1], Shahrzad Oveisi[2], Abbas Efranian[3] and Ioannis Patras[1]
[1]Queen Mary University of London
[2]Azad University
[3]Iran University of Science and Technology
[1]UK
[2,3]Iran

1. Introduction

The electroencephalogram (EEG) is a complex and a periodic time series, which is a sum over a very large number of neuronal membrane potentials. Despite rapid advances of neuroimaging techniques, EEG recordings continue to play an important role in both the diagnosis of neurological diseases and understanding the psychophysiological processes. Recently, many efforts have been done to use the electroencephalogram as a new communication channel between human brain and computer (Lotte & Guan, 2011; Oveisi, 2009; Ortner et al., 2011). This new communication channel is called EEG-based brain-computer interface (BCI). Most of these efforts have been dedicated to the improvement of the accuracy and capacity of this EEG-based communication channel. One of the most important factors about the performance of BCI systems is classification system. A classification system typically consists of both a preprocessor and a classifier. Preprocessors are used to improve the performance of classifier systems. One of the preprocessors can be used to improve the performance of brain–computer interface (BCI) systems is independent component analysis (ICA) (Van et al., 2011; Oveisi, 2009). ICA is a signal processing technique in which observed random data are transformed into components that are statistically independent from each other (Oveisi et al., 2012). ICA is a useful technique for blind separation of independent sources from their mixtures. Sources are usually original, uncorrupted signals or noise sources. Linear ICA was used to separate neural activity from muscle and blink artifacts in spontaneous EEG data (Jung et al., 2000). It was verified that the ICA can separate artifactual, stimulus locked, response-locked, and non-event related background EEG activities into separate components (Jung et al., 2001). Furthermore, ICA would appear to be able to separate task-related potentials from other neural and artifactual EEG sources during hand movement imagination in form of independent components. In (Peterson et al., 2005), it has been showed that the power spectra of the linear ICA transformations provided feature subsets with higher classification accuracy than the power spectra of the original EEG signals. However, there is no guarantee for linear combination of brain sources in EEG signals. Thus the identification of non-linear dynamic of EEG signals should be taken into consideration. For non-linear mixing model, linear ICA algorithms fail to extract original signals and become inapplicable because the assumption of linear mixtures is violated and the linear algorithm cannot compensate for the information distorted by the non-linearity.

ICA is currently a popular method for blind source separation (BSS) of linear mixtures. However, nonlinear ICA does not necessarily lead to nonlinear BSS (Zhang & Chan, 2007). Hyvarinen and Pajunen (1999) showed that solutions to nonlinear ICA always exist, and that they are highly non-unique. In fact, nonlinear BSS is impossible without additional prior knowledge on the mixing model, since the independence assumption is not strong enough in the general nonlinear mixing case (Achard & Jutten, 2005; Singer & Coifman, 2007). If we constrain the nonlinear mixing mapping to have some particular forms, the indeterminacies in the results of nonlinear ICA can be reduced dramatically, and as a consequence, in these cases nonlinear ICA may lead to nonlinear BSS. But sometimes, the form of the nonlinear mixing procedure may be unknown. Consequently, in order to model arbitrary nonlinear mappings, one may need to resort to a flexible nonlinear function approximator, such as the multi-layer perceptron (MLP) (Woo & Sali, 2002; Almeida, 2003) or the radius basis function (RBF) network (Tan et al., 2001), to represent the nonlinear separation system. In this situation, in order to achieve BSS, nonlinear ICA requires extra constraints or regularization. In (Woo & Sali, 2002), a general framework for a demixer based on a feedforward multilayer perceptron (FMLP) employing a class of continuously differentiable nonlinear functions has been explained. In this method, Cost functions based on both maximum entropy (ME) and minimum mutual information (MMI) have been used. In (Almeida, 2003), the MLP has been used to model the separation system and trains the MLP by information maximization (Infomax). Moreover, smoothness provided by the MLP was believed to be a suitable regularization condition to achieve nonlinear BSS. In (Tan et al., 2001), a blind signal separation approach based on an RBF network is developed for the separation of nonlinearly mixed sources by defining a contrast function. This contrast function consists of mutual information and cumulants matching. However, the matching between the relevant moments of the outputs and those of the original sources was expected to guarantee a unique solution. But the moments of the original sources may be unknown.

In this research, a nonlinear ICA has been used to separate task-related potentials from other neural and artifactual EEG sources. The proposed method has been tested on several different subjects. Moreover, the results of proposed method were compared to the results obtained using linear ICA, and original EEG signals.

2. Background

2.1 Mutual information

Mutual information is a non-parametric measure of relevance between two variables. Shannon's information theory provides a suitable formalism for quantifying these concepts. Assume a random variable **X** representing continuous-valued random feature vector, and a discrete-valued random variable C representing the class labels. In accordance with Shannon's information theory, the uncertainty of the class label C can be measured by entropy $H(C)$ as

$$H(C) = -\sum_{c \in C} p(c) \log p(c), \quad (1)$$

where $p(c)$ represents the probability of the discrete random variable C. The uncertainty about C given **X** is measured by the conditional entropy as

$$H(C|X) = -\int p(\mathbf{x}) \left(\sum_{c \in C} p(c|\mathbf{x}) \log p(c|\mathbf{x}) \right) d\mathbf{x}, \tag{2}$$

where $p(c|\mathbf{x})$ is the conditional probability for the variable C given X.

In general, the conditional entropy is less than or equal to the initial entropy. It is equal if and only if the two variables C and X are independent. The amount by which the class uncertainty is decreased is, by definition, the mutual information $I(X,C) = H(C) - H(C|X)$ and after applying the identities $p(c,\mathbf{x}) = p(c|\mathbf{x})p(\mathbf{x})$ and $p(c) = \int p(c,x)dx$ can be expressed as

$$I(X,C) = \sum_{c \in C} \int p(\mathbf{x},c) \log \frac{p(\mathbf{x},c)}{p(c)p(\mathbf{x})} d\mathbf{x} \tag{3}$$

If the mutual information between two random variables is large, it means two variables are closely related. The mutual information is zero if and only if the two random variables are strictly independent. The mutual information and the entropy have the following relation, as shown in Fig. 1:

$$\begin{aligned} I(X;Y) &= H(X) - H(X|Y) \\ I(X;Y) &= H(Y) - H(Y|X) \\ I(X;Y) &= H(X) + H(Y) - H(X,Y) \\ I(X;Y) &= I(Y,X) \\ I(X,X) &= H(X). \end{aligned} \tag{4}$$

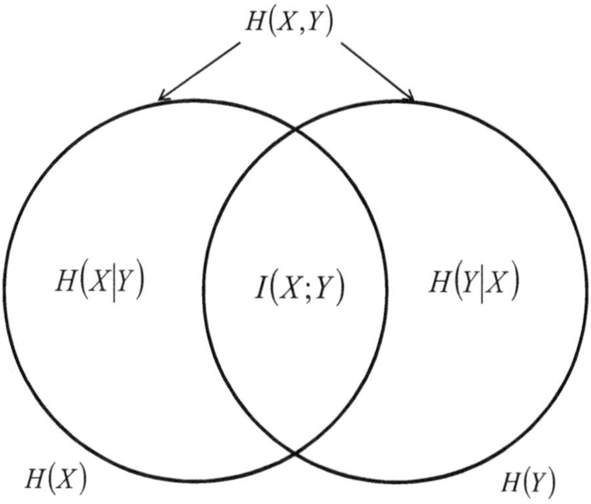

Fig. 1. The relation between the mutual information and the entropy.

2.2 Genetic algorithm

In a genetic algorithm, a population of strings (called chromosomes or the genotype of the genome), which encode candidate solutions (called individuals, creatures, or phenotypes) to an optimization problem, evolves toward better solutions. Traditionally, solutions are represented in binary as strings of 0s and 1s, but other encodings are also possible. The evolution usually starts from a population of randomly generated individuals and happens in generations. In each generation, the fitness of every individual in the population is evaluated, multiple individuals are stochastically selected from the current population (based on their fitness), and modified (recombined and possibly randomly mutated) to form a new population. The new population is then used in the next iteration of the algorithm. Commonly, the algorithm terminates when either a maximum number of generations has been produced, or a satisfactory fitness level has been reached for the population. If the algorithm has terminated due to a maximum number of generations, a satisfactory solution may or may not have been reached.

A standard representation of the solution is as an array of bits. Arrays of other types and structures can be used in essentially the same way. The main property that makes these genetic representations convenient is that their parts are easily aligned due to their fixed size, which facilitates simple crossover operations. Variable length representations may also be used, but crossover implementation is more complex in this case. Tree-like representations are explored in genetic programming and graph-form representations are explored in evolutionary programming.

The fitness function is defined over the genetic representation and measures the quality of the represented solution. The fitness function is always problem dependent. For instance, in the knapsack problem one wants to maximize the total value of objects that can be put in a knapsack of some fixed capacity. A representation of a solution might be an array of bits, where each bit represents a different object, and the value of the bit (0 or 1) represents whether or not the object is in the knapsack. Not every such representation is valid, as the size of objects may exceed the capacity of the knapsack. The fitness of the solution is the sum of values of all objects in the knapsack if the representation is valid or 0 otherwise. In some problems, it is hard or even impossible to define the fitness expression; in these cases, interactive genetic algorithms are used.

Once the genetic representation and the fitness function are defined, a GA proceeds to initialize a population of solutions (usually randomly) and then to improve it through repetitive application of the mutation, crossover, inversion and selection operators.

Initially many individual solutions are (usually) randomly generated to form an initial population. The population size depends on the nature of the problem, but typically contains several hundreds or thousands of possible solutions. Traditionally, the population is generated randomly, allowing the entire range of possible solutions (the search space). Occasionally, the solutions may be "seeded" in areas where optimal solutions are likely to be found (Akbari & Ziarati, 2010).

During each successive generation, a proportion of the existing population is selected to breed a new generation. Individual solutions are selected through a fitness-based process, where fitter solutions (as measured by a fitness function) are typically more likely to be selected. Certain selection methods rate the fitness of each solution and preferentially select the best solutions. Other methods rate only a random sample of the population, as the latter process may be very time-consuming.

The next step is to generate a second generation population of solutions from those selected through genetic operators: crossover (also called recombination), and/or mutation.

For each new solution to be produced, a pair of "parent" solutions is selected for breeding from the pool selected previously. By producing a "child" solution using the above methods of crossover and mutation, a new solution is created which typically shares many of the characteristics of its "parents". New parents are selected for each new child, and the process continues until a new population of solutions of appropriate size is generated. Although reproduction methods that are based on the use of two parents are more "biology inspired", some research suggests more than two "parents" are better to be used to reproduce a good quality chromosome.

These processes ultimately result in the next generation population of chromosomes that is different from the initial generation. Generally the average fitness will have increased by this procedure for the population, since only the best organisms from the first generation are selected for breeding, along with a small proportion of less fit solutions, for reasons already mentioned above.

Although Crossover and Mutation are known as the main genetic operators, it is possible to use other operators such as regrouping, colonization-extinction, or migration in genetic algorithms.

This generational process is repeated until a termination condition has been reached. Common terminating conditions are:

- A solution is found that satisfies minimum criteria
  - Fixed number of generations reached
  - Allocated budget (computation time/money) reached
  - The highest ranking solution's fitness is reaching or has reached a plateau such that successive iterations no longer produce better results
  - Manual inspection
  - Combinations of the above

Simple generational genetic algorithm procedure:

1. Choose the initial population of individuals
2. Evaluate the fitness of each individual in that population
3. Repeat on this generation until termination (time limit, sufficient fitness achieved, etc.):
   1. Select the best-fit individuals for reproduction
   2. Breed new individuals through crossover and mutation operations to give birth to offspring
   3. Evaluate the individual fitness of new individuals
4. Replace least-fit population with new individuals

3. Independent Component Analysis (ICA)

3.1 Linear ICA

We assume that we observe n linear mixtures $x_1, x_2, \ldots, x_n$ of n independent components:

$$x_j = a_{j_1} s_1 + a_{j_2} s_2 + \ldots + a_{j_n} s_n \tag{5}$$

In this equation the time has been ignored. Instead, it was assumed that each mixture x_j as well as each independent component s_i are random variables and $x_j(t)$ and $s_i(t)$ are samples of these random variables. It is also assumed that both the mixture variables and the independent components have zero mean (Oveisi et al., 2008).

If not subtracting the sample mean can always center the observable variables x_i. This procedure reduces the problem to the model zero-mean:

$$\hat{x} = x - E(x) \tag{6}$$

Let x be the random vectors whose elements are the mixtures $x_1, x_2, \ldots, x_n$ and let s be the random vector with the components $s_1, s_2, \ldots, s_n$. Let **A** be the matrix containing the elements a_{ij}. The model can now be written:

$$x = As \text{ or } x = \sum_{i=1}^{n} a_i s_i \tag{7}$$

The above equation is called independent component analysis or ICA. The problem is to determine both the matrix A and the independent components s, knowing only the measured variables x. The only assumption the methods take is that the components s_i are independent. ICA looks a lot like the "blind source separation" (BSS) problem or blind signal separation: a source is in the ICA problem an original signal, so an independent component. In ICA case it is also no information about the independent components, like in BSS problem.

Whitening can be performed via eigenvalue decomposition of the covariance matrix:

$$VDV^T = E\{\hat{x}\hat{x}^T\} \tag{8}$$

where **V** is the matrix of orthogonal eigenvectors and **D** is a diagonal matrix with the corresponding eigenvalues. The whitening is done by multiplication with the transformation matrix **P**:

$$\tilde{x} = P\hat{x} \tag{9}$$

$$P = VD^{\frac{1}{2}}V^T \tag{10}$$

The matrix for extracting the independent components from $\tilde{x}$ is $\tilde{W}$, where $W = \tilde{W}P$

3.2 Nonlinear ICA

Conventional linear ICA approaches assume that the mixture is linear by virtue of its simplicity. However, this assumption is often violated and may not characterize real-life signals accurately. A realistic mixture needs to be non-linear and concurrently capable of treating the linear mixture as a special case (Lappalainen & Honkela, 2000; Gao et al., 2006; Jutten & Karhunen, 2004). Generally, a non-linear ICA problem can be defined as follows: given a set of observations, $x(t)=\left[x_1(t),x_2(t),\ldots,x_n(t)\right]^T$ which are random variables and generated as a mixture of independent components $s(t)=\left[s_1(t),s_2(t),\ldots,s_n(t)\right]^T$ according to

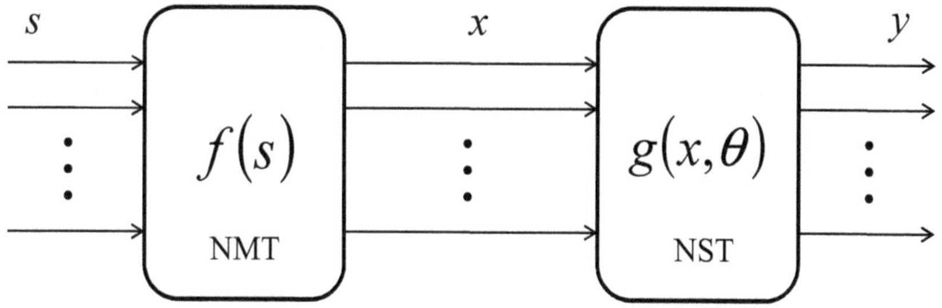

Fig. 2. Nonlinear mixing and separating systems for independent component analysis.

$$x(t)=f\left[s(t)\right] \tag{11}$$

where f is an unknown nonlinear mixing transform (NMT). The block diagram of the nonlinear ICA is shown in Figure 2.

The separating system $g(.,\theta)$ in the right part of Fig. 2, called nonlinear separation transform (NST) is used to recover the original signals $x(t)$ from the nonlinear mixture without the knowledge of the source signals $s(t)$ and the mixing nonlinear function f. However, a fundamental difficulty in nonlinear ICA is that it is highly non-unique without some extra constraints; therefore, finding independent components does not lead us necessarily to the original sources (Achard & Jutten, 2005).

ICA in the nonlinear case is, in general, impossible. In (Rojas et al., 2004), it has been added some extra constraints to the nonlinear mixture so that the nonlinearities are independently applied in each channel after a linear mixture. As figure 3 shows, the proposed algorithm in (Rojas et al., 2004) needs to estimate two different mixtures: a family of nonlinearities g which approximates the inverse of the nonlinear mixtures f and a linear unmixing matrix W which approximates the inverse of the linear mixture A. For the demixing system, first we need to approximate g_i, which is the inverse of the nonlinear function in each channel, and then separate the linear mixing by applying W to the output of the g_i nonlinear function:

$$y_i(t)=\sum_{j=1}^{n} w_{ij} g_i\left(x_j(t)\right) \tag{12}$$

In order to develop a more general and flexible model of the function g_i, it can be used a M th order odd polynomial expression of nonlinear transfer function (g_i):

$$p_j\left(x_j\right)=\sum_{k=1}^{M} p_{jk} x_j^{2k-1} \tag{13}$$

where $p_j = \left[p_{j1}, p_{j2}, \ldots, p_{jM}\right]$ is a parameter vector to be determined. By using relations (12) and (13), we can write the following criterion for the output sources y_i :

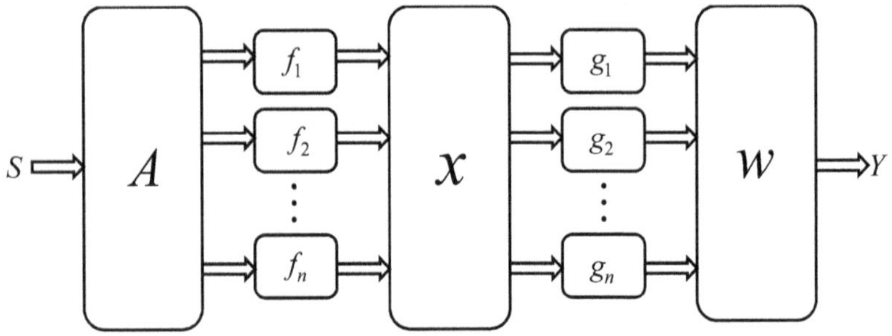

Fig. 3. Post-nonlinear mixing and demixing models for independent component analysis.

$$y_i(t) = \sum_{j=1}^{n} w_{ij} \sum_{k=1}^{M} p_{jk}.x_j^{2k-1} \tag{14}$$

The parameter vector p_j should be determined so that the inverse of the mutual information of the output sources y_i is maximized. To achieve this objective, can be defined the following criterion (Rojas et al., 2004):

$$eval\_function(y) = \frac{1}{I(y)} \tag{15}$$

Nevertheless, computation of the parameter vectors p_j is not easy, as it presents a problem with numerous local minima when the usual BSS cost functions are applied. Thus, we require an algorithm that is capable of avoiding entrapment in such a minimum. As a solution, in this work, a genetic algorithm (GA) (Goldberg, 1989) was used for mutual information optimization. Unlike many classical optimization techniques, GA does not rely on computing local first- or second-order derivatives to guide the search process; GA is a more general and flexible method that is capable of searching wide solution spaces and avoiding local minima (i.e., it provides more possibilities of finding an optimal or near-optimal solution). To implement the GA, we use genetic algorithm and direct search toolbox for use in Matlab (The Mathworks, R2007b).

The linear demixing stage has been performed by the well-known Infomax algorithm (Hyvarinen et al., 2001). To be precise, Infomax has been embedded into the GA in order to approximate the linear mixture.

In this application, the genetic algorithm is run for 30 generations with population size of 20, crossover probability 0.8, and uniform mutation probability of 0.01. The number of individuals that automatically survive to the next generation (i.e., elite individuals) is selected to be 2. The scattered function is used to create the crossover children by creating a random binary vector and selects the genes where the vector is a 1 from the first parent, and the genes where the vector is a 0 from the second parent.

4. Experimental setup

The EEG data of healthy right-handed volunteer subjects were recorded at a sampling rate of 256 from positions Cz, T5, Pz, F3, F4, Fz, and C3 by Ag/AgCl scalp electrodes placed according to the International 10-20 system that has been shown in Fig. 4. The eye blinks were recorded by placing an electrode on the forehead above the left brow line. The signals were referenced to the right earlobe.

Data were recorded for $5\ s$ during each trial experiment and low-pass filtered with a cutoff 45 Hz. There were 100 trails acquired from each subject during each experiment day. At $t = 2\ s$, a cross ("+") was displayed on the monitor of computer as a cue visual stimulus. The subjects were asked to imagine the hand grasping in synchronization with the cue and to not perform a specific mental task before displaying the cue. In the present study, the tasks to be discriminated are the imaginative hand movement and the idle state. The experimental setup has been shown in Fig. 5.

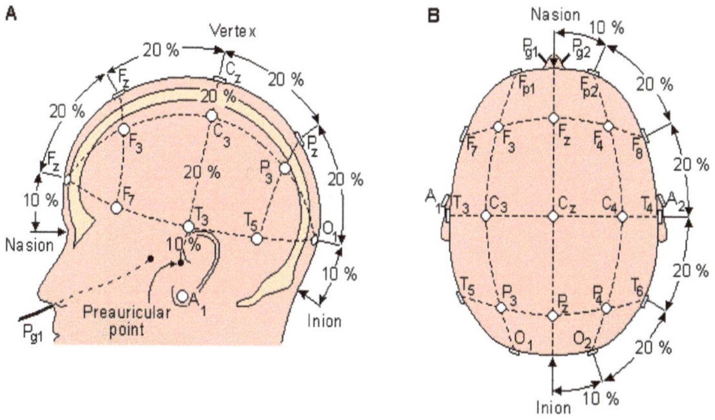

Fig. 4. The international 10-20 system

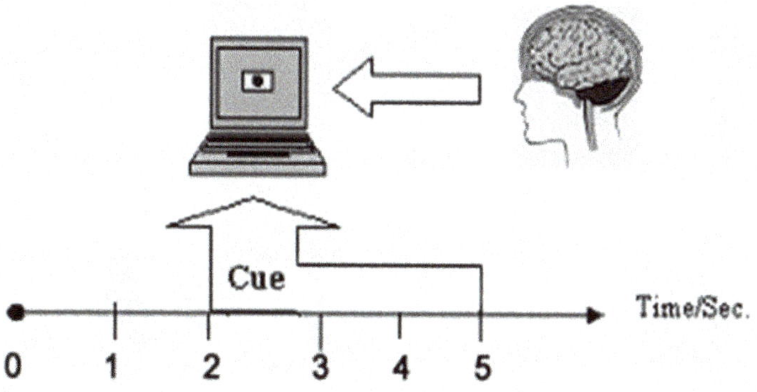

Fig. 5. Experimental Setup

Eye blink artifact was suppressed by using independent component analysis. The artifactual independent components were visually identified and set to zero. This process has been shown in Fig. 6.

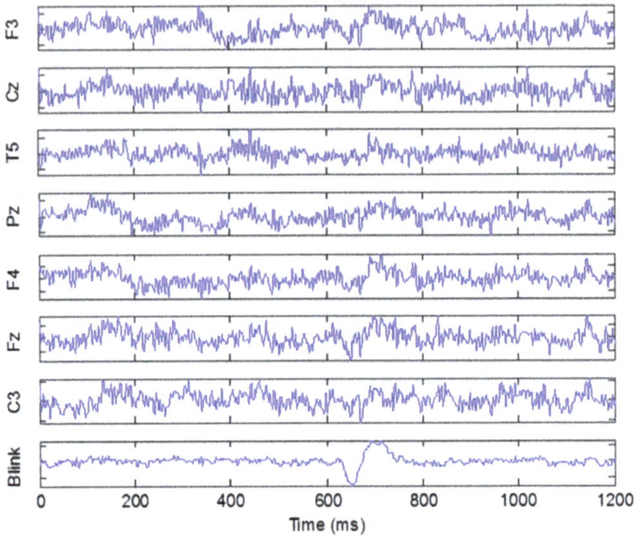

(a)

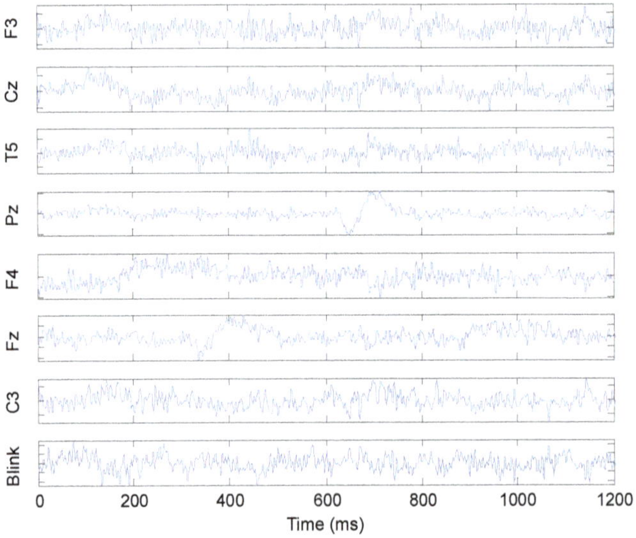

(b)

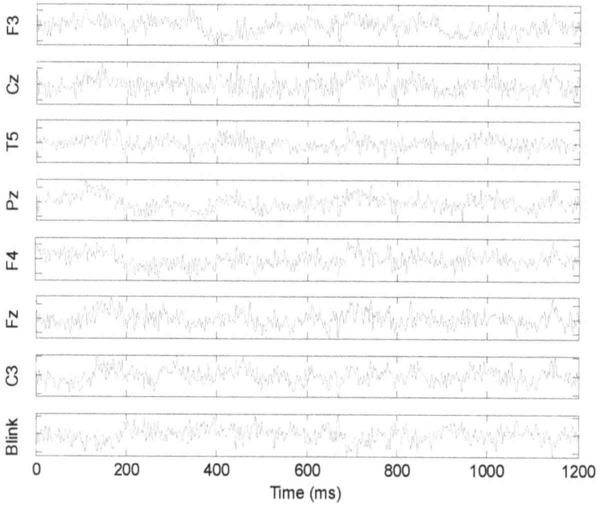

(c)

Fig. 6. (a) Raw EEG signals, (b) ICA components, (c) EEG signals after eye blink removal.

5. Results

The nonlinear ICA algorithm, proposed in (Rojas et al., 2004), was applied to given training 7-channel EEG data sets associated to the hand movement imagination and resting state. Original features are formed from 1 second interval of each component, in the time period 2.3–3.3 seconds, during each trial of experiment. The window starting 0.3 seconds after cue presentation is used for classification. The number of local extrema within interval, zero crossing, 5 AR parameters, variance, the mean absolute value (MAV), and 1Hz frequency components between 1 and 35Hz constitute the full set of features with size 44. The classifier is trained to distinguish between rest state and imaginative hand movement. The imaginative hand movement can be hand closing or hand opening. From 200 data sets, 100 sets are randomly selected for training, while the rest is kept aside for validation purposes. Training and validating procedure is repeated 10 times and the results are averaged.

Multiple classifiers are employed for classification using extracted components obtained by linear and nonlinear ICA. The Multiple Classifiers are used if different sensors are available to give information on one object. Each of the classifiers works independently on its own domain. The single classifiers are built and trained for their specific task. The final decision is made on the results of the individual classifiers. In this work, for each component, separate classifier is trained and the final decision is implemented by a simple logical majority vote function. The desired output of each classifier is −1 or +1. The output of classifiers is added and the *signum function* is used for computing the actual response of the classifier. The diagonal linear discrimination analysis (DLDA) (Krzanowski, 2000) is here considered as the classifier. The classifier is trained to distinguish between rest state and imaginative hand movement. The block diagram of classification process is shown in Fig. 7.

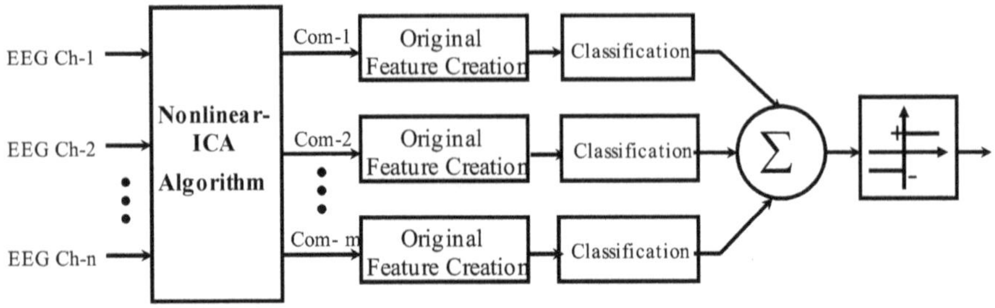

Fig. 7. The block diagram of classification process.

The results have been recorded for four subjects (AE, ME, BM, SN) for different experiment days. Table 1 summarizes the results of classification accuracy of the original EEG signals. The average classification accuracy is 73.84%.

Table 2 summarizes the results of classification accuracy for different subjects by using linear ICA. For these experiments, the Infomax algorithm (Hyvarinen et al., 2001) as a linear ICA has been used. The average classification accuracy over all subjects is 74.61% which 1% better than that obtained original EEG signals. An average classification rate of 77.95% is achieved by using nonlinear ICA. As can be observed, components which are obtained by nonlinear ICA improved the EEG classification accuracy compared to the linear ICA and original EEG signals. These results are 4 percent higher than average classification results by using the raw EEG data. Fig. 8 shows the classification accuracy rate obtained by nonlinear ICA (NICA), linear ICA (LICA), and original EEG signals (channel).

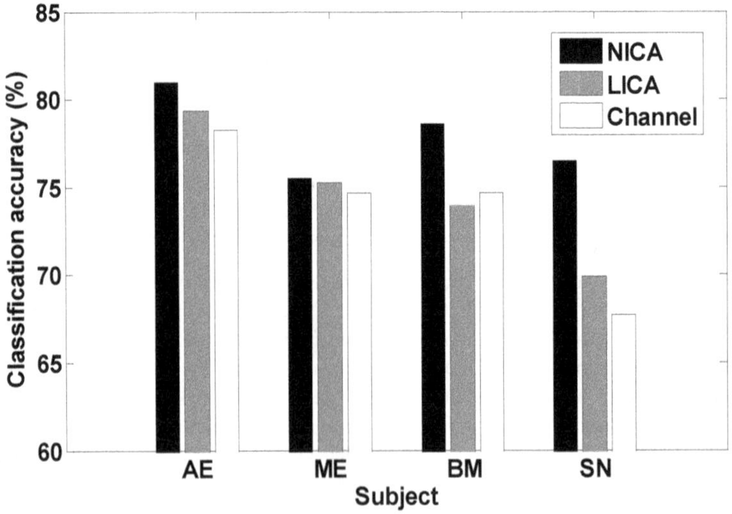

Fig. 8. Mean classification accuracy of EEG patterns for different subjects using nonlinear ICA (NICA), linear ICA (LICA), and original EEG signals (channel).

Subject	DAY1	Day2	Day3	DAY4	DAY5	mean
AE	77.3	76.4	75.5	83.9	-	78.27
ME	65.3	84.9	74.6	73.8	-	74.65
BM	67.2	90.6	66	75.2	-	74.75
SN	77.4	66.1	61.6	69.4	64.1	67.7
mean	71.8	79.5	69.42	75.57	64.1	73.84

Table 1. Classification Accuracy Rate of Original EEG Signals During Hand Movement Imagination.

Subject	DAY1	Day2	Day3	DAY4	DAY5	mean
AE	76.3	81.9	77.9	81.4	-	79.37
ME	68.7	84.1	77.2	71.1	-	75.27
BM	67.1	93.3	63	72.5	-	73.97
SN	78.9	71.1	64.1	67.6	67.6	69.86
mean	72.75	82.6	70.55	73.15	67.6	74.61

Table 2. Classification Accuracy Rate of Extracted Components During Hand Movement Imagination using Linear ICA.

Subject	DAY1	Day2	Day3	DAY4	DAY5	mean
AE	77.6	81	80.1	85.3	-	81
ME	72.8	80.5	76.6	72	-	75.47
BM	76.2	93	69	76.2	-	78.6
SN	78.5	79	81.7	72.5	72	76.74
mean	76.28	83.38	76.85	76.5	72	77.95

Table 3. Accuracy Rate of Extracted Components During Hand Movement Imagination using Nonlinear ICA.

6. Conclusion

Preprocessing plays an important role in the performance of BCI systems. One of the preprocessors can be used to improve the performance of BCI systems is independent component analysis (ICA). ICA would appear to be able to separate task-related potentials from other neural and artifactual EEG sources during hand movement imagination in form of independent components. However, there is no guarantee for linear combination of brain sources in EEG signals. Therefore, in this research a novel method was proposed for EEG signal classification in BCI systems by using non-linear ICA algorithm. The results of applying this method on four subjects have demonstrated that the proposed method in this research has improved the mean classification accuracies in relation to raw EEG data and linear ICA. The analysis of variance (ANOVA) shows that the mean classification accuracies achieved by using non-linear ICA are significantly different ($p < 0.01$).

7. Future directions

There are a number of directions in which the research described in this chapter can be extended. One area currently under investigation is to design a non-linear model for separation of nonlinearly mixed sources in the brain. As clarified in this chapter, a fundamental difficulty in nonlinear ICA is that it is highly non-unique without some extra constraints. Now, the question is which extra constraints can more compensate the information distorted by the non-linear combination of brain sources. Answering to this question will be our future work.

8. References

Achard, S.; Jutten, C. (2005). Identifiability of Post-Nonlinear Mixtures, *IEEE Trans. Signal Processing Letters*, vol. 12, no. 5, pp. 423-426.

Akbari, R. & Ziarati, K. (2010). A multilevel evolutionary algorithm for optimizing numerical functions, International Journal of Industrial Engineering Computations, Vol. 2, no. 2, pp. 419- 430.

Almeida, L.B. (2003). MISEP - Linear and Nonlinear ICA Based on Mutual Information, Journal of Machine Learning Research 4, pp. 1297-1318.

Gao, P.; Woo, W.L. & Dlay, S.S. (2006). Non-Linear Independent Component Analysis Using Series Reversion and Weierstrass Network, IEE Proc., *Vis. Image Process*, Vol. 153, no. 2, pp.115- 131.

Goldberg, D. E. (1989). *Genetic Algorithms in Search , Optimization and Machine Learning*. Addison-Wesley.

Hyvarinen, A.; Karhunen, J. & Oja, E. (2001). *Independent Component Analysis,* John Wiley & Sons.

Hyvarinen, A. & Pajunen, P. (1999). Nonlinear Independent Component Analysis Existence and Uniqueness Results, *Neural Network,* Vol. 12, no. 3, pp.429-439.

Jung, T.; makeig, S.; humphiers, C.; Lee, T.; Mckeown, M. & Sejnowski, T. (2000). Removing Electroencephalographic Arifacts by Blind Source Separation, *Psychophysiology*, pp. 163- 178.

Jung, T.; makeig, S.; Westerfield, M.; Townsend, J. ; Courchesne, E. & Sejnowski, T. (2001). Analysis and Visualizion of Single-Trial Event-Related Potentials, *Human Brain Mapping*, vol. 14, pp. 166-185.

Jutten, C. & Karhunen, J. (2004). Advances in Blind Source Separation (BSS) and Independent Component Analysis (ICA) for Nonlinear Mixtures, International Journal of Neural Sysytems, vol. 14, no. 5, pp. 267-292.

Krzanowski, W. J. (2000). *Principles of Multivariate Analysis: A User's Perspective*, Oxford University Press, Oxford.

Lappalainen, H. & Honkela, A. (2000). Bayesian Nonlinear Independent Component Analysis by Multi-Layer Perceptrons, *In Advances in Independent Component Analysis*, pp. 93-121.

Lotte, F. & Cuntai Guan (2011). Regularizing Common Spatial Patterns to Improve BCI Designs: Unified Theory and New Algorithms, *IEEE Transactions on Biomedical Engineering*, vol. 58, no. 2, pp. 355-362.

Ortner, R.; Allison, B.Z.; Korisek, G.; Gaggl, H. & Pfurtscheller, G. (2011). An SSVEP BCI to Control a Hand Orthosis for Persons With Tetraplegia, *IEEE Transactions on Neural Systems and Rehabilitation Engineering*, vol. 19, no. 1, pp. 1-5.

Oveisi, F. (2009). EEG Signal Classification Using Nonlinear Independent Component Analysis, *in Proceedings of the 34th IEEE International Conference on Acoustics, Speech, and Signal Processing (ICASSP).*

Oveisi, F. (2009). Information Spectrum and Its Application to EEG-Based Brain-Computer Interface, *in Proceedings of the 4th International IEEE EMBS Conference on Neural Engineering*, Antalya, Turkey.

Oveisi, F. & Erfanian, A. (2008). A Minimax Mutual Information Scheme for Supervised Feature Extraction and Its Application to EEG-based brain-computer Interfacing, *EURASIP Journal on Advances in Signal Processing*, vol. 2008.

Oveisi, F.; Oveisi, S.; Erfanian, A.; Patras, I. (2012). Tree-Structured Feature Extraction Using Mutual Information, *IEEE Transactions on Neural Networks and Learning Systems*, Vol. 23, no. 1, pp.127-137.

Peterson, D. A.; Knight, J. N.; Kirby, M. J.; Anderson, C. & Thaut, M. (2005). Feature Selection and Blind Source Separation in an EEG-Based Brain-Computer Interface, *EURASIP Journal on Applied Signal Processing*, vol. 19, pp. 3128–3140.

Rojas, F.; Puntonet, C.; Alvarez, M.; Rojas, I. & Clemente, R. (2004). Blind Source Separation in Post-Nonlinear Mixtures Using Competitive Learning, Simulated Annealing, and a genetic algorithm, *IEEE Trans. Systems, Man, and Cybernetics.*, vol. 34, no. 4, pp. 407-416.

Singer, A. & Coifman, R. R. (2007). Non-linear independent component analysis with diffusion maps, *Applied and Computational Harmonic Analysis*, vol. 25, pp. 226-239.

Tan, Y.; Wang, J. & Zurada, J. M. (2001). Nonlinear Blind Source Separation Using a Radial Basis Function Network, *IEEE Trans. on Neural Networks* , vol. 12, no. 1, pp. 124-134.

Van, L.; Wu, D. & Chen, C. (2011). Energy-Efficient FastICA Implementation for Biomedical Signal Separation, *IEEE Trans. on Biomedical Engineering* , vol. 58, no. 6, pp. 1865-1873.

Woo, W.L. & Sali, S. (2002). General Multilayer Perceptron Demixer Scheme for Nonlinear Blind Signal Separation, IEE Proc., *Vis. Image Process*, Vol.149, no. 5, pp.253-262.

Zhang, K. & Chan, L. (2007). Nonlinear Independent Component Analysis with Minimal Nonlinear Distortion, *24th International Conference on Machine Learning*, New York, USA.

9

Application of Polynomial Spline Independent Component Analysis to fMRI Data

Atsushi Kawaguchi[1], Young K. Truong[2] and Xuemei Huang[3]
[1]*Biostatistics Center, Kurume University, Kureme, Fukuoka*
[2]*Department of Biostatistics, University of North Carolina at Chapel Hill, NC*
[3]*Department of Neurology, Penn State University, PA*
[1]*Japan*
[2,3]*USA*

1. Introduction

In independent component analysis (ICA), it is assumed that the components of the observed k-dimensional random vector $\mathbf{x} = (x_1, \ldots, x_k)$ are linear combinations of the components of a latent k-vector $\mathbf{s} = (s_1, \ldots, s_k)$ such that $s_1, \ldots, s_k$ are mutually independent. This is denoted by

$$\mathbf{x} = \mathbf{A}\mathbf{s}, \tag{1}$$

where $\mathbf{A}$ is a $k \times k$ full-rank non-random mixing matrix. The main objective then is to extract the mixing matrix through a set of observations $\mathbf{x}_1, \mathbf{x}_2, \ldots, \mathbf{x}_n$. For a detailed description of this method, including its motivation, existence and relationship with other well known statistical methods such as principal component analysis, factor analysis, see Hyvärinen et al. (2001).

In signal processing applications, it will be convenient to view the observations as values recorded at k locations over time periods $1, \ldots, n$. The number of locations k varies depending on the application area. For instance, $k = 2$ in a blind source seperation problem to $k \approx 10^5$ in a typical human brain imaging data set. The number of time points n also varies and it ranges from $n \approx 10^2$ to $n \approx 10^6$.

The spatial (location) and temporal (time) description of the data has generated a huge number of biomedical applications such as cognitive or genomic research. In this paper, we will focus on human brain data acquired from functional magnetic resonance imaging (fMRI) technique where $k \approx 10^5$ and $n \approx 10^2$. This imaging technique has been used to effectively study brain activities in a non-invasive manner by detecting the associated changes in blood flow. Typically, fMRI data consists of a 3D grid of voxels; each voxel's response signal over time reflects brain activity. However, response signals are often contaminated by other signals and noise, the magnitude of which may be as large as that of the response signal. Therefore, independent component analysis has been applied to extract the spatial and temporal features of fMRI data (Calhoun & Adali, 2006; McKeown et al., 1998).

For fMRI datasets, we remark that it is theoretically possible to search for signals that are independent over space (spatial ICA) or time (temporal ICA). In fact, the above ICA description involving the spatial k and temporal n scales should be called more precisely as the temporal ICA, while in spatial ICA, k will be treated as time, and n as location. Thus one can see that temporal ICA is just the *transpose* of spatial ICA. However, in practice, it is very difficult to obtain accurate and meaningful results from the temporal ICA of fMRI data because of the correlation among the temporal physiological components. Therefore, the use of spatial ICA is preferred for fMRI analysis (McKeown et al., 1998).

Our ICA on fMRI data is carried out by first reducing the number of independent components (IC) using tools such as principal component analysis (PCA) or singular value decomposition (SVD), followed with an algorithm for determining the ICs. The most commonly used ICA algorithms for analyzing fMRI data are Infomax (Bell & Sejnowski, 1995), FastICA (Hyvärinen & Oja, 1997), and joint approximate diagonalization of eigenmatrices (JADE) (Cardoso & Souloumiac, 1993). Calhoun & Adali (2006) reported that Infomax consistently yielded the most reliable results, followed closely by JADE and FastICA. In this study, we propose a novel ICA algorithm that is a modification of the logspline ICA algorithm (LICA) (Kawaguchi & Truong, 2011) and apply it to fMRI data. In ICA, we employ a likelihood approach to search for ICs by estimating their probability distributions or density functions (pdf). This is equivalent to maximizing the independence among ICs, and it is realized by using polynomial splines to approximate the logarithmic pdf; we call this the logspline model. To account for the sparsity of spatial fMRI maps, we further treat the pdf as a mixture of a logspline and a logistic density function; this approach has proven to be very effective for treating sparse features in data. Using simulated and real data, we compared our method with several well-known methods and demonstrated the relative advantage of our method in extracting ICs.

The remainder of this paper is organized as follows. Section 2 describes the proposed method. Section 3 presents the simulation studies. Section 4 describes the application of the proposed method to real data. Finally, Section 5 presents discussions and concluding remarks of our method.

2. Method

Let $\mathbf{Y}$ denote a $T \times V$ data matrix: each column of this matrix corresponds to a voxel time series, and there are V voxels and T time points. We invoke singular value decomposition (SVD) to yield the approximation $\mathbf{Y} \approx \mathbf{UDX}$, where $\mathbf{U}$ is a $T \times M$ orthogonal matrix, $\mathbf{D} = diag(d_1, d_2, \ldots, d_M)$ with $d_1 \geq d_2 \geq \cdots \geq d_M$, and $\mathbf{X}$ is an $M \times V$ orthogonal matrix. Here, we selected (orthogonal) columns of $\mathbf{U}$ to represent some experimental task functions as well as the physiological components. In addition, the dimension of $\mathbf{D}$ has been reduced by discarding values below a certain threshold; in other words, these values are essentially treated as noise.

We determine the ICs based on the matrix $\mathbf{X}$ so that $\mathbf{X} = \mathbf{AS}$, where $\mathbf{A}$ is an $M \times M$ mixing matrix and S is an $M \times V$ source matrix. That is, the v-th column of $\mathbf{X}$ is equal to $\mathbf{A}$ multiplied by the v-th column of $\mathbf{S}$, where $v = 1, 2, \ldots, V$. Equivalently, each column of $\mathbf{X}$ is a mixture of M independent sources. Let $\mathbf{S}_v$ denote the source vector at voxel v so that $\mathbf{S}_v = (S_1, S_2, \ldots, S_M)$, $v = 1, 2, \ldots, V$. Suppose that each S_j has a density function f_j for $j = 1, 2, \ldots, M$. Then, the density function of $\mathbf{X}$ can be expressed as $f_{\mathbf{X}}(\mathbf{x}) = \det(\mathbf{W}) \prod_{j=1}^{M} f_j(\mathbf{w}_j \mathbf{x})$, where $\mathbf{W} = \mathbf{A}^{-1}$ and $\mathbf{w}_j$ is the j-th row of $\mathbf{W}$.

We now model each source density according to the mixture with unknown probability a:

$$f_j(x) = a f_{1j}(x) + (1 - a) f_{2j}(x), \tag{2}$$

where the logarithm of $f_{1j}(x)$ is modeled by using polynomial splines

$$\log(f_{1j}(x)) = C(\boldsymbol{\beta}_j) + \beta_{01j}x + \sum_{i=1}^{m_j} \beta_{1ij}(x - r_{ij})_+^3,$$

with $\boldsymbol{\beta}_j = (\beta_{01j}, \beta_{11j}, \ldots, \beta_{1m_jj})$ being a vector of coefficients, $C(\boldsymbol{\beta}_j)$ a normalized constant, r_{ij} the knots; and $f_{2j}(x) = sech^2(x)/2$ is a logistic density function. Here $(y)_+ = \max(y, 0)$.

We denote the vector of parameters in the density function by $\boldsymbol{\theta} = (a, \boldsymbol{\beta})$. The maximum likelihood estimate (MLE) of $(\mathbf{W}, \boldsymbol{\theta})$ is obtained by maximizing the likelihood of $\mathbf{X}$ with respect to $(\mathbf{W}, \boldsymbol{\theta})$:

$$\ell(\mathbf{W}, \boldsymbol{\theta}) = \sum_{i=1}^{n}\sum_{j=1}^{k} \log(f_j(\mathbf{w}_j^T\mathbf{x}_i)).$$

We use a profile likelihood procedure to compute the MLE because a direct computation of the estimates is generally not feasible. The iterative algorithm is shown in Table 1. Note that

1. Initialize $\mathbf{W} = \mathbf{I}$.
2. Repeat until the convergence of $\mathbf{W}$, using the Amari metric.
   (a) Given $\mathbf{W}$, estimate the log density $g_j = \log f_j$ for the jth element X_j of $\mathbf{X}$ (separately for each j) by using the stochastic EM algorithm shown in Appendix 7.
   (b) Given g_j $(j = 1, 2, \ldots, p)$,

$$\mathbf{w}_j \leftarrow \text{ave}[\mathbf{X}g_j'(\mathbf{w}_j^T\mathbf{X})] - \text{ave}[g_j''(\mathbf{w}_j^T\mathbf{X})]\mathbf{w}_j$$

   where $\mathbf{w}_j$ is the jth column of $\mathbf{W}$ and ave is a sample average over $\mathbf{X}$.
   (c) Orthogonalize $\mathbf{W}$

Table 1. Algorithm

the Amari metric (Amari et al., 1996) used in the algorithms is defined as

$$d(\mathbf{P}, \mathbf{Q}) = \frac{1}{p(p-1)} \left\{ \sum_{i=1}^{p} \left(\frac{\sum_{j=1}^{p} |a_{ij}|}{\max_j |a_{ij}|} - 1 \right) + \sum_{j=1}^{p} \left(\frac{\sum_{i=1}^{p} |a_{ij}|}{\max_i |a_{ij}|} - 1 \right) \right\},$$

where $a_{ij} = (\mathbf{P}^{-1}\mathbf{Q})_{ij}$, $\mathbf{P}$, and $\mathbf{Q}$ are $p \times p$ matrices. This metric is normalized, and is between 0 and 1.

Several authors have discussed initial guesses for ICA algorithms. Instead of setting several initial guesses, as discussed in Kawaguchi & Truong (2011), $\mathbf{X}$ is multiplied by $\widetilde{\mathbf{W}}$, which is the output of the algorithm when the log density function $g(x)$ is replaced with $g(x) = 1/\{2b^{1/b}\Gamma(1 + 1/b)\}\exp\{-|x|^b/b\}$ with $b = 3$. The final output is obtained in the form $\widehat{\mathbf{W}} = \widetilde{\mathbf{W}}\widehat{\mathbf{W}}_0$, where $\widehat{\mathbf{W}}_0$ is the output of the algorithm shown in Table 1.

The purpose of spatial ICA is to obtain independent spatial maps and the corresponding temporal activation profiles (time courses). By multiplying $\mathbf{X}$ with $\widehat{\mathbf{W}}$, we can obtain the estimates of the spatial map $\widehat{\mathbf{S}}$ as $\widehat{\mathbf{S}} = \widehat{\mathbf{W}}\mathbf{X}$. On the other hand, the corresponding time courses are obtained in the form $\widehat{\mathbf{A}} = \widehat{\mathbf{W}}(\mathbf{UD})^{-1}$.

3. Simulation study

In this section, we conducted a simulation study to compare the proposed method with existing methods such as Infomax (Bell & Sejnowski, 1995), fastICA (Hyvärinen & Oja, 1997), and KDICA (Chen & Bickel, 2006). We designed our comparative study by using data that emulated the properties of fMRI data. The spatial sources **S** consisted of a set of 250 × 250 pixels. These spatial sources were modulated with four corresponding time courses **A** of length 128 to form a 62,500 × 128 dataset. The spatial source images **S** shown in the left-hand side of Figure 1 are created by generating random numbers from normal density functions with mean 0 and standard deviation 0.15 for a non-activation region, and mean 1 and standard deviation 0.15 for an activation region. The activated regions consist of squares of d_i pixels on a side, for $i = 1, 2, 3, 4$, that are located at different corners. We consider two situations: d_i's are the same among the four components ($d_1 = d_2 = d_3 = d_4 = d$) and d_i's are different. For the former, we used $d = 20, 30, 40$, and 50. For the latter, we generated uniform random numbers between 20 and 50 for each d_i. The temporal source signals in the right-hand side of Figure 1 are the stimulus sequences convolved with an ideal hemodynamic response function as a task-related component, and sin curves with frequencies of 2, 17, and 32 as other sources. We generated the task-related component by using the R package `fmri` with onset times (11,75) and a duration of 11. We repeated the above procedure 10 times for the case in which d_i's were the same and 50 times for the case in which d_i's were different.

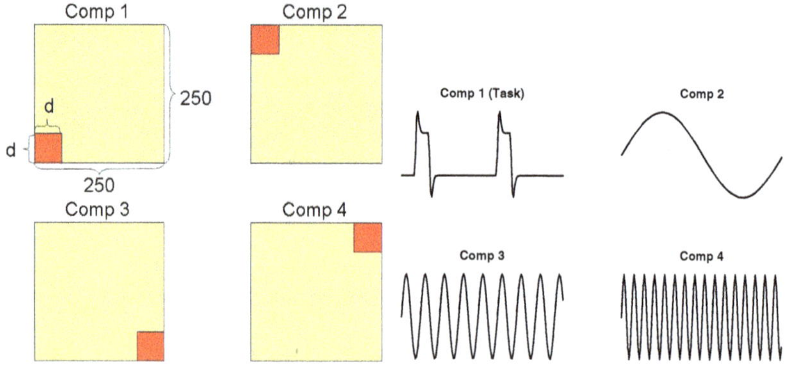

Fig. 1. Spatial and Temporal Simulation Data

Both the spatial and the temporal accuracies of ICA were assessed by R-square fitting of a linear regression model. The evaluation was carried out as follows. For every estimated time course, the R-square is computed from the linear regression model with the response being each of the estimates and the predictor being true, that is, the stimulus sequence (`Comp 1` on the right-hand side in Figure 1). The component that has the maximum R-square is considered to be task-related. We used the R-square value of this component for the comparison with the existing methods with respect to temporal accuracy and to determine the corresponding spatial map. The intensities of the spatial map are vectorized and used in the linear regression model as the response with the vectorized true (`Comp 1` on the left-hand side in Figure 1) as the predictor to compute R-square for the spatial accuracy.

The averaged R-squares over simulations are summarized in Tables 2 and 3 for the temporal and spatial data, respectively. When the sizes of the activation region were the same among all

	Infomax	fastICA	KDICA	PSICA
d=50	0.627	0.852	0.679	0.843
d=40	0.456	0.460	0.472	0.735
d=30	0.408	0.463	0.424	0.586
d=20	0.358	0.270	0.709	0.518
average	0.462	0.511	0.571	0.670
rand	0.623	0.651	0.529	0.699

Table 2. Temporal R-square for simulation data. The mean over d = 20, 30, 40, and 50 is calculated in the row labeled as average. The rand row shows the average over 50 replications when d_i's were chosen randomly from the range 20 to 50.

	Infomax	fastICA	KDICA	PSICA
d=50	0.801	0.765	0.641	0.761
d=40	0.462	0.502	0.545	0.726
d=30	0.409	0.528	0.552	0.680
d=20	0.323	0.478	0.624	0.587
average	0.499	0.568	0.591	0.688
rand	0.537	0.607	0.579	0.643

Table 3. Spatial R-square for simulation data. The mean over d = 20, 30, 40, and 50 is calculated in the row labeled as average. The rand row shows the average over 50 replications when d_i's were chosen randomly from the range 20 to 50.

components, R-squares of the proposed method were significantly larger than those of others for moderate sizes (d = 40 and 30) for both temporal and spatial data. For d = 50, fastICA had the largest R-square for both temporal and spatial data, with the difference from the result of the proposed method being small. For d = 20, KDICA had the largest R-square for both temporal and spatial data, with the difference from the result of the proposed method being significant for temporal data but not for spatial data. With respect to the average for d = 50, 40, 30, and 20, the proposed method had the largest R-square value than the others did. When d_i was determined randomly, which might be more practical, we observed that the largest R-square value in the rand row of the table was achieved by the proposed method.

4. Application

To demonstrate the applicability of the proposed method to real data, we separate fMRI data into independent spatial components that can be used to determine three-dimensional brain maps. To study brain regions that are related to different finger tapping movements, fMRI data were obtained from a twin pair (Twin 1 and Twin 2) performing different tasks alternately.The paradigm shown in Figure 2 consisted of externally guided (EG) or internally guided (IG) movements based on three different finger sequencing movements performed alternately by either the right or the left hand.

The fMRI dataset has 128 scans that were acquired using a modified 3T Siemens MAGNETOM Vision system. Each acquisition consists of 49 contiguous slices. Each slice contains 64$\times$64 voxels. Hence, each scan produces 64$\times$64$\times$49 voxels. The size of each voxel is 3 mm$\times$3 mm$\times$3 mm. Each acquisition took 2.9388 s, with the scan-to-scan repetition time (TR) set to 3 s. The dataset was pre-processed using SPM5 (Friston et al., 1995). The preprocessing

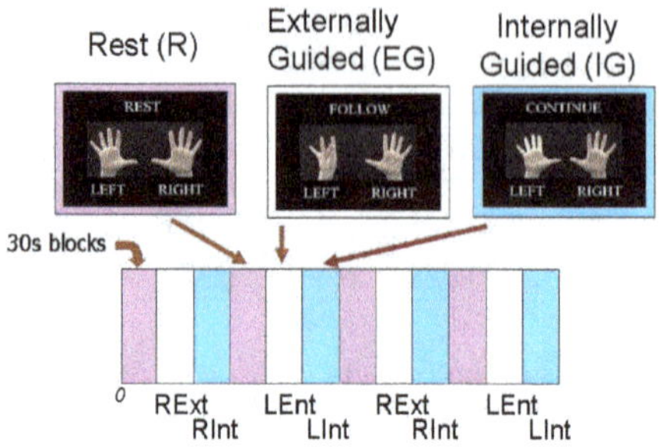

Fig. 2. Experimental Paradigm

included slice timing, realignment, and smoothing. We masked the image outside the human head using the GIFT software package (Group ICA of fMRI Toolbox, Calhoun et al., 2001). We used 21 components for Twin 1 and 30 for Twin 2; these were estimated using the minimum description length (MDL) criteria.

We applied four ICA algorithms—Infomax (Bell & Sejnowski, 1995), fastICA (Hyvärinen & Oja, 1997), KDICA (Chen & Bickel, 2006), and the proposed method (PSICA)—to the twins' data. The R-square statistic was calculated from the fitted multiple linear regression model with the estimated time course as the response. The predictors were the right EG, right IG, left EG, and left IG, which consists of the expected BOLD response for the task indicator function given by the argument as a convolution with the hemodynamic response function modeled by the difference between two gamma functions. Table 4 shows the corresponding R-square statistics. From this table, we can see that the proposed method extracted more correlated components for a task than did the other methods for both twins.

	Infomax	fastICA	KDICA	PSICA
Twin 1	0.640	0.666	0.655	0.680
Twin 2	0.847	0.661	0.805	0.862

Table 4. Temporal R-square statistics for the twin data

Figure 3 shows one of the resulting spatial maps of SPICA for Twins 1 and 2 respectively, in which the right motor area is highly activated and the corresponding time course shows a fit to the left-hand task paradigm.

We mention a few important observations in this real human brain analysis:

1. After the analysis, it was revealed to us that Twin 1 had shown signs and symptoms (tremors and slowed movements) of the Parkinson's disease (PD), while Twin 2 was considered normal at the time the data were collected. This may help to explain why Twin 2 — the normal subject has higher R-squares in three of the four methods (Table 4). In these methods, fastICA shows practically no difference of the twins.
2. In interpreting results from ICA, one should note that ICA is ambiguous about the sign: $\mathbf{x} = \mathbf{A}\mathbf{s} = (-\mathbf{A})(-\mathbf{s})$. This fact has produced different colour scales in the spatial maps (located in the lower right corner). With this in mind, one can say that Twin 2 or the normal subject has a higher intensity or activation level in the right motor area (because of the left-hand task paradigm).

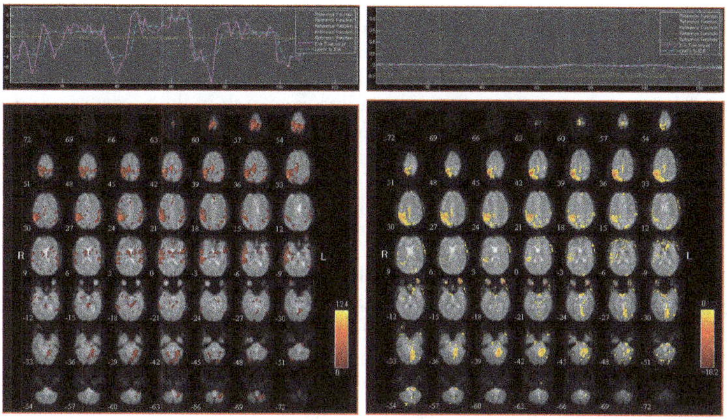

Fig. 3. Spatial Images for Twin 1 (left) and Twin 2 (right)

3. Further examination of the spatial maps indicates that the normal subject (on the right panel) has a more focused location of the motor area, see particularly the red region in slices 51, 48, 45, 42, 39, 36 and 33. The activated motor area of the PD twin (the left panel) is not as sharply defined.

5. Discussion and conclusion

In this study, we developed an ICA algorithm based on a maximum likelihood approach using a mixture of logspline and logistic density models with adaptive knot locations. The first concern about this approach is that its model dimension seems to be much higher than those of its peers. Here model dimensionality is defined as the number of model parameters including possibly the spline knot locations. Depending on how noisy the data are, the built-in model selection procedure (which is based on AIC or BIC) works in a sensible adaptive way: there is constantly a trade-off in balancing the bias and variance of the estimate of the parameter since the optimal strategy is to minimize the mean square error loss at the expense of the model dimension. Moreover, the logistic component is included to reduce the model dimension from the spline part in handling the sparsity of the spatial map. The main issue then is the time required to extract the ICs this way. It is considerably more time consuming, but the accuracy is very rewarding. The improvement over its peers performance was demonstrated numerically in Tables 2 and 3 using the R-square as a criterion.

It is important to point out that we should also provide a sensitivity and specificity analysis of the activated spatial locations as described in Lee et al. (2011), where popular methods such as Infomax and fastICA were shown to have a higher false-positive/nagative rate. This implies that brain activation should be studied more carefully, and one should avoid using methods that tend to yield false activation.

As in our previous approahces to ICA, the key feature has always been the flexibility modeling the source. In Kawaguchi & Truong (2011), the marginal distribution of the temporal source component was modelled by the logspline methodology and we noted the improvement over its peers. The comparatrive study was based on a wide variety of density functions, some are known to be very challenging to estimate. Further details of this approach can be found in

Kawaguchi & Truong (2011). In pursuing spatial ICA for fMRI based on human brain data, we observed that simply taking the transpose of the temporal ICA approach mentioned in the introduction did not always work. This is due to the fact that the spatial activation maps are very sparse: density estimation using the logspline approach in the presence of sparsity has never been investigated before. One of the our findings is that the logspline estimate of the spatial distribution is too noisy, perhaps the model dimension is too high. Thus the logistic component is added to our previous temporal ICA procedure in order to address this issue. The advantage over the simple tranposition of the temporal approach has been clearly shown in this paper.

The mixture modeling has been used previously for the detection of brain activation in fMRI data (Everitt & Bullmore, 1999; Hartvig & Jensen, 2000; Neumann et al., 2008). In fMRI data, the density functions of spatial sources are known to be supergaussion with heavy tails due to the fact that brain activation is sparse and highly localized (McKeown et al., 1998), and often skewed due to larger signal amplitudes in activated regions (Stone et al., 2002). Cordes & Nandy (2007) modeled source densities as improved exponential power family. Our modeling would be more flexible than these approaches.

In addition, the method may have some important extensions. Namely, it has been an important problem as how to assess the variability of ICA, especially how the variance of the spatial map can be best displayed. One way to examine the variation of the mixing coefficient estimates is to use bootstrap method while preserving information about the spatial structure. For example, in spatial ICA, one can generate bootstrap random samples from the logspline density estimates of the source over space. Mix these samples using the estimate mixing coefficients to yield the observed fMRI (BOLD) signals, which will then pass through ICA to produce the so called bootstrapped spatial maps and mixing coefficients. We outline this as an algorithm:

1. $\mathbf{x} \approx \hat{\mathbf{A}}\hat{\mathbf{s}}$ via our ICA algorithm.
2. $\hat{\mathbf{s}} \rightarrow \mathbf{s}^*$ which is a bootstrapped source sample drawnn from the distribution of $\hat{\mathbf{s}}$.
3. $\mathbf{x}^* := \hat{\mathbf{A}}\mathbf{s}^*$ to yield bootstrapped observed samples.
4. $\mathbf{x}^* = \hat{\mathbf{A}}^*\hat{\mathbf{s}}^*$ using our ICA algorithm.
5. Repeat until a desirable number of bootstrap samples is achieved.

Table 5. Bootstrap Algorithm

The bootstrapped sample $\mathbf{s}^*$ can be regarded as a by-product of the adequately modelled spatial map density function. The algorithm can be described similarly for temporal ICA. Thus it is feasible to develop the statistical inference framework for assessing the variability of the estimator of the mixing matrix via the bootstrap method while preserving information about the spatial or temporal structure.

In extending our temporal ICA to spatial ICA, we merely added the logistic component to the logspline piece, which is essentially a one-dimensional density estimation, or marginal density estimation procedure. Alternatively, in order to capture the actual spatial feature of the three dimensional brain, or the two dimensional map, one can incorporate the spatial correlation structure of the spatial map by introducing tensor products of spline functions or the interaction terms in the logspline formulation. For temporal ICA, this can be implemented by using time series models to account for the source serial correlations. Indeed, Lee et al. (2011) has reported that there is noticeable improvement over the marginal density based ICA procedures. It will be important to see if the same will hold for the above spatial ICA approach using tensor products of splines.

Another issue that we have not addressed is how to extend our method to compare groups of subjects. This is known as the group ICA problem. In principle, we can follow the Group ICA of fMRI Toolbox (Calhoun et al., 2001) by simply concatenating the observed data matrices. This will certainly increase the computational complexity and one has to address the efficiency problem as well.

Finally, we recall that prior to applying any of the ICA algorithms, one must carry out a dimension reduction step on the observed data matrix first. In temporal ICA with T and V as time and space scales, V will be reduced by, typically, employing the principal component analysis (PCA), while the time factor T will be reduced in the spatial ICA. We have found that even greater improvement can be achieved by using informative dimension reduction methods such as singular value decomposition (SVD) by choosing the eigen-vectors to relate to the experimental task paradigm closely. This is being referred to as a *supervised SVD dimension reduction* procedure (Bai et al., 2008) and has been used effectively in Lee et al. (2011).

In conclusion, the results presented in this paper can be viewed as a tool for setting up a new framework for addressing some of known issues in applying ICA to fMRI or other brain imaging modalities such as EEG or neural spike sorting problems. We have demonstrated that the key element here is the flexibility in modeling the source distribution and that was achieved by using polynomial splines as an approximation tool. We also used a mixture of distribution approach to account for the spatial distribution in ICA for fMRI data analysis. Although there are still many issues to be addressed, we have illustrated the usefulness of our approach to fMRI brain activation detection in both simulated and real data analysis.

6. Acknowledgment

We are grateful to Dr. Aiyou Chen for providing the KDICA programming code. We are also deeply grateful to Dr. Mechelle Lewis for her insight about the Twin data set, and her fruitful discussion on our analysis. This research was supported in part by the Banyu Fellowship Program sponsored by the Banyu Life Science Foundation International and by Grants-in-Aid from the Ministry of Education, Culture, Sport, Science and Technology of Japan (21700312) to AK, and by NSF DMS-0707090 to YT.

7. Appendix

A. Stochastic EM algorithm for mixture density estimation

In statistics, an expectation-maximization (EM) algorithm is an iterative method for finding maximum likelihood estimates (MLE) of parameters in statistical models. Typically these models involve latent variables in addition to unknown parameters and known data observations (Dempster et al., 1977). In our mixture model (2), parameter a is associated with the latent variable of the number of non-activated voxels in a given sample, and the unknown parameter β is related to the distribution of the fMRI intensity, coming from the logspline component.

The EM algorithm is particularly useful when the score function cannot be solved directly. The algorithm iteration alternates between performing an expectation (E) step, which computes the expectation of the log-likelihood evaluated using the current estimate for the parameters, and a maximization (M) step, which locates parameters maximizing the expected log-likelihood resulted in the E step. A version of this algorithm called a stochastic EM algorithm was introduced in (Celeux & Diebolt, 1992) to avoid stabilization on saddle points in parametric mixture models by incorporating the stochastic step (S-step) into the EM algorithm. (Bordes et al., 2007) generalized it to semiparametric mixture models by using kernel density estimation.

Suppose we have observations $x_1, x_2, \ldots, x_n$, and the observations are grouped by the k-means clustering method with k being the integer part of $n/10$. Let us denote the number of members in each group by n_g ($g = 1, 2, \ldots, k$) and $\boldsymbol{x}_g = (x_{i_{g1}}, x_{i_{g2}}, \ldots, x_{i_{gn_g}})$. The algorithm used in this paper is given below.

(1) E-step: Compute $\tau(j|\boldsymbol{x}_g)$ ($g = 1, 2, \ldots, k$, $j = 1, 2$) using

$$\tau(j|\boldsymbol{x}_g) = \frac{1}{n_g}\sum_{h=1}^{n_g} \tilde{\tau}(j|x_{i_{gh}})$$

where $\tilde{\tau}(j|x) = af_j(x)/f(x)$.

(2) S-step: Draw $z(\boldsymbol{x}_g)$ randomly from a Bernoulli distribution with probability of $\tau(1|\boldsymbol{x}_g)$ and define $z(\boldsymbol{x}_{i_{gh}}) = 1$ if $z(\boldsymbol{x}_g) = 1$ and $z(\boldsymbol{x}_{i_{gh}}) = 1$ otherwise for $g = 1, 2, \ldots, k$ and $h = 1, 2, \ldots, n_g$.

(3) M-step: The estimator of a is given by

$$\hat{a} = \frac{1}{n}\sum_{i=1}^{n} z(x_i).$$

f_1 is estimated by maximizing the likelihood described in Appendix 7 based on x_i for $i \in \{i;\, z(x_i) = 1\}$.

These steps are repeated until convergence. For the log spline density f_1, the maximum likelihood estimation is applied. The data-driven knot locations in f_1 are optimized as described in Appendix 7. We use the k-means method to initialize a and f_1. From the observations that are separated by the k-means method, those having a larger mean are used to initialize f_1. It is possible that the stochastic EM algorithm may not converge but be stable (Bordes et al., 2007; Celeux & Diebolt, 1992). Therefore, we use a large number of iterations so as to stabilize the estimate of f. We then select $\hat{f}$ as the final estimate, whose likelihood $\sum_{i=1}^{n} \log \hat{f}(x_i)$ is the maximum among the iterations.

B. Logspline density estimation

Let X be a random variable having a continuous and positive density function. The log density of X is modeled by

$$g(x) = \log(f(x)) = C(\boldsymbol{\beta}) + \beta_{01}x + \sum_{i=1}^{m} \beta_{1i}(x - r_i)_+^3,$$

where $\boldsymbol{\beta} = (\beta_{01}, \beta_{11}, \ldots, \beta_{1m})$ is a vector of coefficients, $C(\boldsymbol{\beta})$ is a normalized constant, r_{ji} are the knots, and $(a)_+ = \max(a, 0)$. Let $X_1, \ldots, X_n$ be independent random variables having the same distribution as X. The log-likelihood function corresponding to the logspline family is given by $\ell(\boldsymbol{\beta}) = \sum_{i=1}^{n} g(X_i)$. The maximum likelihood estimate $\hat{\boldsymbol{\beta}}$ is obtained by maximizing the log-likelihood function. This methodology was introduced by Stone (1990) and the software was implemented by Kooperberg & Stone (1991). An ICA algorithm based on the logspline density estimation was initiated by Kawaguchi & Truong (2011).

The knot selection methodology involves initial knot placement, stepwise knot addition, stepwise knot deletion, and final model selection based on the information criterion. We set the initial knot placement to be the minimum, median, and maximum values of the distribution of data. At each addition step, we first find a good location for a new knot in each of the intervals $(L, r_1), (r_1, r_2), \dots, (r_{K-1}, r_K), (r_K, U)$ determined by the existing knots $r_1, r_2, \dots, r_K$ and some constants L and U. Let $X_{(1)}, \dots, X_{(n)}$ be the data written in nondecreasing order. Set $l_1 = 0$ and $u_K = n$. Define l_i and u_i by

$$l_i = d_{\min} + \max\{j : 1 \leq j \leq n \text{ and } X_{(j)} \leq r_i\}, \quad i = 2, \dots, K$$

and

$$u_i = -d_{\min} + \max\{j : 1 \leq j \leq n \text{ and } X_{(j)} \geq r_i\}, \quad i = 1, \dots, K-1,$$

where $d_{\min}$ is the minimum distance between consecutive knots in order statistics.

For $i = 0, \dots, K$ and for the model with X_{j_i} as a new knot where $j_i = [(l_i + u_i)/2]$ with $[x]$ being the integer part of x, we compute the Rao statistics R_i defined by

$$R_i = \frac{[\boldsymbol{S}(\hat{\boldsymbol{\beta}})]_i}{\sqrt{[\boldsymbol{I}^{-1}(\hat{\boldsymbol{\beta}})]_{ii}}},$$

where $\boldsymbol{S}(\hat{\boldsymbol{\beta}})$ is the score function, that is, the vector with entries $\partial\ell(\hat{\boldsymbol{\beta}})/\partial\beta_j$, and $\boldsymbol{I}(\hat{\boldsymbol{\beta}})$ is the matrix whose entry in row j and column k is given by $-\partial^2\ell(\hat{\boldsymbol{\beta}})/\partial\beta_j\partial\beta_k$. We place the potential new knot in the interval $[X_{l_{i^*}}, X_{u_{i^*}}]$ where $i^* = \operatorname{argmax} R_i$. Within this interval, we further optimize the location of the new knot. To do this, we proceed by computing the Rao statistics R_l for the model with $X_{(l)}$ as the knot with $l = [(l_{i^*} + j_{i^*})/2]$ and R_u for the model with $X_{(u)}$ as the knot with $u = [(j_{i^*} + u_{i^*})/2]$. If $R_{i^*} \geq R_l$ and $R_{i^*} \geq R_u$, we place the new knot at $X_{(i^*)}$; if $R_{i^*} < R_l$ and $R_l \geq R_u$, we continue searching for a knot location in the interval $[X_{(l_{i^*})}, X_{(j_{i^*})}]$; and if $R_{i^*} < R_u$ and $R_l < R_u$, we continue searching for a knot location in the interval $[X_{(j_{i^*})}, X_{(u_{i^*})}]$.

After a maximum number of knots $K_{\max} = \min(4n^{1/5}, n/4, N, 30)$, where N is the number of distinct X_i's, we continue with stepwise knot deletion. During knot deletion, we successively remove the knot that has minimum Wald statistics, defined by

$$W_i = \frac{\hat{\beta}_i}{\sqrt{[\boldsymbol{I}^{-1}(\hat{\boldsymbol{\beta}})]_{ii}}}$$

of the existing knots.

Among all the models that are fit during the sequence of knot addition and knot deletion, we choose the model that minimizes the Bayesian information criterion (BIC) defined by $BIC = -2\ell(\hat{\boldsymbol{\beta}}) + m\log(n)$.

8. References

Amari, S., Cichocki, A. & Yang, H. H. (1996). A new learning algorithm for blind signal separation, *Advances in Neural Information Processing Systems* 8: 757–763.

Bai, P., Shen, H., Huang, X. & Truong, Y. (2008). A Supervised Singular Value Decomposition for Independent Component Analysis of fMRI, *Statistica Sinica* 18: 1233–1252.

Bell, A. J. & Sejnowski, T. J. (1995). An information maximisation approach to blind separation and blind deconvolution, *Neural Computation* 7: 1129–1159.

Bordes, L., Chauveau, D. & Vandekerkhove, P. (2007). A stochastic em algorithm for a semiparametric mixture model, *Comput. Stat. Data Anal.* 51: 5429–5443.

Calhoun, V. D. & Adali, T. (2006). Unmixing fmri with independent component analysis, *Engineering in Medicine and Biology Magazine, IEEE* 25: 79–90.

Calhoun, V. D., Adali, T., Pearlson, G. D. & Pekar, J. J. (2001). A method for making group inferences from functional mri data using independent component analysis, *Human Brain Mapping* 14: 140–151.

Cardoso, J. F. & Souloumiac, A. (1993). Blind beamforming for non gaussian signals, *IEE-Proc.-F* 140: 362–370.

Celeux, G. & Diebolt, J. (1992). A stochastic approximation type EM algorithm for the mixture problem, *Stochastics Stochastics Rep.* 41(1-2): 119–134.

Chen, A. & Bickel, P. J. (2006). Efficient independent component analysis, *Annals of Statistics* 34: 2825–2855.

Cordes, D. & Nandy, R. (2007). Independent component analysis in the presence of noise in fmri, *Magnetic Resonance Imaging* 25(9): 1237–1248.

Dempster, A., Laird, N. & Rubin, D. (1977). Maximum likelihood from incomplete data via the em algorithm, *Journal of the Royal Statistical Society. Series B (Methodological)* pp. 1–38.

Everitt, B. S. & Bullmore, E. T. (1999). Mixture model mapping of brain activation in functional magnetic resonance images, *Human Brain Mapping* 7: 1–14.

Friston, K., Holmes, A., Worsley, K., Poline, J., Frith, C. & Frackowiak, R. (1995). Statistical parametric maps in functional imaging: A general linear approach, *Human Brain Mapping* 2: 189–210.

Hartvig, N. V. & Jensen, J. L. (2000). Spatial mixture modeling of fmri data, *Human Brain Mapping* 11: 233–248.

Hyvärinen, A., Karhunen, J. & Oja, E. (2001). *Independent Component Analysis*, John Wiley & Sons.

Hyvärinen, A. & Oja, E. (1997). A fast fixed point algorithm for independent component analysis, *Neural Computation* 9: 1483–1492.

Kawaguchi, A. & Truong, K. Y. (2011). Logspline independent component analysis, *Bulletin of Informatics and Cybernetics* 43: 83–94.

Kooperberg, C. & Stone, C. (1991). A study of logspline density estimation, *Computational Statistics & Data Analysis* 12(3): 327–347.

Lee, S., Shen, H., Truong, Y., Lewis, M. & Huang, X. (2011). Independent component analysis involving autocorrelated sources with an application to functional magnetic resonance imaging, *Journal of the American Statistical Association* 106(495): 1009–1024.

McKeown, M. J., Makeig, S., Brown, G., Jung, T.-P., Kindermann, S., Bell, T., Iragui, V. & Sejnowski, T. J. (1998). Analysis of fmri by blind separation into independent spatial components, *Human Brain Mapping* 6: 160–188.

Neumann, J., von Cramon, D. Y. & Lohmann, G. (2008). Model-based clustering of meta-analytic functional imaging data, *Human Brain Mapping* 29(2): 177–192.

Stone, C. (1990). Large-sample inference for log-spline models, *The Annals of Statistics* 18(2): 717–741.

Stone, J. V., Porrill, J., Porter, N. R. & Wilkinson, I. D. (2002). Spatiotemporal independent component analysis of event-related fmri data using skewed probability density functions, *Neuroimage* 15: 407–421.

10

Associative Memory Model Based in ICA Approach to Human Faces Recognition

Celso Hilario, Josue-Rafael Montes, Teresa Hernández, Leonardo Barriga and Hugo Jiménez
CIDESI- Centro de Ingeniería y Desarrollo Industrial
México

1. Introduction

The human-like activities have been representing a research topic in several areas, which try to understand the internal processes involved. However, the complexity and the diversity of this situation has allowed to propose approaches of different areas. These approaches pretend to imitate via simulation/emulation particular behaviors. Human perception as information of an acquiring process and brain data manipulation process approach represent two open tasks in pattern recognition analysis. Human Face Recognition (HFR) involves two approaches. The main overcome is given by a set of adequate features to characterize a human face image and multiple situations involved in the recognition process.

HFR is typically performed by the use of previous well-known features and a set of classifiers; both of them used to define a criterion for clustering and classifying each set of features. However, these features are not completely invariant on different environment conditions as well as changes of perspective, luminance conditions and shadows generation. The majority of existent approaches are limited and conditioned to specific scenario conditions, where face features are well-behaved. In this sense, there are approximations as Santini & Jain (1999); Zhang & Zhang (2010); Zhao et al. (2003) ,where the authors show several criteria, focused on the human face features and its invariance to different scenario conditions.

On the other hand, the classifiers used for clustering human face characteristics are deeply dependable on the feature behavior; i.e. over the space of features, faces with high degree of similarity are spread in cumulus; then these cumulus are feasible for clustering via any grouping criterion. In this scenario, the most representative approaches include Abdelkader et al. (2001); Aggarwal & Cai (1999); Collins et al. (2002); Duda et al. (2000), where different clustering criteria for human face recognition are shown. Note the foundations and paradigms used are different, and consequently the results obtained are distinct in similar scenarios. Different approaches are similar in the point of each one proposed a new way to discriminate information that results independent[1] one of the other. One well-accepted classifier described above are the associative memories . The associative memory approach is a connective one; which uses the linearity expressed in data set as well as a linear transformation. This approach usually is refereed as a kind of Neural Network. In several scenarios it could represent a robust

[1] Independent term is referred to the disposition of different measures that results different among them under certain well-defined operators

approach because it supports the interference of noise in the data Duda et al. (2000); Minsky & Papert (1987). Some of the most distinctive works done around associative memories include Hopfield (1982); Kosko (1998); Polyn & Kahana (2008). The majority of current Associative Memories approaches use a binary representation or some discrete scenarios, where the coding process consists on symmetric representation models. This situation might be a limit to express robust classifiers.

Finally, other approaches make emphasis in the mixture of a set of features and clustering approaches. Some of the most significant works are Ben et al. (2002); Bray (2005); Giese & Poggio (2000), where the authors explicitly show criteria to define human features and a cluster approach to group each class.

In this work, we present a different way to figure out the problem of face recognition. The proposal consists on consider the problem of HFR as an associative task. Face variations represent the data mixed with noise. Then, the proposal consists of an heterogeneous process, where each different codification face represents a class and its variations the data versions mixed with noise. The associative process expands traditional approaches, due to this it uses two ways of independence: the linear independence and the probabilistic independence. Both of them are deeply based on the superposition property of signal analysis Books (1991). Finally, the proposal has been tested with a data base of faces. This data base considers different face position, luminance conditions and face gesticulation.

2. Foundations

In this chapter, we describe main concepts, where our proposal is based. In the first stage a description of the independence data concept viewed on different areas and which characteristics are important to take care of. In second part is introduced the concept of linear independence and its main properties. Finally, in the third part, the statistical independence is introduce and its main properties.

2.1 Data independence

The concept of independence in several areas is related to the idea of analyzing when certain events or objects are exclusive; i.e. there is no affectation in the behavior of any interacting objects. Typically, the independence, according to the area of research is oriented of the theoretical foundations used to express and measure it. The independence is close related with its opposite, the dependence concept. Both definitions are completely dependable of the information representation and the way of operate it. This fact implies, the way of measure independence/dependence is strictly related of what characteristic or form to manipulate information is used. For instance, two algebraic expression may be linear dependentTomasi (2004); but it does not imply being statistical dependentDuda et al. (2000), nor grammatically dependent too. Analyzing several dependence definitions, there are common characteristics surrounding the concept. These characteristics are:

1. A well-defined domain , which is usually mapped to an order/semiorden relationship[2].
2. A representation of data structure.
3. An operator to determine the independence/dependence.

[2] It could be discrete or continuous.

4. An operator to mix independent data.

The first one is achieved with the aim to declare an explicit order in the work space. This order establishes basic foundations to other operators. The second property is focused to any data needs of a representation to figure out certain behaviors, which defines the interdependence and dependency of the data.The next property consists on an explicit operator or set of properties which define, for a particular space, which is the criteria of dependence/independence using the last two points above mentioned. Finally, the last property represents an operator as a Cartesian product like, which defines the rules of mixing two independent datum.

2.2 Linear independence

Firstly, a first kind of independence is discussed. It is linear independence. Linear algebra is a branch that studies vector spaces (also called linear spaces) along with linear maps, mappings between vector spaces that preserve the linear structure. Because vector spaces have bases, matrices can be used to represent both vectors and linear transformations, this facilitates computation and makes vector spaces more concrete. Linear algebra is commonly restricted to the case of finite-dimensional vector spaces, while the peculiarities of the infinite-dimensional case are traditionally covered in linear functional analysis.

In linear algebra, a family of vectors is linearly independent if none of them can be written as a linear combination of finitely many other vectors in the collection. A family of vectors which is not linearly independent is called linearly dependent. This is, two or more functions, equations or vectors $f_1, f_2, \ldots, f_n$ which are not linearly dependent can not be expressed in the form

$$a_1 f1_{+} a_2 f_2 + \ldots + a_n f_n = 0 \tag{1}$$

with $a_1, \ldots, a_n$ constants which are not zero values.

The linear dependence of vectors is defined from basic operators in the algebra: Summation of two vectors and scalar product. Both of them in combination of basic structure elements derive in concepts (like rank, determinant, inverse, Gauss diagonalizing), used to test the dependence of two vectors.

Associative Memories encode information as matrices; which make emphasis in linear dependent methods to group classes, and linear independent to discriminate among vectors.

2.3 Probabilistic independence

Probability independence is the second independence criterion. Probabilistic theory is the branch of mathematics concerned with probability, the analysis of random phenomena. The central objects of the probability theory are random variables, stochastic processes, and events: mathematical abstractions of non-deterministic events or measured quantities that may either be single occurrences or evolve over time in an apparently random fashion.

In probability theory, to say that two events are independent intuitively means that the occurrence of one event makes it neither more nor less probable that the other occurs. Similarly, two random variables are independent if the conditional probability distribution of either given the observed value of the other is the same as if the other's value had not been observed. The concept of independence extends to dealing with collections of more than two events or random variables. Formally, this definition says: Two elements A and B are independent if only if

$$Pr(A \bigcap B) = Pr(A)Pr(B) \quad (2)$$

Here $A \bigcap B$ is the intersection of A and B, that is, it is the event that both events A and B.

This kind of independence is oriented to the data probability of occurrence in the data domain. Note the probabilistic independence analyzes the form and behavior of the probabilistic density function (pdf) of a given data; which results totally different to the linear assumption of a linear combination before described. This kind of independence is focused on match the range domain distribution. In pattern recognition it usually becomes useful, because a pdf of an event represent the data variation representation of events with small affectations.

3. The proposal

The actual work is focused in the case of Human Face Recognition (HFR), where a new way of classification and recognition based on the concept of independence is proposed. In this section is described the process of the information coding for distinctive feature identification and the associative model used to classify the different faces.

3.1 Information coding

Decision process given a set of evidence of a cluster of classes is dependable of data coding. The capabilities of clustering are deeply dependable of the information coding process and the expressiveness of information encoded. Several authors has proposed different methods for classifying the information Chaitin (2004); Shannon & Weaver (1949). This classification usually is based on a numerical criterion to define an order relationship over a descriptor space; which is conformed with the characteristic measured. Typically the clustering consists on define any distance function and the establishment of a radius-like criterion; to choose which elements belongs to a particular class Duda et al. (2000). However they are limited by the distribution of the information coding and the expressiveness of information coding.

The problem of face recognition should be viewed as a pattern recognition process; however as it was comment above, it consists on select a previously well-known descriptor, carrying the limitations described in above paragraphs. Then, generalizing, we need to define a clustering criterion without explicit descriptors. Consequently, the descriptors must be located without explicit knowledge of scenario. According to our proposes, we use a set of descriptors which contains the normalized distances of features. These features result of estimate the derivatives of order n of the distance matrix as is described as follows.

Given an image $I(\mathbf{x})$, indexed by the vector position $\mathbf{x}$, such that it contains a face. Image will be operated with gradient operator $\nabla_k^{(n)} I(\mathbf{x})$. The k parameter denotes the parameters of the derivative approach used to estimate it, and (n) is the order of the operator. Furthermore, the derivative is normalized from $m \times n$ size to $m' \times n'$, which is represented by $I'(\mathbf{x}) = \nabla_k^{(n)} I(\mathbf{x})$. The dimension $m' \times n'$ will be fixed by subsequent images to be analyzed. Using $I'(\mathbf{x})$, a distance matrix is built, representing the derivatives as a long vector of $m'n'$ dimension by linking of each row of the image derivative as follows

$$M_d(i,j) = d_k(\mathbf{I}'(i), \mathbf{I}'(j)) \tag{3}$$

for all i, j positions in $\mathbf{I}'(x)$ which is the version as vector of $I'(\mathbf{x})$; d_k is any given distance function defined and M_d is a square matrix of $m'n' \times m'n'$. Matrix M_d is used as a set of descriptors of each face.

Gradient operator $\nabla_k^{(n)}$ provides information of pixel's intensities variations, which they indicates pixel the degree of texture and border information in the images. Note this operator results invariant at diffuse light sources. Additionally, matrix distance M_d is dependable of d_k distance function based on L_k norm; i.e. values of k less than 1 increases the sparseness the data on M_d, and values greater than 1 for k decreases the sparseness of data.

3.2 Associative memory

In this section we describe the proposal of a new kind of associative memory based in linear and statistical independence.

3.2.1 The principles

Associative models consist on build a model $\mathcal{M}$ such that for a pair of set $\mathcal{A}$ and $\mathcal{B}$ create a relationship $\mathcal{R} : \mathcal{A} \to \mathcal{B}$. Being strictly the $\mathcal{R}$ relation has a property where a pair of elements $(a, b) \in \mathcal{A} \times \mathcal{B}$, and the elements with almost a distance criteria d_a in $\mathcal{A}$ and d_b in $\mathcal{B}$ are related too; i.e. elements with a small similarity with a pair $(a, b) \in \mathcal{A} \times \times \mathcal{B}$ are related in the same way. Typically the memories are classified according to the nature of associated sets: when $\mathcal{A} = \mathcal{B}$ memory is named as auto associative; when $\mathcal{A} \neq \mathcal{B}$ is named as hetero associative; when $|\mathcal{A}| > |\mathcal{B}|$ is considered as a classifier and finally, when $|\mathcal{A}| \leq |\mathcal{B}|$ is considered as a transducer Knuth (1998) or codifier Shannon & Weaver (1949).

Model $\mathcal{M}$ is build usually with a few samples (commonly named as learning or training samples). Being strictly, there is no particular expressions which decide over related elements; instead, the process is well-based in theoretical foundations. $\mathcal{M}$ is build attending the main theoretical foundations which define the class of the memory used. The nature of the majority models $\mathcal{M}$ are based in connective approaches, and consequently, it express a lineal mixture among inputs and outputs. The quality of learning process depends on the capabilities of associate with fewer errors the training samples and it will be used as good estimators for non-considered pair of elements in the relationship Minsky & Papert (1987); Trucco & Verri (1998).

However, even the learning process results robust, there are situations where the linearity expressed by the inputs and outputs results insufficient for establish a relationship between them. In sections described above we speak about the independence concept. Then, define a theoretical framework which uses several independence criteria should be beneficial to develop better models $\mathcal{M}$ and consequently better associative models.

The aim contribution of this work consists on a new model of Associative Memory based on real domain and the mixing of two different approaches of independence: the lineal independence and statistic independence. The proposal works under the assumption that two signal can be mixed/unmixed if we know the structure of the distribution of each signal. Our approach works with large vectors, such that the distribution of the data inputs that conform the event encoded can be estimated.

An associative Memory have at least two operators: similarity operator, which indicates the radio of match with some class previously learned; and belonging operator, which verifies the data previously encoded in the memory. Additionally, an scheme to learn and estimate each class is needed. In further paragraphs the proposal is showed, which one mixes two kind of independences: the linear and statistic. Both of them are used to define a framework which associates inputs and outputs.

Given a set of information sources $\{S_1, \ldots, S_n\}$, the data contained in all signals should be mixed. This fact becomes true whenever sensing process is related and should be affected by the same external variables. A first consideration is, the *true* variables of the system are not perceived directly; but we can assume that they are the result of any combination of the measured signals. For simplicity, this combination is viewed as a linear combination. Then, from the sources $\{S_1, \ldots, S_n\}$, we can estimate each *true* variable as any linear combination as follows

$$U_i = \sum_{j=1}^{n} w_{ij} S_j \tag{4}$$

Consequently, to obtain simplicity the expression above can be rewriting as the dot product of $\mathbf{w}_i$ vector with a vector $\mathbf{S} = [S_1, S_2, \ldots, S_n]$, as $U_i = \mathbf{w}_i \mathbf{S}^T$. After, as it is appreciated, in n sources, there are a maximum of U_n variables, it leads to $\mathbf{U} = W_{n \times n} \mathbf{S}^T$, for a particular time stamp t.

Then, unmixing and mapping this sources to well-behaved space becomes an overcome that could be estimated as Independent Component Analysis problem; i.e. the real source measurement must be considered as linear and statistically independent among each component. Under these assumptions, one way to estimate the independent variables could be done with and ICA approach. The approach consists on estimate a W matrix such that mix/unmix the sources to orthogonal variables as follows

$$U = W_{n \times n} X_{n \times m} \tag{5}$$

where $X = [S_1, S_2, \ldots, S_n]^T$ is a matrix composed with all information sources; W a square matrix with the form of $W = [\mathbf{w_1}, \ldots, \mathbf{w_n}]$ for unmixing the sources in X and U represents a set of linear and statistical independent variables. The values of W are estimated iteratively via fast-ICA algorithm as it is showed in Table 1. Fast-ICA algorithm consists on detect the orthogonal protections which maximize the information. The last one is measured via neg-entropy, as an approach to the real measurement result from the calculus of system entropy. As it can be appreciated the algorithm is non-deterministic starting from random values for each one $\mathbf{w}_i$ projection; which each iteration is tunning in direction of maximum information.

Unfortunately, one of the greatest disadvantages consists on the transformation W separates the mixed data, but the output are not sorted. This cause, the use of any component sort of returned by ICA is totally dependable of the phenomenon nature. However, for our purposes

Pseudo-Code
Estimate for each component $\mathbf{w}_i$ in W as follows
1. Initialize $\mathbf{w}_i$ with random numbers.
2. Let $\mathbf{w}_i^+ \leftarrow E\{\mathbf{x}g(\mathbf{w}_i^T\mathbf{x})\} - E\{\mathbf{x}g'(\mathbf{w}_i^T\mathbf{x})\}\mathbf{w}$.
3. Let $\mathbf{w}_i \leftarrow \frac{\mathbf{w}_i^+}{\lVert\mathbf{w}_i^+\rVert}$
4. If the convergence is not achieved, go back to 2.

Table 1. Pseudo-code of W estimation

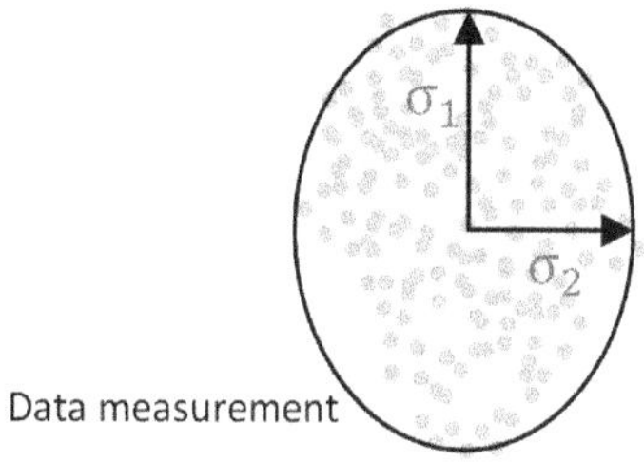

Fig. 1. Data Measurement Sparseness; as it is appreciated the eigen-values are located in the principal orthogonal axis.

this property could be used for developing a similarity criterion as is described in further paragraphs.

3.2.2 Similarity criterion

The similarity criterion for the proposal is based on the information given in the matrix W. The orthogonal components expressed in W can not be sorted, but it is possible to weight the contribution of each orthogonal component.

The weight process is done by analyzing the eigen-values of the matrix W. The relative magnitude of each eingen-value is used as a contribution of each orthogonal axis. This is, factorizing W as $U\Sigma V^T$, where Σ is an square matrix and the diagonal vector $[\lambda_1, \ldots, \lambda_n]$ has the singular values of W. The eigen-values Σ of W are related with the rank and the sparseness of each orthogonal component in the source space (see Figure 1). The most simple case of similarity using ICA is defined for two different signals as the proportion between each one of different eigen-values of the unmixing matrix W; i.e. formally is defined as

$$d(s_1, s_2) = \frac{\lambda_2}{\lambda_1} \tag{6}$$

where $W = U \begin{bmatrix} \lambda_1 & 0 \\ 0 & \lambda_2 \end{bmatrix} V^T$, and W is estimated with the algorithm described in Table 1. This similarity criterion is given for a pair of signals and can be interpreted as a degree of orthogonality existent in two components lineal and statistical independent; i.e. for a pair of signals s_1 and s_2 both of them are *similar* if and only if both s_1 and s_2 are linear and statistical dependent. The degree of independence is measured with the proportion of the second and the first eigen-value. If the proportion is near to 0 means, means s_1 and s_2 are linear and statistical dependent; i.e. they are similar. The proportion $\frac{\lambda_2}{\lambda_1}$ provides a normalized measure, and λ_1 is a normalized factor. In other scenarios we can use directly λ_2 as the degree of similarity between s_1 and s_2. Note, for a not normalized distance $d(s_1, d_2) = \lambda_2$, the expression is a metric Santini & Jain (1999).

Next,the belonging operator must be defined using a similarity function (see Equation 6), as a function $\epsilon_\lambda : \mathbb{R}^n \times \mathbb{R}^n \to \{true, false\}$, and its equation is defined as

$$\epsilon_\lambda(s_1, s_2) = d(s_1, s_2) \geq \lambda \tag{7}$$

where s_1 and s_2 represent the data information of encoded event. This operator is performed with the aim to decide when a given reference data s_1 and testing data s_2, correspond to the same information encoded.

3.2.3 Memory architecture

In last section the similarity criterion and the belonging operator has been defined. In this section we describe the architecture of an Associative Model based on last definition. The focus consists on use a set of representative data vectors, which represent the different classes of the memory. These vectors are matched with the data for a classification process using the belonging operator. Learning and discovering process are discussed in follows section. After this, we make the assumption that the representative classes are known.

The architecture of the Associative Memory proposed consists on given k classes denoted by $\Psi_1, \Psi_2, \ldots, \Psi_n$. Each class Ψ_i is denoted by a set of signals with the form $\Psi_i = \{s_1, s_2, \ldots, s_k\}$. Each signal s_i represents the events coding as a large vector fixed vectors. The process of encoding must warranty the best descriptors are encoded. For HFR purposes, it is assumed, the coding process described above has been used. We point out, the quality and properties of encoding process will affect the association capabilities. To get adequate results we emphasize that the information representation follows the properties of linear and statistical independence in the created relationships in the memory. The most representative information, of each class, is used as reference to match for discovering which class corresponds. For each Ψ_i there is a representative element denoted by the operator $E_i(\Psi_i)$. This operator is defined in general terms as the expected information of class Ψ_i. The nature of this operator is stochastic, but it might change, according to the data nature and distribution of coding data.

Then, the Associative Memory is defined by a matrix $\Xi = \begin{bmatrix} E(\Psi_1) \\ \vdots \\ E(\Psi_k) \end{bmatrix}$, such that it is conformed by all expected elements of each class Ψ_i. The matrix Ξ represent the Associative Memory.

Next, the memory needs an operator to verify when it contains a particular data. The operator of belongings aforementioned, is extended to verify if a given data s_i expressed as vector, is contained in any class of a memory Ξ. The operator is extended in two ways: the first variation return which elements has a certain degree of similarity with the testing data s_i. This operator will be useful when someone need to know what classes have closed similarity to the data s_i.

The operator is expressed as

$$\in_\Lambda(\Xi, s_j) = \{\epsilon_\Lambda(E(\Upsilon_i), s_j) = true\} \quad (8)$$
$$\text{for } j = 1,2,\ldots,k.$$

Note that it return a set with the most similar classes. Second variation return the most similar class to our testing data s_j. Then a variation of last operator consists on returning the class which has the minimum distance to the representative information of the class. Finally, the operator is defined as

$$\in^*_\Lambda(\Xi, s_j) = \min\{d_i(E(\Upsilon_i), s_j)\} \quad (9)$$
$$\text{for all } E(\Upsilon_i) \in \epsilon_\Lambda(E(\Upsilon_i), s_j). \quad (10)$$

Both operators are used to determine when a given data s_j belongs to any learned class. As final comments, the value of Λ is a threshold of belongings, and it is a parameter dependable of the scenario.

3.2.4 Learning process

Learning process, in an Associative Memories, consists on discover the relationships between $\mathcal{A}$ and $\mathcal{B}$ sets, and it usually is viewed as a class discovering process under a metric criterion defined by the belonging operator (see Equation 7).

The class discovering process, in the associative memory, is achieved with a learning process that needs a belongings threshold Λ. Note that the signal used has been normalized in frequency and length. Both assumptions help to operate them. For general purposes, it has been considered only the automatic learning process, which results useful in several scenarios. For a given set of signals $\{S_1, \ldots, S_n\}$, the process of discovering the classes is performed as follows.

The simplest case is whenever the process involves only one class. At the beginning, class Υ_1 is an empty set. First signal S_1 is added to the first class $\Upsilon_1 \leftarrow \Upsilon_1 \bigcup \{S_1\}$ and consequently, $\Xi = [\Upsilon_1]^T$. For further signals S_i, they are added to set Υ_1 if and only if $\in^*_\Lambda (\Xi, S_i) = 1$.

Generalizing for k classes, whenever $\in^*_\Lambda (\Xi, S_i) = 0$, a new class Υ_k is created, such that $\Xi \leftarrow [\Xi, \Upsilon_k]^T$, where Λ is a distance threshold among classes. The number of classes Υ_k added represents the different orthogonal concepts in the memory. Note that, if we had previously information of k classes, it became as a supervised learning process. Analogous case, if we had not previously information, it became as dynamic learning process. Additional to mention, if learning process is always computed, the expected value $E(\Upsilon_i)$ for each class should be changed, adapting dynamically to new data evidence.

In this point we need to define the $E(\Upsilon_i)$ operator. In several cases the expected operator is defined as the average; which is computed as the average of each component of all elements in Υ_i. Note that we make the assumption of Υ_i elements are sparse over the feature space uniformly. However, when the data distribution does not follows an uniform distribution, it will be insufficient. In this sense $E(\Upsilon_i)$ becomes as the expected value. The computation can be easily computed and estimated for each component. In this situation, the expected value is computed for each vector position. Note for the estimation of expected value, we need to have enough evidence to approximate component distribution and estimate global maximum. The expected value needs that the learning data must be sparse uniformly in the feature space.

Sample	1	2	3	4	5	6	7	8	9	10	11	12	13	14	**Total**
# Pictures	24	11	23	10	25	26	18	15	30	10	24	40	14	32	302

Table 2. Number of postures per sample of data base used.

The Two last approaches discussed above would result useful; however data in Ψ could result affected by noise or might not being sampled uniformly. In this cases is needed a method to dismiss this variations. One approach consists on eliminate the least significant orthogonal components in Ψ_i. Then, the filter process would be performed via PCA (Principal Component Analysis) approach. PCA approach consists on reconstruct the original information only taking most significant linear components. The advantage of eliminate small components is that eliminate several redundancies making more compact and better sparseness the learning evidence in the feature space.

For Ψ_i signal set is constructed a matrix D with each data vector transposed as follows

$$D = [S_1^T S_2^T \dots S_n^T] \text{ for } S_1, S_2, \dots, S_n \text{ signals in } \Psi_i \tag{11}$$

For dismissing the noise effects and sparse better the feature space a matrix D^* is build factorizing D as $U\Sigma V^T$ and build a matrix D^* without least significant data. D^* is reconstructed with as

$$D^* = U\Sigma^* V \tag{12}$$

where Σ^* is equal to Σ, but with the difference that the lasted singular values $\sigma_l^*, \sigma_{l+1}^* \dots \sigma_n^*$ are zero values and l value is estimated considering a percentage of original information. Proportion between summation of l principal component and summation of all components represents noise/signal ratio. Then election of l must define a percentage function of total information used in data. The percentage of information represented with l singular values is computed with $\% = \frac{\sum_{i=1}^{l} \sigma_i}{\sum_{j=1}^{n} \sigma_j}$. The election of l value, must be defined covering at least α percentage of information as follows

$$I(\alpha) = \max \arg\{l | 1, 2, \dots, n\} \tag{13}$$

$$\text{such that } \frac{\sum_{i=1}^{l} \sigma_i}{\sum_{j=1}^{n} \sigma_j} \leq \alpha$$

Finally $D^* = [(S_1^*)^T (S_2^*)^T \dots (S_n^*)^T]$ represents filtered data and it will be applied any scheme to estimate the calculus of expected value for Ψ_i.

4. Experimental results

In this section, we describe an experimental method for validating the proposal. The validation process consists on developing an Associative Memory to classify and to recognize Human Faces. The implementation details are given and described in follows sections.

4.1 Experimental model

Our proposal was tested with the development of Human Face recognition. The information consists on a data base of Human Faces. The data base includes different faces and different faces gesticulations. Figure 2 (a) shows samples of the data base. Each face in data base has several poses and gesticulation. The variation of each face is important, because it is used in

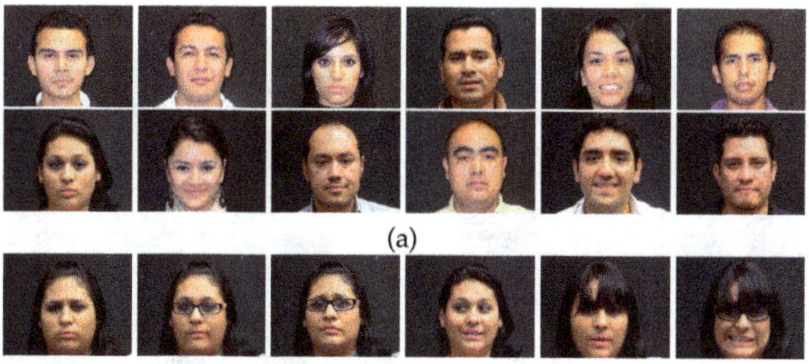

(a)

(b)

Fig. 2. Face data base: (a) some different samples contained in the data base; where (b) for each sample, there is different face gesticulation.

the learning process, for extraction of the main features which characterize each face. Pictures were taken with a Panasonic Lumix Camera at 7.1 Megapixels (3072 × 2304) in RAW format. The number of photos and persons involved are appreciated in Table 2.

In Section 3.1, a scheme for encoding was presented. This process is applied to the different faces in the data base. For practical purposes, derivative of order 1 has been used for encoding and characterizing each face. Derivative has been implemented with a symmetrical mask $d = [-10+1]$, which has assumed Gaussian. The parameters of Gaussian were fixed to $\mu = 0$ and $\sigma = 1$ with length 9, which define a Gaussian from -4 to 4 in normalized dimensions. Finally, the derivative filter was defined as $F = [-G(0,1)0 + G(0,1)]_{9\times19}$. The Gaussian results beneficial due to, it allows dismiss noise effects in images.

Next, features descriptors were estimated with Equation (3). This equation needs a normalized version of ∇I with $m' \times n'$ dimensions. The amount of information encoded in a pattern is directly affected by dimension of normalized image. A first view, a fixed value is used to compute these patterns (32 × 32 pixels). In a second stage, dimensions of normalized version has changed in follows dimensions: $8, 16, 24, 32, 40, 48, 56, 64, 96, 128, 160, 192$ and 224. Then a matrix M of descriptors has been created as comment in Section 3.2.4. Image patterns represent the relationships among all pixels in the image border.

Process learning uses a set of patterns, as input. This process computes distinctive face patterns. In Section 3.2.4 were described two approaches for estimating these patterns. In our implementation, we only test with a single average. This consideration might be strictly simple; but for our approach the average results enough for implementations purposes. This is, superposition principle, states that, for all linear systems, the net response at a given place and time caused by two or more stimuli is the sum of the responses which would have been caused by each stimulus individually. So that if input A produces response X and input B produces response Y then input $(A + B)$ produces response $(X + Y)$; which for a simple average is the same, factorizing by $\frac{1}{n}$ each term involved in sum operator Books (1991).

To illustrate this process, in Figure 3 are shown some faces patterns. This patterns were computed with normalized images of 16 × 16. As is appreciated, vertical and horizontal lines provided information of distribution and behavior data for several faces. Note, image descriptor has a resolution of $16^2 \times 16^2$ resulted of distance image.

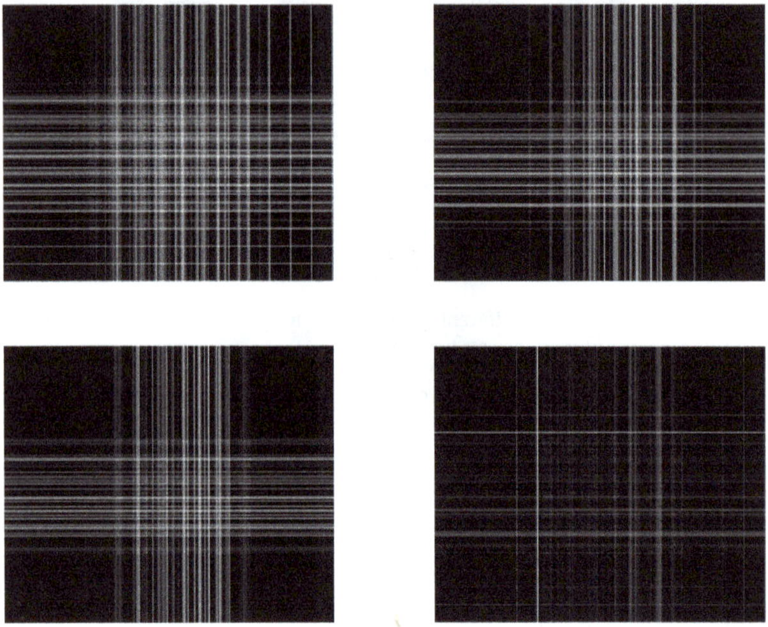

Fig. 3. Samples of patterns estimated in learning process. Visual patterns represent inner relations among features borders of each face.

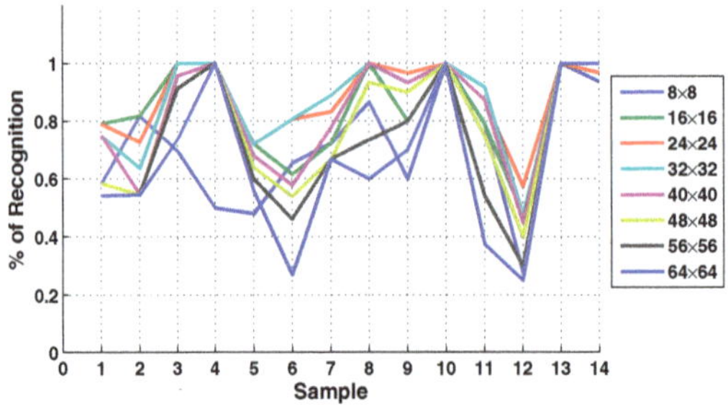

Fig. 4. Efficiency of recognition changing dimensions of normalized image.

4.2 Results and discussion

After computing a face pattern from each different sample, a memory is created as follows;

$$\mathcal{M} = \begin{bmatrix} E(\Psi_1) \\ \vdots \\ E(\Psi_n) \end{bmatrix}$$

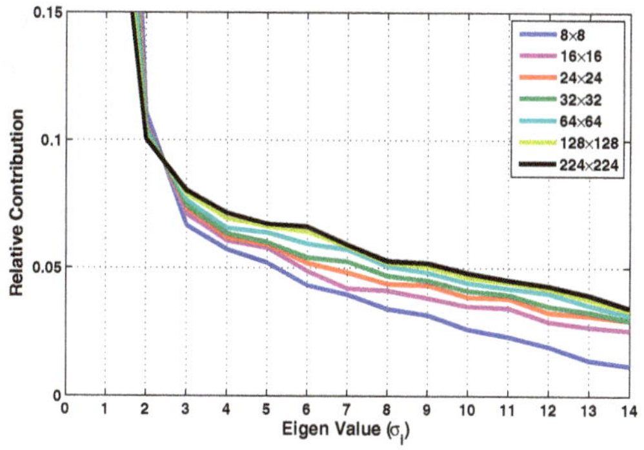

Fig. 5. Samples of patterns estimated in learning process. Visual patterns represent inner relations among features borders of each face.

where $E(\Psi_1)$ corresponds to a distinctive pattern of Ψ. Hence, an associative memory is created. For putting the memory on line, it only need to verify an encoded face and infer, which class encoded in $\mathcal{M}$ has less similarity and estimate its class. Note, an associative model is developed with the aim to infer which class has less similarity, so one must be careful because the class with less similarity additionally, need to follow a threshold similarity criterion too. This is, even there is no exist a similar class, memory return most similar.

Then, the validation of memory is performed measuring the degree of accuracy of recognition any face, to corresponding class. To perform it, all faces $s_1, s_2, \ldots, s_n$ in data base are tested. Normalized image dimension has changed. Results are shown in Figure 4.

The level of recognition of our approach is over 85% with pattern sizes of 24×24 and 32×32. Note, for small sizes (8 and 16) of image patterns, and for considerable high dimensions, the memory miss classify (64 or more). This is, in the first case there is not enough information for choosing which class is more similar. To justify this fact, a matrix of similarity has been created with patterns in memory. The matrix has 14×14 dimensions and represent in pseudo color the similarity degree. White color represents high degree of similarity and black color represents no similarity. In fact, the face patterns are analyzed; hence, we expect the matrix similarity has only similitudes in the diagonal. Then, Figure 6 (a) shows that degree of similarity among classes is small and it might cause miss classification problems. But when dimensions become higher, the degree of independence usually increases too, as is appreciated in Figure 6 (b) and Figure 6 (c), for 32×32 and 224×224 respectively. Additionally this can be verified with Figure 5, where principal components of patterns are computed and its relative contribution. Note, in higher sizes, they become more linear.

Intuitively, someone expect to be more accurate with high dimensions. However, note, the face recognition approach is based on a distance matrix, and associative memory approach is founded in two kind of independences: linear and statistical. Whenever data become higher, distribution of face patterns becomes similar, being not possible to classify. This point is illustrated in Figure 7, where two pdf's of different classes are estimated with 8×8, 32×32 and 224×224 pattern size. Note all of them, are dissimilar at the beginning and become similar when dimension of pattern is increased. This is, this approach is suitable for classes well differentiated.

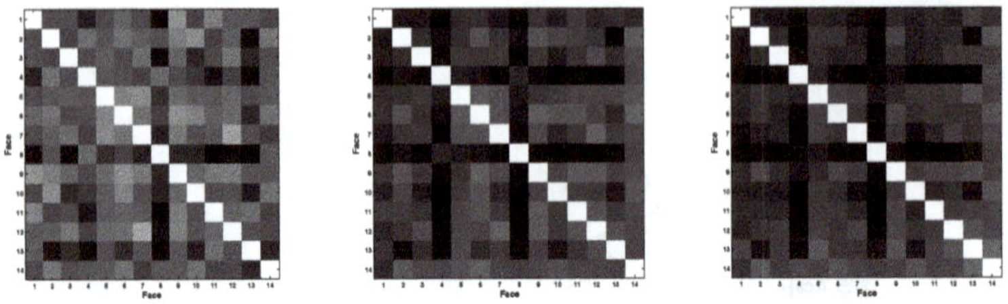

(a) (b) (c)

Fig. 6. Similarity matrix among classes for (a) normalized image of 8×8; 32×32 and 224×224.

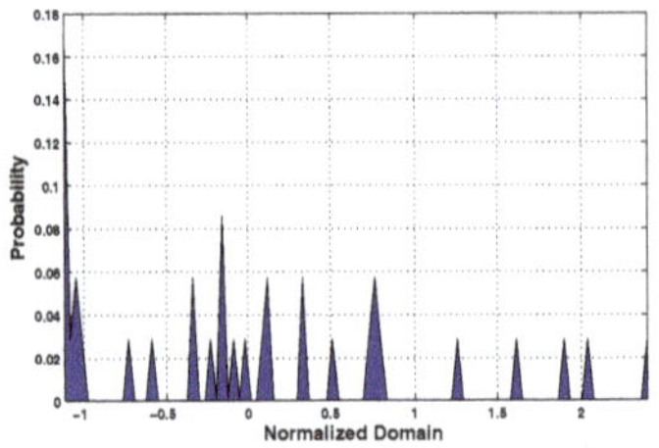

(a)

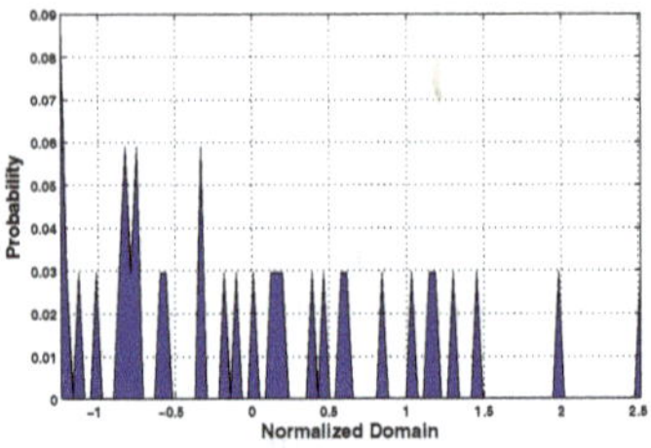

(b)

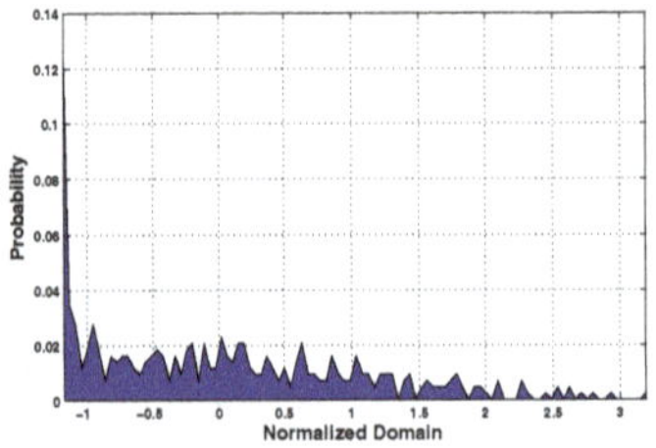

(c)

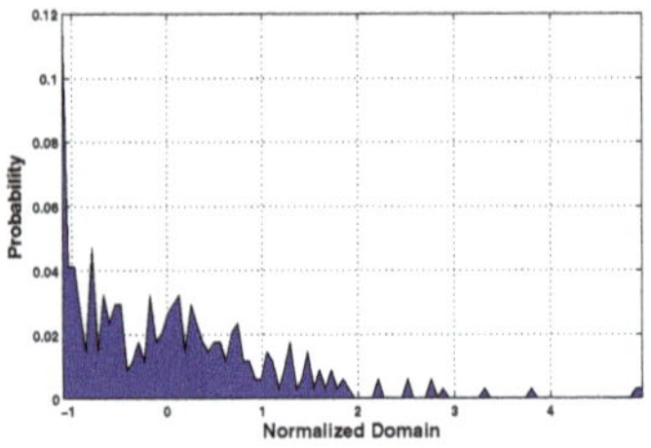

(d)

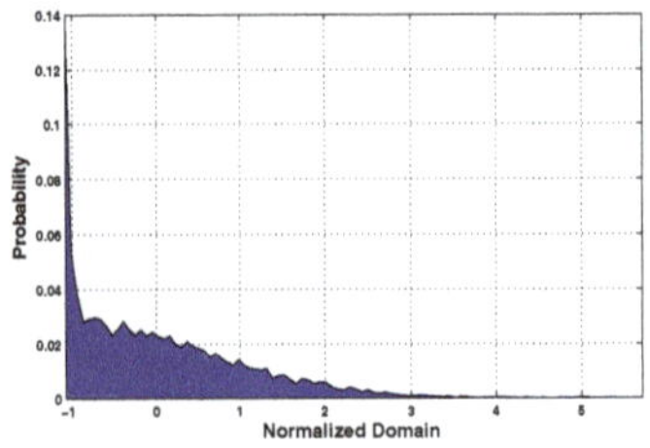

(e)

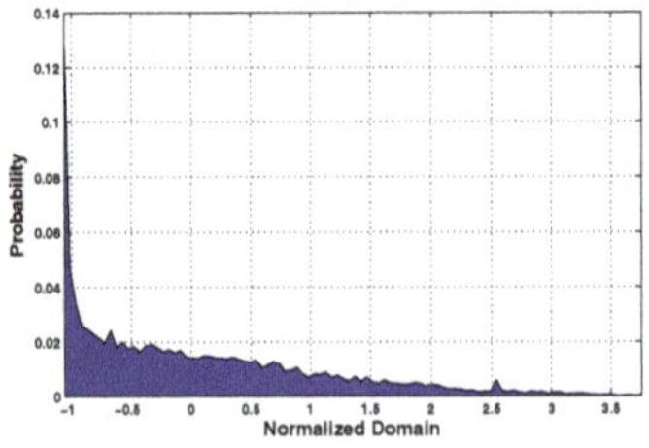

(f)

Fig. 7. Probabilistic density functions of two image patterns (a),(c),(e) and (b),(d),(f) with dimension 8×8; 32×32; and 224×224.

At this point, the miss classification effects when data become increased might be considered as inconvenient; however, it is not true, because framework is operated under its basic assumptions, and it shows the limits of encoding process; i.e. we need to define a better scheme to extract the most suitable features for face recognition which warrant linear independence and probabilistic independence. Then, to define an optimums classifier we need to consider the amount of information and its distribution associated to these data; however this point escape of the scope of this paper.

Summarizing, through this work, we propose a scheme for face recognition based on simple texture features, which uses a new paradigm of associative machines based on ICA approach. Its main characteristics include the use of real domain for characterizing data, including two kinds of independence.

5. Conclusion

In this chapter, we discussed about the independence concept, as an approach to characterize information by different features properties. Additionally, we pointed out the importance of different features when they are used to classify or discriminate certain events encoded. This fact results important to the well-development of classifier and coding process. Typically, common approaches are based on linear independence or statistical independence. Both approaches are distinct and measure different structure over data.

In this chapter, we proposed a new approach for classifying which take the advantage of both kind of linearity. It becomes relevant because we offer a new approach that characterize more strictly the information features used for data identification and data grouping. In this order, this work proposes a new family of associative memories, based on the ICA approach. The main characteristics include a real domain in the data, tolerance to certain data variations, and the association is given by the possibility of express each class as an orthogonal component and probabilistic independent. Preliminary tests show the viability of use this approach as general classifier being applicable in several research areas.

To test and validate the proposal, we implement a face recognizer application. It uses a basic coding data based on the derivative of the image. This encoding analyses differences between face border and its texture. In our tested scenario, the proposal is capable to discover and define a correct correspondence between the face variations and its corresponding face. In tested scenarios and information used the conditions were varied with the aim to show the robustness of the approach.

6. Future directions

As comment above paragraphs, a new framework based on linear independence and statistical independence is presented. Its main contributions are focused to the development of new kind of classifiers. This work exhibits the basis for new classifiers and recognizers. Further researches are focused to the application of this framework on different areas as signal analysis, image analysis and areas where the amount of information and lack of predefined features might difficult its analysis. This involves tasks as information coding, which extract good features and make more feasible the data manipulation.

Parallel, other areas of interest are related with the characterization of accuracy and efficiency of this approach, and the theoretical foundations that express new operators based on this proposal to manage in a better way the data.

Finally, one last direction is focused on the analysis of the feasibility to implement this approach in hardware, making more efficient time process and its application in real time systems.

7. References

Abdelkader, C., Cutler, R., Nanda, H. & Davis, L. (2001). EigenGait: Motion-Based Recognition of People using Image Self-Similarity, *Audio- and Video-Based Biometric Person Authentication* pp. 284–290.

Aggarwal, J. K. & Cai, Q. (1999). Human Motion Analysis: A Review, *IEEE Journal on Computer Vision and Image Understanding* 73(3): 428–440.
URL: *citeseer.ist.psu.edu/aggarwal99human.html*

Ben, J., Wang, Z., Pandit, P. & Rajaram, S. (2002). Human Activity Recognition using Multidimensional Indexing, *IEEE Transactions on Pattern Analysis and Machine Intelligence* 24(8): 1091–1104.

Books, P. (1991). *The Penguin Dictionary of Physics*, Penguin Books, Valerie Illingworth, London.

Bray, J. (2005). Markerless Based Human Motion Capture: A Survey, *Technical report*, Brunel University, Department System Engineering.

Chaitin, G. (2004). *Algorithmic Information Theory*, Cambridge University Press. IBM Research Center.

Collins, R. T., Gross, R. & Shi, J. (2002). Silhouette-Based Human Identification from Body Shape and Gait, *IEEE International Conference on Automatic Face and Gesture Recognition* pp. 351–356.

Duda, R. O., Hart, P. E. & Stork, D. G. (2000). *Pattern Classification*, 2nd. edition edn, Wiley-Interscience.

Giese, M. & Poggio, T. (2000). Morphable Models for the Analysis and Synthesis of Complex Motion Patterns, *International Journal of Computer Vision* 38(1): 59–73.

Hopfield, J. J. (1982). Neural networks and physical systems with emergent collective computational abilities, *Proceedings of the National Academy of Sciences of the USA* 79(8): 2554–2558.

Knuth, D. (1998). *Art of Computer Programming*, Vol. 1, 2 edn, Addison-Wesley Professional.

Kosko, B. (1998). Bidirectional associative memories, *IEEE Transactions on Systems, Man, and Cybernetics* 8(11): 40–46.

Minsky, M. L. & Papert, S. A. (1987). *Perceptrons - Expanded Edition: An Introduction to Computational Geometry*, second edition edn, The MIT Press; Expanded edition.

Polyn, S. & Kahana, M. (2008). Memory search and the neural representation of context., *Trends in Cognitive Sciences* 12: 24–30.

Santini, S. & Jain, R. (1999). Similarity Measures, *IEEE Transactions on Pattern Analysis and Machine Intelligence* 21(9): 871.

Shannon, C. & Weaver, W. (1949). *The Mathematical Theory of Communication*, University of Illinois Press.

Tomasi, C. (2004). *Mathematical Modelling of Continuos Systems*, Duke University.

Trucco, E. & Verri, A. (1998). *Introductory Techniques for 3-D Computer Vision*, 1 edn, Prentice Hall.

Zhang, C. & Zhang, Z. (2010). A survey of recent advances in face detection, *Technical Report MSR-TR-2010-66*, Microsoft Research Microsoft Corporation.

Zhao, W., Chellappa, R., Phillips, P. & Rosenfeld, A. (2003). Face recognition: A literature survey, *ACM Computing Surveys* 35(4): 399–458.

11

Preservation of Localization Cues in BSS-Based Noise Reduction: Application in Binaural Hearing Aids

Jorge I. Marin-Hurtado[1] and David V. Anderson[2]
[1]*Universidad del Quindio, Department of Electronics Engineering, Armenia, Q.*
[2]*Georgia Institute of Technology, School of Electrical and Computer Engineering, Atlanta, GA*
[1]*Colombia*
[2]*USA*

1. Introduction

For speech applications, blind source separation provides an efficient strategy to enhance the target signal and to reduce the background noise in a noisy environment. Most ICA-based blind source separation (BSS) algorithms are designed under the assumption that the target and interfering signals are spatially located. When the number of interfering signals is small, one of the BSS outputs is expected to provide an excellent estimation of the target signal. Hence, the overall algorithm behaves as an "ideal" noise-reduction algorithm. However, when the number of interfering signals increases, problem known as the cocktail party effect, or when the background noise is diffusive (i.e., non-point-source noise), this BSS output is no longer a good estimation of the target signal. (Takahashi et al., 2009) showed that in a two-output ICA-based BSS algorithm under these adverse environments, one BSS output includes a mixture of the target signal and residual noise related to the interfering signals, while the other output provides an accurate estimation of the background noise. This particular property validates the experimental results achieved by different post processing strategies to enhance the BSS output associated to the target signal (Noohi & Kahaei, 2010; Parikh et al., 2010; Parikh & Anderson, 2011; Park et al., 2006). These methods are based on Wiener filtering (Kocinski, 2008; Noohi & Kahaei, 2010; Park et al., 2006), spectral subtraction (Kocinski, 2008), least-square (LS) minimization (Parikh et al., 2010), and perceptual post processing (Parikh & Anderson, 2011). All these methods take advantage of a reliable background noise estimator obtained at one of the BSS outputs.

The above BSS-based noise-reduction methods provide a single output, which means that the direction of arrival of the target signal (also known as binaural cue or localization cue) is lost in the enhanced signal. There are some applications such as the new generation of hearing aids, called binaural hearing aids, which demands noise-reduction algorithms that preserve localization cues. These binaural hearing aids are targeted for hearing-impaired people who suffer from hearing losses at both ears. A binaural hearing aid consists of two hearing devices, one per each ear, and a wireless link to exchange information between both hearing devices.

This wireless link can be used to synchronize the processing performed by both hearing aids or to exchange the signals received at each side. The latter allows the use of multi-microphone noise-reduction algorithms such as BSS-based noise reduction algorithms. The perceptual advantages of a binaural processing over independent non-synchronized hearing aids have been extensively documented by (Moore, 2007; Smith et al., 2008; Van den Bogaert et al., 2006). These perceptual studies showed subject preference for those algorithms that preserve the direction of arrival (localization cues) of the target and interfering signals. Hence, this chapter addresses the problem about the preservation of the localization cues in noise-reduction algorithms based on BSS, whose main target application is a binaural hearing aid.

This chapter includes an overview of the state-of-the-art BSS-based noise-reduction algorithms that preserve localization cues. This overview describes in detail five BSS algorithms to recover the localization cues: BSS constrained optimization (Aichner et al., 2007; Takatani et al., 2005), spatial-placement filter (Wehr et al., 2008; 2006), post processing based on adaptive filters (Aichner et al., 2007), post processing based on a Wiener filter (Reindl et al., 2010), and perceptually-inspired post processing (Marin-Hurtado et al., 2011; 2012). This chapter also discusses the advantages and limitations of each method, and presents the results of a comparative study conducted under different kinds of simple and adverse scenarios: multi-talker scenario, diffusive noise, babble noise. Performance of these algorithms is evaluated in terms of signal-to-noise ratio (SNR) improvement, subjective sound quality, and computational cost. The comparative study concludes that the perceptually-inspired post processing outperforms the adaptive-filter-based and the Wiener-filter-based post processing in terms of SNR improvement, noise reduction, and computational cost. Therefore, the perceptually-inspired post processing is outlined as a good candidate for the implementation of a binaural hearing aid. A discussion about the proposed future work and improvements in the proposed methods are also addressed at the end of this chapter.

2. The problem of preservation of localization cues in blind source separation

This section presents a general overview of a blind source separation (BSS) process, and its problem with respect to the spatial placement of the separated sources in the output of the BSS algorithm.

Suppose a BSS system with P sensors. In the frequency domain, a source signal $s_1(\omega)$ is perceived at the sensor array as

$$\boldsymbol{x}(\omega) = \begin{bmatrix} x_1(\omega) \\ \vdots \\ x_P(\omega) \end{bmatrix} = \boldsymbol{h}_1(\omega) s_1(\omega) \tag{1}$$

where $x_p(\omega)$, $p = 1, ..., P$, are the signals at each sensor, and $\boldsymbol{h}_1(\omega)$ is a vector that describes the propagation from the point source to each sensor. In particular for a hearing aid with one microphone per hearing device, i.e., $P = 2$, this vector is called the head-related transfer function (HRTF). In a binaural system, the preservation of these HRTFs is critical since they provide information to the human auditory system about the direction of arrival of the target signals.

When Q sources are present in the environment, the input vector $\boldsymbol{x}(\omega)$ at the sensor array is given by

$$\boldsymbol{x}(\omega) = \sum_{q=1}^{Q} \boldsymbol{h}_q(\omega)s_q(\omega) = \begin{bmatrix} h_{11}(\omega) & \cdots & h_{1Q}(\omega) \\ \vdots & \ddots & \vdots \\ h_{P1}(\omega) & \cdots & h_{PQ}(\omega) \end{bmatrix} \begin{bmatrix} s_{1(\omega)} \\ \vdots \\ s_Q(\omega) \end{bmatrix} = \boldsymbol{H}(\omega)\boldsymbol{s}(\omega), \tag{2}$$

where $\boldsymbol{H}(\omega) = [\boldsymbol{h}_1(\omega) \cdots \boldsymbol{h}_Q(\omega)]$, is called the mixing matrix, and the vector $\boldsymbol{s}(\omega)$ holds the frequency components of each source signal. For BSS-based noise-reduction applications, the source s_1 is typically assigned to the target signal, and the sources s_q, $q = 2, \ldots, Q$, are related to the interfering signals.

The purpose of any blind source separation algorithm is to recover the source signals $s_q(\omega)$ from the mixture $\boldsymbol{x}(\omega)$ by means of a linear operation denoted by the unmixing matrix $\boldsymbol{W}(\omega)$,

$$\boldsymbol{y}(\omega) = \begin{bmatrix} y_{1(\omega)} \\ \vdots \\ y_P(\omega) \end{bmatrix} = \begin{bmatrix} w_{11}(\omega) & \cdots & w_{1Q}(\omega) \\ \vdots & \ddots & \vdots \\ w_{P1}(\omega) & \cdots & w_{PQ}(\omega) \end{bmatrix} \begin{bmatrix} x_{1(\omega)} \\ \vdots \\ x_P(\omega) \end{bmatrix} = \boldsymbol{W}(\omega)\boldsymbol{x}(\omega), \tag{3}$$

where the elements of the matrix $\boldsymbol{W}(\omega)$ denote FIR filters designed to separate the source signals (Fig. 1). These filter weights are designed by an optimization process, where the minimization of the mutual information between the source signals is one of the most successful methods to derive these filter weights (Haykin, 2000). This chapter does not include a detailed description about the methods to estimate the unmixing matrix $\boldsymbol{W}$, except those to recover the localization cues in the BSS filter (Section 3.1).

The whole process can be described by

$$\boldsymbol{y}(\omega) = \boldsymbol{W}(\omega)\boldsymbol{H}(\omega)\boldsymbol{s}(\omega) = \boldsymbol{C}(\omega)\boldsymbol{s}(\omega), \tag{4}$$

where $\boldsymbol{C}(\omega) = \boldsymbol{W}(\omega)\boldsymbol{H}(\omega)$. When the number of the sources and sensors is identical, i.e., $P = Q$, the problem is well-posed, and the matrix $\boldsymbol{C}(\omega)$ becomes diagonal. In this case, $\boldsymbol{y}(\omega) \approx \boldsymbol{s}(\omega)$ or equivalently $y_p(\omega) = \hat{s}_p(\omega)$, $p = 1, \ldots, P$, and $\hat{s}_p(\omega)$ is an estimate of the source signal. Hence, the localization cues of each source signal are lost after the blind source separation. For example, if a binaural hearing aid with one microphone per hearing device, i.e., $P = 2$, is used to cancel out the interfering signal in an environment with one target and one interfering signal, i.e., $Q = 2$, the BSS outputs are expected to be $y_1(\omega) = \hat{s}_1(\omega)$ and $y_2(\omega) = \hat{s}_2(\omega)$. Then, the output $y_1(\omega)$ holds an estimate of the target signal. If the signal $y_1(\omega)$ is applied simultaneously to the left and the right ear, the signal is heard coming always from the front. To avoid this issue, a spatial placement of the estimate $\hat{s}_1$ is required at the output of the entire process. This recovery of the localization cues is described by

$$\boldsymbol{z}(\omega) = \begin{bmatrix} z_1(\omega) \\ z_2(\omega) \end{bmatrix} = \boldsymbol{h}_1(\omega)\hat{s}_1(\omega) \tag{5}$$

where z_1 and z_2 are the signals to deliver to the left and right channel, respectively, and $\boldsymbol{h}_1$ denotes the HRTF for the target signal. The above process can be performed by different approaches. A first approach is to modify the derivation of the BSS filter weights, $\tilde{\boldsymbol{W}}$, such as the output of the BSS algorithm, $\boldsymbol{z}(\omega) = \tilde{\boldsymbol{W}}(\omega)\boldsymbol{H}(\omega)\boldsymbol{s}(\omega)$ is constrained to be $\boldsymbol{z}(\omega) \approx$

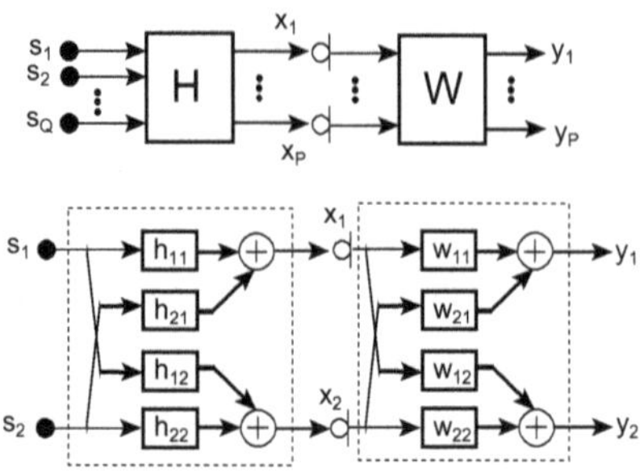

Fig. 1. General description of the blind source separation process for $P \neq Q$ (top); and two-sources and two-sensors $P = Q = 2$ (bottom).

$\boldsymbol{h}_1(\omega)s_1(\omega)$. These methods are discussed in the Section 3.1. Another approach, is to use a BSS post processing such as the output $y_1(\omega)$ is placed spatially by means of a filter $\boldsymbol{b}(\omega)$ such as

$$\boldsymbol{z}(\omega) = \boldsymbol{b}(\omega)y_1(\omega) \tag{6}$$

that ensures the fulfillment of the condition (5). This filter, called spatial-placement filter, is addressed in the Section 3.2. Another approach to recover the localization cues is to estimate a set of noise-suppression gains from the BSS outputs, and to apply these noise-suppression gains to the unprocessed signals. These methods are presented in the Sections 3.3 through 3.5, and are shown to provide more advantages than the BSS constrained optimization or the spatial-placement filter.

Up to this point the problem about recovery of the localization cues has been discussed for the case when $P = Q$ but in many practical applications, such as noise reduction, this condition cannot be met in adverse environments. In these environments, the number of interfering signals is larger than the number of sources, $Q > P$. This situation, called the undetermined case, leads to an ill-conditioned problem. Although the performance of the BSS algorithm is degraded for an undetermined case, the strategies to recover the localization cues in the undetermined case are exactly the same as described above. The main difference between both cases is regarding to the preservation of the localization cues for the interfering signals. These issues are described in detail in the next section.

3. BSS-based binaural noise-reduction algorithms

Most BSS algorithms are designed to support more than two sensors. As a general rule, increasing the number of sensors can separate more interfering sources but at expenses of increasing the computational complexity. In speech enhancement applications, for some adverse environments, the number of interfering signals is typically larger than the number of sources, $Q > P$, or even worse, the interfering signals are non-point noise sources, e.g., babble noise. Hence, increasing the number of sensors cannot improve significantly the quality of the source separation performed by the BSS algorithm. For this reason, a wide range of BSS-based speech enhancement algorithms are proposed for two-output BSS systems. Using two-output BSS algorithms provides additional advantages for some applications such as binaural hearing aids since the computational complexity and the wireless-link bandwidth can be reduced.

When two-output BSS algorithms are used in noise-reduction applications, the primary BSS output provides an estimate of the target signal, and the secondary BSS output, an estimate of the interfering signals. However, the estimate of the target signal does not provide information about the direction of arrival, and additional strategies are required to recover these localization cues. The approaches proposed in (Takatani et al., 2005) and (Aichner et al., 2007) employ a constrained optimization to derive the BSS filter weights. Unfortunately, these methods have shown a poor performance based on subjective tests (Aichner et al., 2007). More recent approaches use a BSS post processing stage to recover the localization cues and to enhance the target signal (Aichner et al., 2007; Marin-Hurtado et al., 2011; Reindl et al., 2010; Wehr et al., 2006). In these post-processing methods, the BSS outputs are used to compute noise-suppression gains that enhance the target signal. These post-processing methods have shown to be successful in the recovery of the localization cues and the reduction of the background noise, which is explained by the theoretical analysis conducted in (Takahashi et al., 2009) for two-output ICA-based BSS algorithms. In (Takahashi et al., 2009), authors showed that the estimate of the interfering signals is close to the true value whereas the estimate of the target signal includes a large amount of residual noise. Hence, when the estimate of the interfering signals is used in the post processing stage to compute the noise-suppression gains, the background noise can be significantly reduced.

In the BSS post-processing methods, depending on how these noise-suppression gains are applied to obtain the enhanced signal, it is possible to distinguish two groups. In the first group, these gains are applied to enhance the BSS output corresponding to the estimate of the target signal (Fig. 2a). In the second group, these gains are applied directly to the unprocessed signals (Fig. 2b). In BSS-based binaural speech enhancement applications, these noise-suppression gains are used not only to enhance the speech signal but also to recover the direction of arrival (or localization cues) of the speech signal. Although both groups of post processing are successful to recover the localization cues of the target signal, experimental and theoretical analysis show that the first group, in which BSS noise-suppression gains are applied to the BSS outputs, cannot recover the localization cues for the interfering signals (Aichner et al., 2007; Marin-Hurtado et al., 2012; Reindl et al., 2010; Wehr et al., 2008). In this case, the interfering signals are usually mapped to the direction of arrival of the target signal. This effect is not a desirable feature for binaural hearing aids, in which the displacement of the localization cues is identified as annoying through perceptual experiments (Moore, 2007; Smith et al., 2008; Sockalingam et al., 2009). On the other hand, the BSS post-processing methods that apply the noise-reduction gains to the unprocessed signals are shown to be successful in the recovery of the localization cues for both target and interfering signals simultaneously (Marin-Hurtado et al., 2012; Reindl et al., 2010).

3.1 BSS constrained optimization

As mentioned in the Section 2, localization cues can be recovered by using a constrained optimization in the derivation of the BSS filter weights, $\tilde{\boldsymbol{W}}$, such as the BSS output $\boldsymbol{z}(\omega) = \tilde{\boldsymbol{W}}(\omega)\boldsymbol{H}(\omega)\boldsymbol{s}(\omega)$ is constrained to be $\boldsymbol{z}(\omega) \approx \boldsymbol{h}_1(\omega)s_1(\omega)$, where s_1 and $\boldsymbol{h}_1$ are the target signal and its HRTF.

In (Takatani et al., 2005), authors proposed a BSS algorithm using the structure shown in Fig. 3, which uses a cost function that involves two terms,

$$\mathcal{J}(n) = \mathcal{J}_{\boldsymbol{y}}(n) + \beta \mathcal{J}_{\tilde{\boldsymbol{y}}}(n). \tag{7}$$

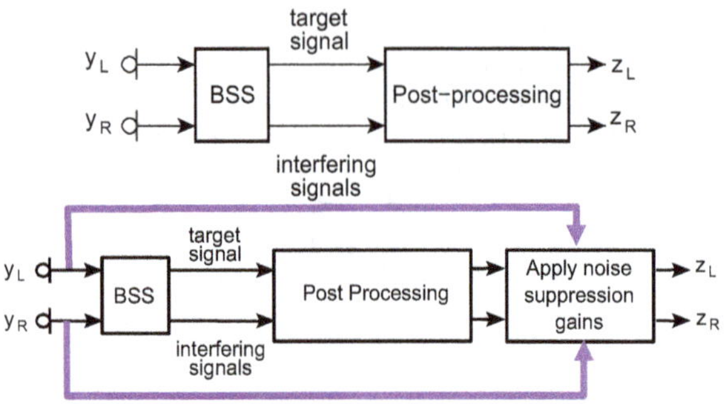

Fig. 2. BSS post processing to recover the localization cues: Post processing that enhances the BSS output (top), and post processing that enhances the unprocessed signals (bottom).

The first term, $\mathcal{J}_{\boldsymbol{y}}(n)$, is related to the classical source separation algorithms by minimization of the mutual information between the output channels y_1 and y_2, $\boldsymbol{y}(n) = [y_1(n)\, y_2(n)]^T$, and the second term, $\mathcal{J}_{\tilde{\boldsymbol{y}}}(n)$, is the minimization of the mutual information between the combination of the channels, $\tilde{y}(n) = [\tilde{y}_1(n)\, \tilde{y}_2(n)]^T$,

$$\begin{aligned} \tilde{y}_1(n) &= x_1(n-l) - y_1(n) \\ \tilde{y}_2(n) &= x_2(n-l) - y_2(n), \end{aligned}$$

where l is a time delay to compensate the processing delay introduced by the unmixing filters $\boldsymbol{w}$, and the parameter β controls a trade-off between both cost functions. The cost functions $\mathcal{J}_{\boldsymbol{y}}(n)$ and $\mathcal{J}_{\tilde{\boldsymbol{y}}}(n)$ are based on the statistical independence measurement given by the Kullback-Leibler divergence (KLD) or relative entropy, (Takatani et al., 2005)

$$\mathcal{J}_{\boldsymbol{y}}(n) = \mathcal{E}\left\{\log \frac{\hat{p}_{y,P}(\boldsymbol{y}(n))}{\prod_{q=1}^{P} \hat{p}_{y,1}(y_q(n))}\right\} \tag{8}$$

and

$$\mathcal{J}_{\tilde{\boldsymbol{y}}}(n) = \mathcal{E}\left\{\log \frac{\hat{p}_{y,P}(\tilde{\boldsymbol{y}}(n))}{\prod_{q=1}^{P} \hat{p}_{y,1}(\tilde{y}_q(n))}\right\} \tag{9}$$

where $\hat{p}_{y,P}(.)$ is the estimate of the P-dimensional joint probability density function (pdf) of all channels, $\hat{p}_{y,1}(.)$ is the estimate of the the uni-variate pdfs, and $\mathcal{E}\{.\}$ is the expected value.

A disadvantage of the Takatani *et al.*'s method is the huge computational cost and the slow convergence. An alternative solution proposed by (Aichner et al., 2007) replaces the minimization of the mutual information of the combined channels, $\tilde{\boldsymbol{y}}$, by a minimization of the minimum mean-square error (MMSE) of the localization cues,

$$\mathcal{J}(n) = \mathcal{J}_{\boldsymbol{y}}(n) + \gamma \mathcal{E}\left\|\boldsymbol{x}(n-l) - \boldsymbol{y}(n)\right\|^2, \tag{10}$$

where $\mathcal{J}_{\boldsymbol{y}}(n)$ is given by (8), γ is a trade-off parameter, and l is a time delay to compensate the processing delay introduced by the BSS algorithm (Fig. 4). The rationale behind the above method is that localization cues of the target signal in BSS inputs, $\boldsymbol{x}(n)$, must be kept in the BSS outputs, $\boldsymbol{y}(n)$, which is equivalent to minimize the MMSE between the input and output.

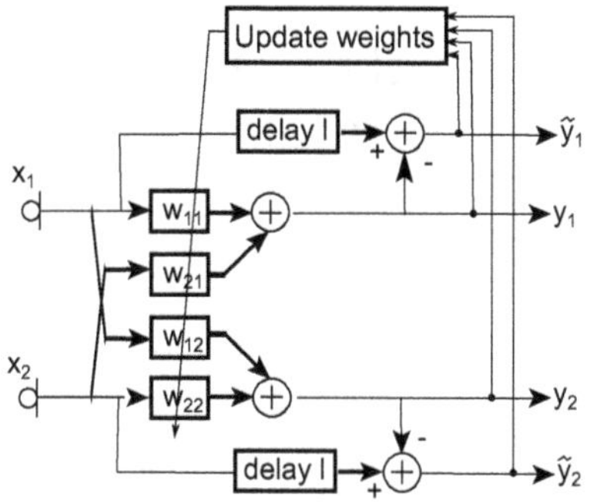

Fig. 3. Block diagram of the BSS constrained optimization to recover the localization cues proposed by (Takatani et al., 2005).

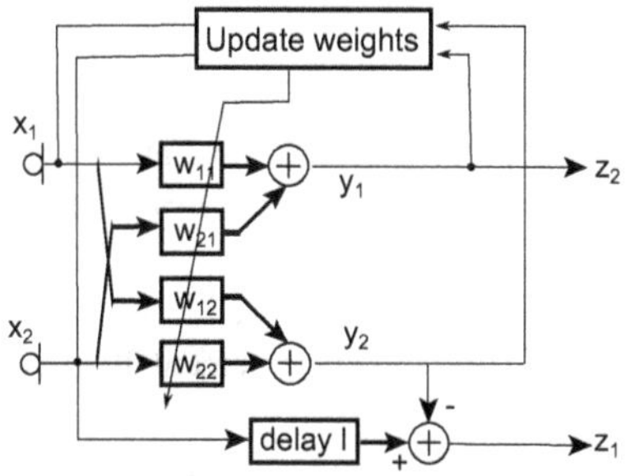

Fig. 4. Block diagram of the BSS constrained optimization to recover the localization cues proposed by (Aichner et al., 2007).

Although the subjective test conducted in (Aichner et al., 2007) showed that both methods can preserve the localization cues of the target signal, both methods cannot preserve the localization cues of the suppressed interfering signals, and the interfering signals are heard strongly distorted. In terms of noise reduction, the BSS constrained optimization method proposed in (Aichner et al., 2007) provides better performance than (Takatani et al., 2005).

3.2 Post processing based on spatial-placement filter

The main disadvantage of the BSS constrained optimization algorithms is their high computational cost. This issue can be solved by the spatial-placement filter introduced in (6), Section 2. A block diagram of the spatial-placement filter is shown in Fig. 5. The purpose of this filter is to recover the localization cues that are lost in the BSS output related to the target signal. If the BSS output holding the estimate of the target signal is $y_1(\omega)$, the spatial-placement filter, $\boldsymbol{b}(\omega)$, $\boldsymbol{z}(\omega) = \boldsymbol{b}(\omega)y_1(\omega)$ must satisfy (5), i.e., in the ideal case,

$$\boldsymbol{b}(\omega)y_1(\omega) = \boldsymbol{h}_1(\omega)s_1(\omega). \tag{11}$$

According to (2), the HRTF $\boldsymbol{h}_1(\omega)$ corresponds to the first column of the mixing matrix $\boldsymbol{H}(\omega)$,

$$\boldsymbol{h}_1(\omega) = \boldsymbol{H}(\omega)\boldsymbol{e}_1 \tag{12}$$

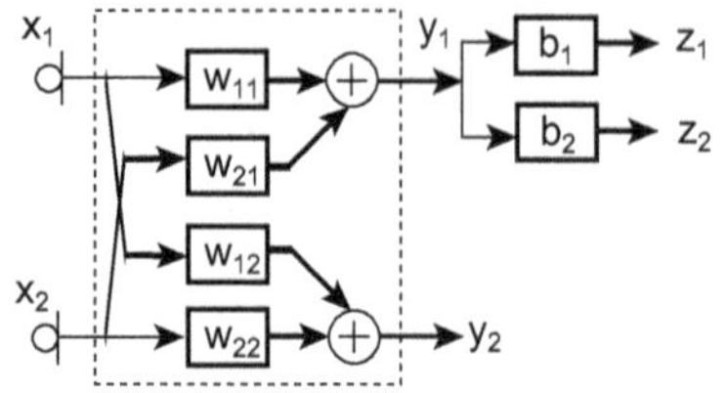

Fig. 5. Block diagram of the spatial-placement filter to recover the localization cues.

with $\mathbf{e}_1 = [1\,0 \ldots 0]^T$. From (4),

$$\boldsymbol{H}(\omega) = \boldsymbol{W}^{-1}(\omega)\boldsymbol{C}(\omega), \tag{13}$$

Thus, replacing (12) and (13) in (11),

$$\boldsymbol{b}(\omega)y_1(\omega) = \boldsymbol{W}^{-1}(\omega)\left[\boldsymbol{C}(\omega)\boldsymbol{e}_1 s_1(\omega)\right] \tag{14}$$

where the term in brackets $\boldsymbol{C}(\omega)\boldsymbol{e}_1 s_1(\omega) = \boldsymbol{W}(\omega)\boldsymbol{H}(\omega)\boldsymbol{e}_1 s_1(\omega) = \boldsymbol{W}(\omega)\boldsymbol{h}_1(\omega)s_1(\omega)$ is the output of the BSS algorithm when only the target signal is present in the environment. In other terms, the term in brackets becomes $\boldsymbol{e}_1 y_1(\omega)$. Thus,

$$\boldsymbol{b}(\omega) = \boldsymbol{W}^{-1}(\omega)\boldsymbol{e}_1 \tag{15}$$

or in other words, the coefficients of the spatial-placement filter correspond to the first column of the inverse matrix $\boldsymbol{W}^{-1}$.

A practical implementation of (15) requires the regularization of the inverse matrix to avoid an unstable algorithm. However, even using this regularization, the method in (15) is impractical for the recovery of the localization cues (Wehr et al., 2006). For example, suppose an environment with two sources, one target signal, s_1, and one interfering signal, s_2. In this environment, the signals perceived in the sensor array are given by

$$\boldsymbol{x} = \boldsymbol{h}_1 s_1 + \boldsymbol{h}_2 s_2 = \begin{bmatrix} h_{11} \\ h_{21} \end{bmatrix} s_1 + \begin{bmatrix} h_{12} \\ h_{22} \end{bmatrix} s_2 \tag{16}$$

Hence, in an ideal binaural noise-reduction system, the spatial-placement filter is expected to provide an output with structure similar as (16) but scaling down the term related to the interfering signal.

If a two-output BSS algorithm is used to cancel out the interfering signal, the output of the spatial-placement filter,

$$\boldsymbol{z}(\omega) = \boldsymbol{W}^{-1}(\omega)\boldsymbol{e}_1 y_1(\omega) = \boldsymbol{W}^{-1}(\omega)\boldsymbol{e}_1 \sum_{j=1}^{P} c_{1j} s_j\ , \tag{17}$$

is described in terms of the matrix elements c_{ij} and h_{ij} as

$$\boldsymbol{z} = \begin{bmatrix} h_{11} \\ h_{21} \end{bmatrix} s_1 - \frac{c_{21}}{c_{22}} \begin{bmatrix} h_{12} \\ h_{22} \end{bmatrix} s_1 + \frac{c_{12}}{c_{11}} \begin{bmatrix} h_{11} \\ h_{21} \end{bmatrix} s_2 \tag{18}$$

where the above derivation used the facts that $\boldsymbol{W}^{-1}(\omega) = \boldsymbol{H}(\omega)\boldsymbol{C}^{-1}(\omega)$, and $\boldsymbol{C}$ becomes a diagonal matrix in the determined case, i.e., $\|c_{11}c_{22}\|^2 \gg \|c_{12}c_{21}\|^2$. In the above equations, the variable ω is omitted for mathematical convenience. In (18) is clear that the target signal, s_1, is mapped to the desired direction of arrival, $\boldsymbol{h}_1 s_1 = [h_{11}\, h_{21}]^H s_1$. On the other hand, the interfering signal, s_2, is scaled by a factor c_{12}/c_{11} but it is also mapped to the direction of arrival of the target signal, which suggests that the localization cues for s_2 are not preserved. Another critical problem of this spatial-placement filter arises from the second term in (18). This term suggests that the target signal, s_1, is also mapped to the direction of arrival of the interfering signal and scaled by a factor c_{21}/c_{22}.

To avoid the regularization of the inverse matrix $\boldsymbol{W}^{-1}$ and the mapping of the target signal into the direction of arrival of the interfering signal, (Wehr et al., 2008; 2006) proposed to use the adjoint of the mixing matrix, $\boldsymbol{H}$, as unmixing matrix, i.e., $\boldsymbol{W}(\omega) = adj\{\boldsymbol{H}(\omega)\}$. Under this assumption, the spatial-placement filter that satisfies (11) is given by

$$\boldsymbol{b}(\omega) = adj\{\boldsymbol{W}(\omega)\}\,\boldsymbol{e}_1 \tag{19}$$

Then the output of the spatial-placement filter is given by

$$\boldsymbol{z}(\omega) = adj\{\boldsymbol{W}(\omega)\}\,\boldsymbol{e}_1 \sum_{j=1}^{P} c_{1j}s_j \tag{20}$$

Again, for an environment with one target and one interfering signal, the output of the spatial-placement filter of a two-output BSS algorithm is given by (Wehr et al., 2006)

$$\boldsymbol{z} = det\{\boldsymbol{W}(\omega)\}\left(\begin{bmatrix} h_{11} \\ h_{21} \end{bmatrix} s_1 + \frac{c_{12}}{c_{11}}\begin{bmatrix} h_{11} \\ h_{21} \end{bmatrix} s_2\right). \tag{21}$$

This equation shows that localization cues of the target signal, s_1, can be recovered correctly. However, the localization cues of the interfering signal are lost since the interfering signal is mapped to the direction of arrival of the target signal. The effect of this displacement in the localization cues for the interfering signal was evaluated in (Aichner et al., 2007) by a subjective test. Results showed that the post processing based on spatial-placement filter can be outperformed by a post processing based on adaptive filter, which is discussed in the next section.

3.3 Post processing based on adaptive filter

Up to this point the approaches discussed to recover the localization cues, BSS constrained optimization and BSS post processing using spatial-placement filter, fail to recover the localization cues of the interfering signals even under the determined case, i.e., when the number of source signals and sensors is the same, $P = Q$. In these methods, the localization cues of the interfering signals are usually mapped to the direction of arrival of the target signal.

To avoid the displacement of the localization cues for the interfering signals, different authors have reported the use of noise-suppression gains applied to the unprocessed signals rather than apply noise-suppression gains to the BSS outputs as in the spatial-placement filter. The first approach proposed to recover efficiently the localization cues was reported by (Aichner et al., 2007), which uses adaptive filters to cancel out the background noise. A block diagram of the method proposed in (Aichner et al., 2007) is shown in Fig. 6. In this approach, a BSS

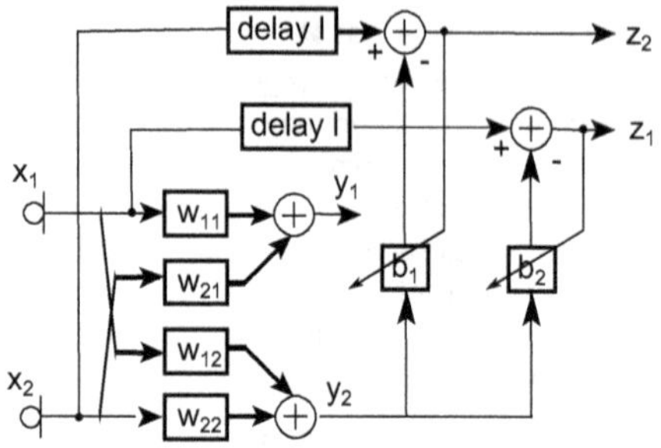

Fig. 6. BSS post processing based on adaptive filters. In this figure y_2 provides an estimate of the interfering signals $\hat{u}(n)$.

algorithm provides an estimate of the interfering signals, $\hat{u}(n)$, and this estimate is used as input for two adaptive filters, one for each side. The desired inputs for these adaptive filters are the unprocessed signals at the left and right channel. Then, the error signals provide enhanced signals in which the localization cues can be preserved.

This post processing can be used together any BSS algorithm. The original description of this algorithm uses the BSS algorithm described in (Aichner et al., 2006). On the other hand, the adaptive filters are designed in the DFT domain to minimize the time-averaged error (Aichner et al., 2007):

$$\mathcal{J}_{AF}(n) = (1-\lambda)\sum_{i=0}^{n}\lambda^{n-i}\sum_{k=0}^{R-1}\left|z_p(k,i)\right|^2 \tag{22}$$

where $z_p(k,i)$, $p \in \{1,2\}$, represents the DFT of the output of the algorithm at the frequency bin k and time index i; $0 < \lambda < 1$ is a forgetting factor; and R is the DFT length. The filter coefficients derived from (22) are given by

$$b_p(k,n) = \frac{r_{ux}(k,n)}{r_{uu}(k,n)} \tag{23}$$

where

$$\begin{aligned} r_{ux}(k,n) &= \lambda r_{ux}(k,n-1) + x_p(k,n)\,u(k,n) \\ r_{uu}(k,n) &= \lambda r_{uu}(k,n-1) + \left|u(k,n)\right|^2 \ ; \end{aligned}$$

$x_p(k,n)$ is the DFT of the input signal at the frequency bin k, time index n, and microphone p; and $u(k,n)$ is the DFT of the BSS output related to the interfering signals.

3.3.1 Limitations

In (Aichner et al., 2007), authors compared the BSS constrained optimizations given in (7) and (10), the spatial-placement filter given in (20), and the post processing based on adaptive filters given in (23), concluding that the post processing based on adaptive filters outperforms the other methods and preserves efficiently the localization cues.

The experimental results of the Aichner's study were conducted only for environments with two sources. Further research identified some problems in the BSS post processing based on adaptive filters. In (Reindl et al., 2010), a theoretical analysis of the adaptive-filter-based post processing shows that the noise reduction can be performed efficiently only under the determined case, i.e., when $P \geq Q$. In the undetermined case, $P < Q$, the noise reduction is possible only if the interfering signals are located at the same position.

To show the above statements lets consider a two-input BSS algorithm. In this algorithm we assume that the BSS output y_2 holds the estimate of the interfering signals, $\hat{u}(\omega) = y_2(\omega)$. This estimate in the frequency domain is given by

$$\hat{u}(\omega) = w_{11}(\omega)x_1(\omega) + w_{21}(\omega)x_2(\omega) = \sum_{p=1}^{2} w_{p1}(\omega)x_p(\omega). \tag{24}$$

In the general case, $x_p(\omega)$ is described by (2),

$$x_p(\omega) = \boldsymbol{e}_p^T \sum_{q=1}^{Q} \boldsymbol{h}_q(\omega)s_q(\omega) = \sum_{q=1}^{Q} h_{pq}(\omega)s_q(\omega). \tag{25}$$

Independent on the algorithm selected for the BSS algorithm, the target signal, s_1, can be assumed to be perfectly canceled out in $\hat{u}(\omega)$, which is expressed through

$$\hat{u}(\omega) = \sum_{p=1}^{2} w_{p1}(\omega) \sum_{q=2}^{Q} h_{pq}(\omega)s_q(\omega) \tag{26}$$

The output of the adaptive filters can be obtained by means of

$$z_p(\omega) = x_p(\omega) - b_p(\omega)\hat{u}(\omega) \quad p \in \{1,2\}. \tag{27}$$

Thus, replacing (25) and (26) in (27),

$$z_p(\omega) = h_{1p}(\omega)s_1(\omega) + \sum_{q=2}^{Q} \left[h_{qp}(\omega) - b_p(\omega)c_q(\omega)\right] s_q(\omega) \tag{28}$$

where

$$c_q(\omega) = w_{11}(\omega)h_{q1}(\omega) + w_{21}(\omega)h_{q2}(\omega). \tag{29}$$

From (28), to cancel out all interfering point sources, the frequency response of the adaptive filters must satisfy the condition

$$\sum_{q=2}^{Q} \left[h_{qp}(\omega) - b_p(\omega)c_q(\omega)\right] s_q(\omega) = 0 \tag{30}$$

In the determined case, $P = Q = 2$, the above equation can be satisfied if $b_p(\omega) = \frac{h_{2p}(\omega)}{c_2(\omega)}$. In the non-determined case, $Q > P$, it is necessary to satisfy the following simultaneous conditions,

$$b_p(\omega) = \frac{h_{2p}(\omega)}{c_2(\omega)} \cap b_p(\omega) = \frac{h_{3p}(\omega)}{c_3(\omega)} \cap \cdots \cap b_p(\omega) = \frac{h_{Qp}(\omega)}{c_Q(\omega)} \tag{31}$$

or equivalently,

$$\frac{h_{2p}(\omega)}{c_2(\omega)} = \frac{h_{3p}(\omega)}{c_3(\omega)} = \cdots = \frac{h_{Qp}(\omega)}{c_Q(\omega)}$$

For the particular case of two interfering sources, $Q = 3$, and two microphones, $P = 2$,

$$\frac{h_{2p}(\omega)}{c_2(\omega)} = \frac{h_{3p}(\omega)}{c_3(\omega)}$$

which leads to

$$w_{21}(\omega)\left[h_{21}(\omega)h_{32}(\omega) - h_{22}(\omega)h_{31}(\omega)\right] = 0$$
$$w_{11}(\omega)\left[h_{22}(\omega)h_{31}(\omega) - h_{21}(\omega)h_{32}(\omega)\right] = 0$$

Avoiding the trivial solution, these equations are true if $h_{21}(\omega)h_{32}(\omega) - h_{22}(\omega)h_{31}(\omega) = 0$, i.e., only if the interfering sources are located at the same position since $h_{21}(\omega) = h_{31}(\omega)$ and $h_{32}(\omega) = h_{22}(\omega)$. Hence, the performance of this post-processing method is fair in multiple-source environments such as babble noise.

Furthermore, a subjective evaluation in (Marin-Hurtado et al., 2011) showed that the adaptive-filter-based post processing cannot preserve the localization cues in the undetermined case. In this case, the interfering signals are mapped to the direction of arrival of the target signal. These experimental findings are explained by a mathematical derivation in (Marin-Hurtado et al., 2012), which is based on an analysis of the interaural transfer function (ITF). The magnitude of the ITF is called interaural level differences (ILD), and its phase is called interaural time differences (ITD). To preserve the localization cues, any post-processing method should ensure an output ITF similar to the input ITF for all frequencies, i.e., $ITF^{in}(\omega) = ITF^{out}(\omega)\ \forall\omega$. These ITFs are defined by the ratios

$$ITF^{in}(\omega) = \frac{x_1(\omega)}{x_2(\omega)}\ ;\ ITF^{out}(\omega) = \frac{z_1(\omega)}{z_2(\omega)} \tag{32}$$

In the post processing based on adaptive filters, the input and output ITF for every interfering signal are defined as

$$ITF_q^{in}(\omega) \triangleq \frac{h_{q1}(\omega)}{h_{q2}(\omega)}\ ;\ ITF_q^{out}(\omega) \triangleq \frac{y_{q1}(\omega)}{y_{q2}(\omega)} \tag{33}$$

where

$$y_{qp}(\omega) = \left[h_{qp}(\omega) - b_p(\omega)c_q(\omega)\right]s_q(\omega).$$

Thus,

$$ITF_q^{out}(\omega) = ITF_q^{in}(\omega) + D_q(\omega)$$

where $q = 2, ..., Q$ and

$$D_q(\omega) = \frac{\left[b_2(\omega)h_{q1}(\omega) - b_1(\omega)h_{q2}(\omega)\right]c_q(\omega)}{\left[h_{q2}(\omega) - b_2(\omega)c_q(\omega)\right]h_{q2}(\omega)}$$

is the ITF displacement. In other words, the perceived direction of arrival for each interfering signal is shifted from its original position. In the determined case, the conditions given by (31) are satisfied, which leads to an ITF displacement $D_q(\omega) = 0$. On the other hand, an ITF displacement $D_q(\omega) \neq 0$ is obtained in the undetermined case since the conditions (31) are not met.

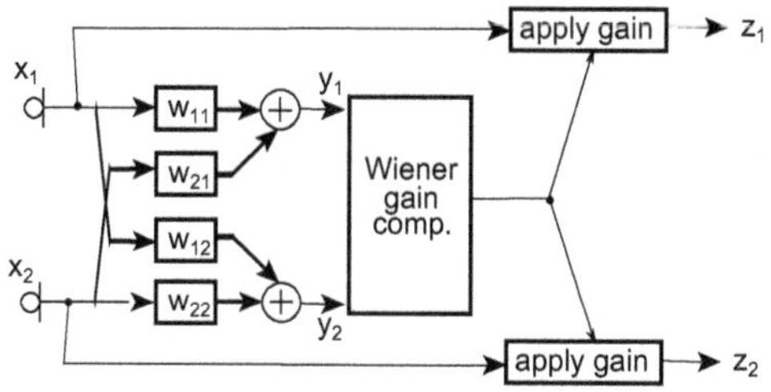

Fig. 7. Post processing based on Wiener filter.

3.4 Post processing based on Wiener filter

The methods described in the previous sections cannot preserve the localization cues for both target and interfering signals simultaneously in the undetermined case. In most of the cases, the interfering signals are mapped to the direction of arrival of the target signal. For other algorithms, such as the post processing based on adaptive filters, the localization cues can only be preserved under certain conditions in which the number of source signals is equal or lower than the number of sensors (determined case). From the perceptual viewpoint, these methods are impractical for binaural hearing aids since the displacement of the localization cues has been identified as annoying for hearing-impaired subjects.

In (Reindl et al., 2010), authors proposed an alternative post-processing stage based on Wiener filter to recover the localization cues. In this method, the BSS outputs are used to compute the Wiener filter gains, and these gains are applied simultaneously to the unprocessed signals (Fig. 7). This method is based on the fact that an ICA-based BSS algorithm provides a good estimate for the interfering signals, i.e., the BSS algorithm provides a good noise estimator. Since the Wiener filter gains are applied symmetrically to both sides, this method is ensured to preserve the localization cues for both target and interfering signals simultaneously.

The Wiener filter gains are computed by (Reindl et al., 2010)

$$g_{Reindl}(\omega) = \max\left\{1 - \alpha_\omega \frac{S_{\hat{n}\hat{n}}(\omega)}{S_{v_1v_1}(\omega)S_{v_2v_2}(\omega)}, 1\right\} \tag{34}$$

where $S_{\hat{n}\hat{n}}(\omega)$, $S_{v_1v_1}(\omega)$, and $S_{v_2v_2}(\omega)$ are the power spectral densities (PSD) of the estimate of the interfering signals (26), and the outputs of the intermediate unmixing filters $v_1(\omega) = w_{11}(\omega)x_1(\omega)$ and $v_2(\omega) = w_{21}(\omega)x_2(\omega)$. If the BSS output that holds the noise estimate $\hat{n}(\omega)$ is $y_1(\omega)$, the signals $v_1(\omega)$ and $v_2(\omega)$ take the forms $v_1(\omega) = w_{12}(\omega)x_1(\omega)$ and $v_2(\omega) = w_{22}(\omega)x_2(\omega)$. These PSDs can be updated by means of a first order estimator,

$$\begin{aligned} S_{\hat{n}\hat{n}}(\omega,n) &= \lambda S_{\hat{n}\hat{n}}(\omega,n-1) + (1-\lambda)\left|\hat{n}(\omega,n)\right|^2 \\ S_{v_1v_1}(\omega,n) &= \lambda S_{v_1v_1}(\omega,n-1) + (1-\lambda)\left|w_{11}(\omega)x_1(\omega)\right|^2 \\ S_{v_2v_2}(\omega,n) &= \lambda S_{v_2v_2}(\omega,n-1) + (1-\lambda)\left|w_{21}(\omega)x_2(\omega)\right|^2 \end{aligned}$$

where λ is a time constant to smooth the estimator, and α_ω is a frequency-dependent trade-off parameter to control the roll-off of the noise reduction. Finally, the enhanced outputs are obtained by

$$\begin{aligned} z_1(\omega) &= g_{Reindl}(\omega)x_1(\omega) \\ z_2(\omega) &= g_{Reindl}(\omega)x_2(\omega) \end{aligned}$$

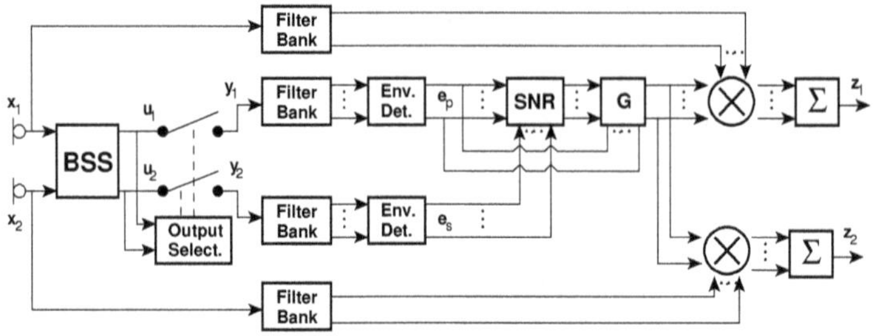

Fig. 8. Perceptually-inspired post processing to preserve the localization cues.

Experimental results in (Reindl et al., 2010) and (Marin-Hurtado et al., 2012) showed that this method can preserve the localization cues for both target and interfering signals simultaneously; however, the performance of this method is slightly below the performance of the post processing based on adaptive filters (Marin-Hurtado et al., 2012).

3.5 Perceptually-inspired post processing

In the previous sections, different BSS post-processing methods were discussed to recover the localization cues. All the above methods can preserve the localization cues of the target signal efficiently. However, only the BSS post-processing method based on Wiener filter can preserve the localization cues for both target and interfering signals simultaneously. This section discusses an alternative BSS post-processing method that preserves both localization cues. In this case, a perceptually-inspired post processing (BSS-PP) is used to compute a set of time-domain gains from the BSS outputs, and these gains are applied to the unprocessed signals (Fig. 8) (Marin-Hurtado et al., 2011; 2012). The BSS post processing used in (Marin-Hurtado et al., 2011; 2012) is an adaptation of the method in (Parikh & Anderson, 2011). This post processing is selected since it outperforms other BSS post processing for monaural speech enhancement applications. This post processing is modified so that it can be used for a binaural hearing aid (Marin-Hurtado et al., 2012):

1. To preserve the localization cues, the gains obtained by the BSS and perceptual post-processing algorithm described in (Parikh & Anderson, 2011) are applied to the unprocessed signals received at each side (Figure 8).
2. To achieve low processing delay, the system is implemented assuming real-time operating constraints, with the envelopes (e_p and e_s), SNR estimates, and gain parameters updated in the frame-by-frame basis, while the gains and outputs are computed in the sample-by-sample basis. In (Parikh & Anderson, 2011), gains are computed assuming an entire knowledge of the signal.
3. To minimize artifacts and to achieve more quality outputs, it is necessary to hold a long-term history for the maximum values of the primary envelope (e_p). Different tests show that the length of this memory should be at least one second.
4. To estimate the SNR, first-order estimators of the signal and noise PSD are used, and the SNR is computed as the ratio of these PSDs.

This perceptually-inspired BSS post processing is shown in Figure 8. Signals received at the left, x_1, and right, x_2, microphones are passed through a BSS algorithm to get u_1 and u_2. An output selection algorithm identifies which BSS output contains the separated target signal (y_1), or primary channel, and the separated interfering signal (y_2), or secondary channel. These outputs, y_1 and y_2, are analyzed using an auditory filter bank, and then, the envelope in each sub-band is extracted. These envelopes are used to estimate the SNR and to compute the noise-suppression gains. The SNR and gains are computed separately for each sub-band. These noise-suppression gains expand the dynamic range of each sub-band by lowering the noise floor. These gains are finally applied simultaneously to the unprocessed signals by time-domain multiplication, and the outputs from each sub-band are summed together to produce the enhanced signals for the left and right ear.

To reduce computational complexity and processing delay in the BSS stage, an info-max BSS algorithm that uses adaptive filters to minimize the mutual information of the system outputs is used. This algorithm is described by the following set of equations (Marin-Hurtado et al., 2012):

$$u_1(n+1) = x_1(n) + \boldsymbol{w}_{12}^T(n)\boldsymbol{u}_2(n) \quad (35)$$

$$u_2(n+1) = x_2(n) + \boldsymbol{w}_{21}^T(n)\boldsymbol{u}_1(n) \quad (36)$$

$$\boldsymbol{w}_{12}(n+1) = \boldsymbol{w}_{12}(n) - 2\mu \tanh(u_1(n+1))\boldsymbol{u}_2(n) \quad (37)$$

$$\boldsymbol{w}_{21}(n+1) = \boldsymbol{w}_{21}(n) - 2\mu \tanh(u_2(n+1))\boldsymbol{u}_1(n), \quad (38)$$

where x_1 and x_2 are the signals received at the left and right microphones, $\boldsymbol{w}_{12}$ and $\boldsymbol{w}_{21}$ are vectors of length N_w describing the unmixing filter coefficients, and $\boldsymbol{u}_1(n)$ and $\boldsymbol{u}_2$ are vectors of length N_w whose elements are the previous outputs of the BSS algorithm, $\boldsymbol{u}_j(n) = [u_j(n)\, u_j(n-1) \cdots u_j(n-N_w+1)]^T$, $j = 1,2$, and n is the time index. To determine which BSS output contains the target signal, the time-average energy of the envelopes of the signals u_1 and u_2 are compared, and then, the output with higher time-average energy is selected as primary channel y_1. This time-average energy is computed by

$$u_j^{env}(n) = \eta_{env} u_j^{env}(n-1) + (1-\eta_{env})u_j^2(n) \quad (39)$$

where η_{env} is a time constant. This update takes place every N samples.

The outputs of the BSS algorithm, $\boldsymbol{y}_1$ and $\boldsymbol{y}_2$, as well as the unprocessed input signals at the left and right microphones, $\boldsymbol{x}_1$ and $\boldsymbol{x}_2$, are passed through a filter bank that resembles the auditory system. This filter bank was implemented using forth-order Butterworth filters. At 22 kHz sampling rate, each filter bank provides 24 sub-bands. At the output of the filter banks, the vectors $\boldsymbol{x}_j(l,k)$ and $\boldsymbol{y}_j(l,k)$ of length N, $j = 1,2$, are obtained, where l corresponds to the frame index and k to the sub-band number. Although the signals x and y are obtained in the sample-by-sample basis, they are analyzed in non-overlapped frames of length N to compute the gain parameters as we will show next.

For each output $\boldsymbol{y}_j(l,k)$, the envelope is extracted using a full-wave rectifier followed by a low-pass filter. In particular, the primary envelope vector $\boldsymbol{e}_p(l,k)$ is extracted from $\boldsymbol{y}_1(l,k)$, and the secondary envelope vector $\boldsymbol{e}_s(l,k)$ from $\boldsymbol{y}_2(l,k)$. The low-pass filters are implemented using a first-order IIR filter whose cutoff frequency is selected to be a fraction of the corresponding bandwidth of the band (Parikh & Anderson, 2011). These cutoff frequencies are set to $1/5$, $1/8$ and $1/15$ of the bandwidth of low, medium and high-frequency bands, respectively. These fractions ensure that the envelope tracks the signal closely but at the same time does not change too rapidly to cause abrupt gain changes that introduce modulation.

The final outputs at the left, z_1, and the right, z_2, side are computed using the time-domain gains $\boldsymbol{g}_{l,k}$ produced by the perceptual post-processing stage:

$$\boldsymbol{z}_j(l) = \sum_k \boldsymbol{g}_{l,k} \circ \boldsymbol{x}_j(l,k) \tag{40}$$

where $\circ$ denotes the element-wise product. The vector form emphasizes that the gains are computed using parameters updated on a frame-by-frame basis. However, these outputs can be computed on a sample-by-sample, reducing the processing delay.

In (Parikh & Anderson, 2011), inspired by a perceptual modeling, these gains modify the envelope of each sub-band $e_k(t)$ such that $\hat{e}_k(t) = \beta e_k^{\alpha}(t)$. To provide noise reduction, the maximum envelope value is preserved (i.e., $\hat{e}_{k_{max}} = e_{k_{max}}$) while the minimum envelope value is lowered (i.e., $\hat{e}_{k_{min}} = K e_{k_{min}}$, where K is an expansion coefficient). Using the previous ideas, (Parikh & Anderson, 2011) developed a method to estimate α and β from the entire signal. To provide a realistic implementation, equations in (Parikh & Anderson, 2011) are modified to a vector form to state the update of α and β is the frame-by-frame basis every N samples (Marin-Hurtado et al., 2012):

$$\boldsymbol{g}_{k,l} = \beta_{l,k} \boldsymbol{e}_p(l,k)^{(\alpha_{l,k}-1)}. \tag{41}$$

The factors α and β are computed as

$$\beta_{l,k} = \max(\boldsymbol{e}_{pmax}(k))^{(1-\alpha_{k,l})} \tag{42}$$

$$\alpha_{k,l} = 1 - \log K / \log M_{l,k}\,, \tag{43}$$

where $M_{l,k}$ is the SNR at k-th sub-band and l-th frame, and $\boldsymbol{e}_{pmax}(k)$, a vector that holds the maximum values of the primary envelopes, is obtained from the previous N_{max} frames:

$$\boldsymbol{e}_{pmax}(k) = [\max(\boldsymbol{e}_p(l,k)) \ldots \max(\boldsymbol{e}_p(l - N_{max}, k))] \tag{44}$$

To avoid computational overflow and preserve the binaural cues, the value of α is constrained in the range $\alpha = [0,5]$. To minimize artifacts and achieve better quality outputs, the history stored in the vector $\boldsymbol{e}_{pmax}$ should hold at least one second, but two-seconds memory, i.e. $N_{max} = \lceil 2f_s/N \rceil$, is recommended. Since α and β are fixed for a given frame, these gains can also be computed in the sample-by-sample basis.

To estimate the SNR at the given sub-band and frame, the signal and noise power are obtained from the envelopes of the primary and secondary channel. This approach reduces miss-classification errors in the SNR estimation when the input SNR is low. To obtain a reliable noise estimate, the noise power is updated using a rule derived from the noise PSD estimator proposed in (Ris & Dupont, 2001):

$$\begin{aligned}
& P_e = \|e_s(l,k)\|^2 \\
& if\ |P_e - P_v(l-1,k)| < \epsilon\sqrt{\sigma_v(l-1,k)} \\
& \quad P_v(l,k) = \lambda_v P_v(l-1,k) + (1-\lambda_v) P_e \\
& \quad \sigma_v(l,k) = \delta\sigma_v(l-1,k) + (1-\delta)\,|P_e - P_v(l-1,k)|^2 \\
& else \\
& \quad P_v(l,k) = P_v(l-1,k) \\
& \quad \sigma_v(l,k) = \sigma_v(l-1,k) \\
& end
\end{aligned} \tag{45}$$

where $P_v(l,k)$ is the noise power at the k-th sub-band and l-th frame, $\sigma_v(l,k)$ is an estimate of the variance of P_v, λ and δ are time constants to smooth the estimation, and ϵ is a threshold coefficient. Finally, the frame SNR is estimated by

$$M_{l,k} = \max\left(\frac{P_x(l,k)}{P_v(l,k)} - 1, 1\right) \tag{46}$$

where P_x is the power of the primary channel estimated by

$$P_x(l,k) = \lambda_x P_x(l-1,k) + (1-\lambda_x)\left\|e_p(l,k)\right\|^2 \tag{47}$$

The values $\lambda_v = 0.95$, $\lambda_x = 0.9$, $\delta = 0.9$, and $\epsilon = 5$ are selected in (Marin-Hurtado et al., 2012) to achieve good performance.

The performance of the BSS-PP depends on the tuning of two parameters: K and N. Whereas K controls the expansion of the dynamic range, N defines how often the parameters to compute the noise-suppression gains are updated. A detailed analysis of the effect of these parameters on the SNR improvement and sound quality is presented in (Marin-Hurtado et al., 2012). In summary, $K = 0.01$ and $N = 8192$ show to be suitable for all scenarios. The mathematical proof that localization cues are preserved in the BSS-PP algorithm is included in (Marin-Hurtado et al., 2012).

3.5.1 Advantages and limitations

In the BSS-PP method, the noise-suppression gains are computed to expand the dynamic range of the noisy signal, in such a way that the maximum signal level is maintained while the noise level is pushed down. The maximum signal level is estimated from the primary channel, and the noise level from the secondary channel. Theoretical analysis conducted in (Takahashi et al., 2009) show that ICA-based BSS algorithms such as the algorithm used in the BSS-PP method provides an accurate noise estimate under non-point-source noise scenarios (e.g., diffusive or babble noise). Therefore, the performance of this method under these scenarios is expected to be high. Since BSS-PP tracks the envelopes of the target speech and noise level simultaneously, it is expected a good performance under highly non-stationary environments. On the other hand, when the interfering signals are few point sources, the BSS algorithm can provide accurate noise estimation only if the target signal is dominant. Thus, the performance of the BSS-PP algorithm is expected to be low under these scenarios at very low input SNR. Fortunately, these kind of scenarios are uncommon. All the above statements are verified through experiments discussed in the next section. In general, the BSS-PP method shows to be efficient in the removal of background noise, provides an acceptable speech quality, preserves the localization cues for both target and interfering signals, and outperforms existing BSS-based methods in terms of SNR improvement and noise reduction (Marin-Hurtado et al., 2012).

4. Comparative study

This chapter discussed different methods to preserve the localization cues in a binaural noise-reduction system based on BSS. These methods are summarized in the Table 1. Based on common features of the algorithms, these methods can be classified in three categories: BSS

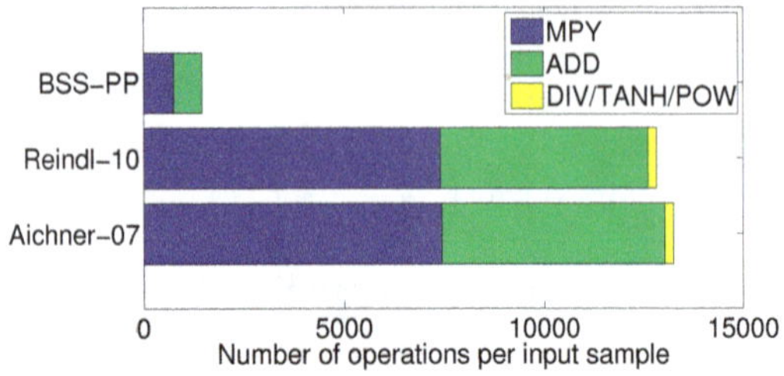

Fig. 9. Number of operations for BSS-PP, Reindl-10, and Aichner-07 per input sample grouped into additions (ADD), multiplications (MPY), divisions (DIV), hyperbolic tangent (TANH), and power raise (POW).

constrained optimization, spatial-placement filters, and BSS post processing to enhance the unprocessed signals. In the first category, the BSS filter weights, w_{qp}, are designed to perform source separation as well as to preserve the localization cues. In the second category, the BSS output corresponding to the estimate of the target signal is enhanced by a FIR filter that restores the localization cues. In the third category, the BSS outputs are used to compute noise-suppression gains that enhance the unprocessed signals. Under the third category, we can include the post-processing methods based on adaptive filters, Wiener filter, and perceptually-inspired processing.

Different reports have been shown that methods based on BSS constrained optimization and spatial-placement filters are unable to provide simultaneous preservation of localization cues for the target and interfering signals. In addition, most of these methods perform a mapping of the direction of arrival of the interfering signals to the direction of arrival of the target signal, which may be perceptually annoying. On the contrary, most methods belonging to the third category, BSS post processing to enhance the unprocessed signals, can preserve the localization cues for both target and interfering signals simultaneously under certain conditions. In particular, among the different methods analyzed, the BSS post-processing method based on Wiener filter and the perceptually-inspired post processing are the only methods able to preserve these localization cues simultaneously.

Since the gains and outputs are computed in the sample-by-sample basis, the processing delay is very small (< 1 ms) in the BSS-PP method compared to other BSS-based post-processing methods such as the method based on adaptive filters, Aichner-07, (Section 3.3), and the method based on Wiener filter, Reindl-10, (Section 3.4). In Aichner-07 and Reindl-10 methods, the processing delay is around 6 ms. In addition, the computational complexity of BSS-PP is significantly smaller than Aichner-07 and Reindl-10 (Fig. 9).

4.1 Experiment

Among the different methods discussed in this chapter, only Aichner-07, Reindl-10, and BSS-PP are evaluated in this experiment. This selection takes into account only the BSS post-processing methods capable of preserving the localization cues for the target and interfering signals simultaneously under certain environmental conditions (Table 1). These methods are implemented in Matlab and tested under different scenarios. Simulations to discern the performance of these techniques are conducted under the following scenarios:

Method	Strategy	Preserve Target Cues	Preserve Noise Cues	Comp. Cost	Processing Delay	Ref.
Takatani-05	BSS constrained optimization	Yes	No	High	?	(Takatani et al., 2005)
Aichner-07B	BSS constrained optimization	Yes	No	High	?	(Aichner et al., 2007)
Wehr-06A	Spatial-placement filter	No	Mapped to target DoA	Medium	?	(Wehr et al., 2006)
Wehr-06B	Spatial-placement filter	Yes	Mapped to target DoA	Medium	?	(Wehr et al., 2006)
Aichner-07	Post processing based on adaptive filters	Yes	Under certain conditions	Medium	~ 6 ms	(Aichner et al., 2007)
Reindl-10	Post processing based on Wiener filter	Yes	Yes	Medium	~ 6 ms	(Reindl et al., 2010)
BSS-PP	Perceptually-inspired post processing	Yes	Yes	Low	~ 1 ms	(Marin-Hurtado et al., 2012)

Table 1. Summary of the binaural noise-reduction methods based on BSS. Processing delay is estimated for a system working at 16 kHz sampling frequency. Question mark is included for the methods not analyzed in the comparative study.

1. Single source under constant-SNR diffusive noise. This scenario is widely used to test virtually all binaural noise-reduction techniques. This background noise is generated by playing uncorrelated pink noise sources simultaneously at 18 different spatial locations.
2. Single source under babble (or cafeteria) noise. The background noise corresponds to a real recording in a cafeteria.
3. Multi-talker. In this scenario, four distinguishable speakers are placed at different azimuthal positions: $40^o, 80^o, 200^o$ and 260°.

The above scenarios are generated by filtering the target signal with the HRTF measured for a KEMAR manikin in absence of reverberation (Gardner & Martin, 1994). The target signal is placed at eight different azimuthal angles: $0^o, 30^o, 90^o, 120^o, 180^o, 240^o, 270^o$ and 330°, where 0° corresponds to the front of the KEMAR, 90° corresponds to the right ear, and 270° to the left ear. Target signals are speech recordings of ten different speakers and sentences taken from the IEEE sentence database (IEEE Subcommittee, 1969). For all scenarios, the interfering signals are added to the target signal at different SNR.

Since the HRTF database in (Gardner & Martin, 1994) is for non-reverberant environments, a secondary database using reverberant conditions is created using the HRTF recordings described in (Jeub et al., 2009; RWTH Aachen University, 2010). This database is included since it is widely known that the performance of the majority of the noise-reduction algorithms is degraded significantly when reverberation is present. This database assumes a babble noise scenario and the following rooms: studio ($RT_{60} = 0.12$s), meeting room ($RT_{60} = 0.23$s), office ($RT_{60} = 0.43$s), and lecture room ($RT_{60} = 0.78$s).

The performance of these techniques is analyzed using the broadband intelligibility-weighted SNR improvement (ΔSNR-SII) (Greenberg et al., 1993). For the subjective test, a MUSHRA (multiple stimulus test with hidden reference and anchor) test is used to assess the overall sound quality. The protocol in (ITU-R, 2003) is used for the subjective test.

4.2 Performance evaluation

SNR improvement for diffusive, babble, and multi-talker scenarios is plotted in Figures 10-12. In general, the perceptually-inspired post-processing method (BSS-PP) outperforms the other BSS-based noise-reduction methods in most scenarios.

The poor performance of BSS-PP in the multi-talker scenario at low input SNR is explained by the errors introduced by a wrong selection of the primary output. When an ideal output selection algorithm is used (dashed line in Fig. 12), the performance of BSS-PP is similar or better than that of the other BSS-based methods. The output selection algorithm can be made more robust by using a direction-of-arrival-estimation algorithm or a permutation algorithm at expenses of increasing the computational complexity. However, scenarios with very few interfering signals at input SNR < 0 dB such as the multi-talker scenario of Fig. 12 are very uncommon, and they are not challenging for the auditory system without any hearing aid. Likewise, binaural noise-reduction methods are useful for challenging scenarios such as babble noise at low input SNR. Since BSS-PP provides an excellent performance under these scenarios (Fig. 11), the output-selection algorithm used by BSS-PP is enough for a large set of practical applications.

Up to this point the performance of all methods has been verified under non-reverberant scenarios. For reverberant scenarios, Fig. 13 shows that for a large reverberant room

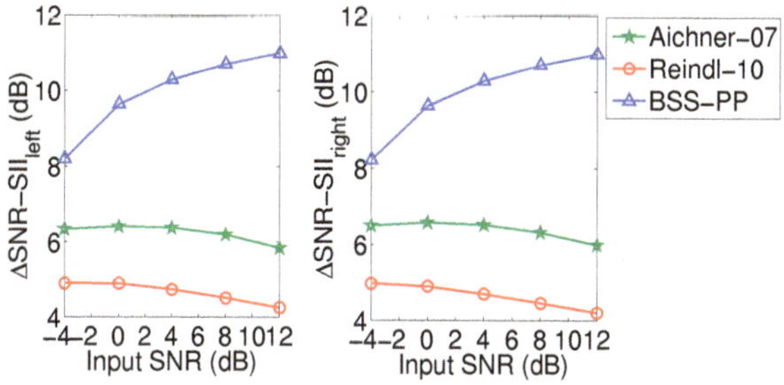

Fig. 10. SNR improvement under diffusive noise scenario.

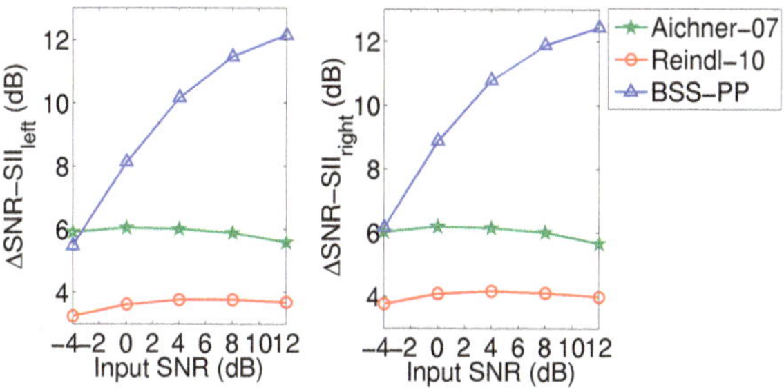

Fig. 11. SNR improvement under babble noise scenario.

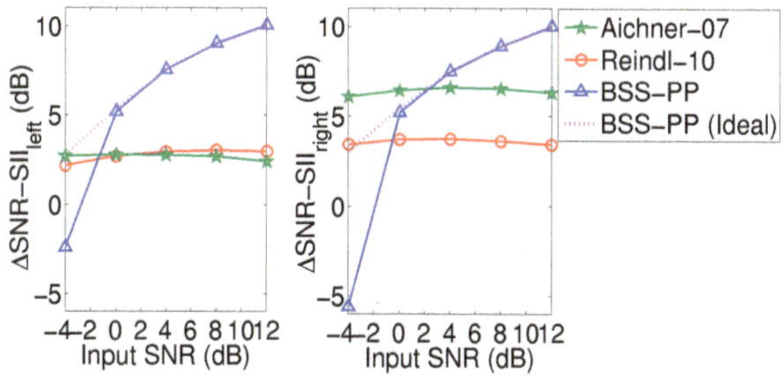

Fig. 12. SNR improvement under multi-talker scenario. The dashed line is the performance for an ideal output-selection algorithm.

($RT_{60} = 0.78$s), BSS-PP provides an acceptable SNR improvement and outperforms the other existing methods for input SNR ≥ 0 dB. Results for other reverberant rooms are included in (Marin-Hurtado et al., 2012).

A subjective test is conducted to assess the subjective sound quality of the methods under study. These results are summarized in the Fig. 14. Sound quality is graded in the scale $[0, 100]$, with 100 the highest value corresponding to a clean signal. To perform the grading, the subject listened to the samples that included clean speech, unprocessed speech in babble noise at an input SNR of 0 dB, and enhanced speech processed by Aichner-07, Reindl-10, and BSS-PP methods. The reference and hidden reference signals are unprocessed noisy

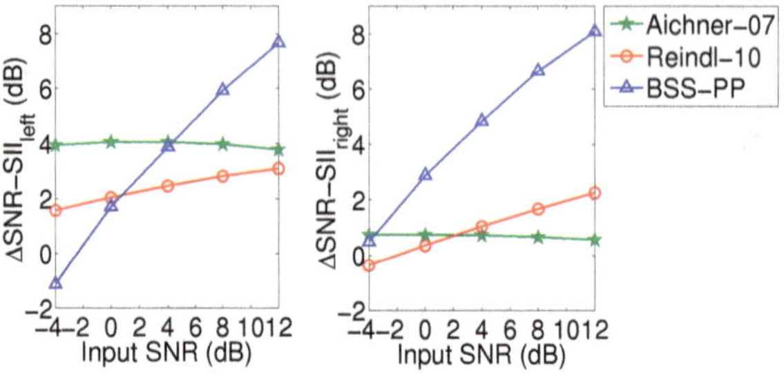

Fig. 13. SNR improvement under babble noise scenario in a lecture room (reverberant condition $RT_{60} = 0.78\text{s}$).

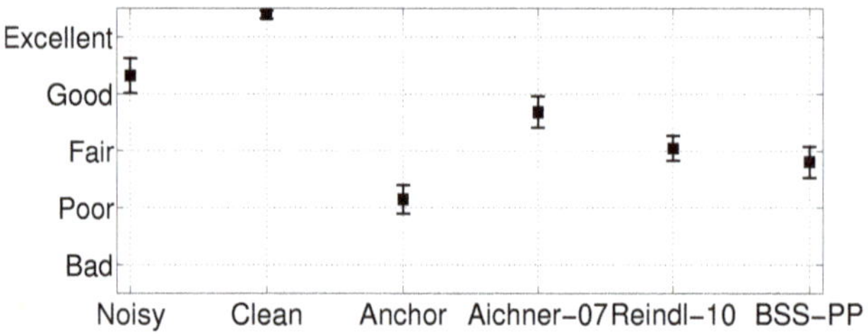

Fig. 14. Subjective test results for speech quality. Reference: speech in babble noise; anchor: noisy speech distorted according to (ITU-R, 2003).

speech while the anchor signal is noisy speech distorted according to (ITU-R, 2003). All samples are five-seconds long, and they are presented randomly to the subject. A total of 20 normal-hearing subjects participated in the experiment. Results show that there is a distortion in the speech quality for all methods, and a subject preference for the speech quality of the unprocessed noisy signal. The methods providing the lowest noise reduction (Aichner-07 and Reindl-10) achieved the best speech quality, and the methods with the highest noise reduction (BSS-PP), the lowest speech quality. However, the speech quality of BSS-PP is higher than the speech quality of the anchor signal (an artificially-distorted speech signal).

5. Conclusions

This chapter described different binaural BSS-based noise-reduction algorithms that are promising for the reduction of the background noise and the preservation of the direction of arrival of the target and interfering signals. The preservation of the direction of arrival, also known as localization cues, is an important issue for some applications such as binaural hearing aids. In these devices, the displacement or lost of these localization cues is reported as perceptually annoying by hearing-impaired users.

The methods reported in the literature to preserve the localization cues in a binaural BSS-based noise-reduction algorithm can be classified into three categories: a) BSS algorithms based on contrained optimization to preserve the localization cues (Section 3.1); b) restoration of the localization cues by means of post processing applied to the BSS output related to the target signal (e.g., spatial-placement filter on Section 3.2); and c) enhancement of the unprocessed inputs by noise-reduction gains computed from the BSS outputs (e.g., adaptive filters, Wiener filter, and perceptual post-processing methods described on Sections 3.3, 3.4, and 3.5). All methods proposed in the literature can preserve the localization cues for the target signal. However, the methods belonging to the first and second category, BSS constrained optimization and spatial-placement filters, cannot preserve the localization cues for the interfering signals. In most cases, these localization cues are mapped to the direction of arrival of the target signal, which suggests that these algorithms are not practical for binaural hearing aids. On the contrary, binaural BSS-based noise-reduction algorithms belonging to the third category, i.e., those methods that compute noise-suppression gains from the BSS outputs and apply these gains to the unprocessed signals, can preserve the localization cues for both target and interfering signals simultaneously. This preservation is identified through subjective and theoretical analysis. This chapter described three methods belonging to the third category: post processing with adaptive filters (Aichner-07), post processing with Wiener filter (Reindl-10), and perceptually-inspired post processing (BSS-PP). An experimental evidence, confirmed through mathematical analysis, showed the post processing based on adaptive filters (Aichner-07) works only in the determined case, i.e., when the number of source signals is equal o lower than the number of sensors. On the contrary, the methods based on Wiener-filter post processing (Reindl-10) and perceptually-inspired post processing (BSS-PP) preserve the localization cues even in the undetermined case.

A comparative study conducted with the Aichner-07, Reindl-10, and BSS-PP methods under different environments showed that BSS-PP outperforms the other methods in terms of SNR improvement and noise reduction. In addition, the BSS-PP method provides a significant reduction in the number of operations compared to the other two methods, and its processing delay is very small. Hence, the BSS-PP turns out a feasible solution for a binaural hearing aid. However, there are two limitations in the BSS-PP method. First, the subjective sound quality is acceptable, with a subjective sound quality graded slightly below the subjective sound quality of the Aichner-07 and Reindl-10 methods. Second, the BSS algorithm demands wireless transmission at full rate. This issue is also present in the Aichner-07 and Reindl-10 methods.

6. Future work

Although the BSS-PP method is a promising binaural noise-reduction algorithm, it is necessary to solve two issues to obtain a practical implementation for a binaural hearing aid. First, to improve the sound quality, the dynamic range expansion performed by the post processing stage must include additional information to take into account a sound quality criteria, or use another perceptual model. Second, to reduce the transmission bandwidth, it is necessary to develop distributive or reduced bandwidth BSS algorithms, or to employ strategies other than BSS to estimate the target and interfering signals.

Most processing in the BSS-PP method can be easily replaced by an analog processing except the BSS algorithm. A mixed-signal solution may reduce computational complexity and power consumption. To obtain a full-analog solution, analog BSS algorithms have to be developed.

Although most BSS-based noise-reduction algorithms such as Reindl-10 and BSS-PP were not initially designed to deal with reverberant conditions, their performance under these environments is acceptable. Hence, their performance could be improvement by modifications in the mathematical framework to take into account the effect of reverberation.

Finally, it is known that the speech intelligibility in noise-reduction applications can be improved by applying a binary mask to the unprocessed signal (Loizou & Kim, 2011). Hence, binary masking can be combined with a BSS algorithm in order to obtain a source separation algorithm that reduces the background noise and improves the speech intelligibility simultaneously. Although some attempts have been explored in (Han et al., 2009; Jan et al., 2011; Mori et al., 2007; Takafuji et al., 2008), these methods are unable to preserve the localization cues for both target and interfering signals simultaneously. Hence, it is necessary to develop post processing algorithms to preserve the localization cues in BSS-based binary masking algorithms.

7. Acknowledges

This work is supported in part by Texas Instruments Inc., formerly National Semiconductor Corporation. Jorge Marin wants to thank Georgia Institute of Technology–USA, Universidad del Quindio–Colombia and Colciencias–Colombia for their financial support.

8. References

Aichner, R., Buchner, H., & Kellermann, W. (2006). A novel normalization and regularization scheme for broadband convolutive blind source separation, *Proc. Int. Symp. Independent Component Analysis and Blind Signal Separation, ICA 2006*, Vol. 3889, pp. 527–535.

Aichner, R., Buchner, H., Zourub, M. & Kellermann, W. (2007). Multi-channel source separation preserving spatial information, *Proc. IEEE Int. Conf. Acoust., Speech Signal Process., ICASSP 2007*, Vol. 1, pp. I–5–I–8.

Gardner, B. & Martin, K. (1994). HRTF measurements of a KEMAR dummy-head microphone, *Technical Report 280*, MIT Media Lab Perceptual Computing. http://sound.media.mit.edu/KEMAR.html.

Greenberg, J. E., Peterson, P. M. & Zurek, P. M. (1993). Intelligibility-weighted measures of speech-to-interference ratio and speech system performance, *J. Acoust. Soc. Amer.* 94(5): 3009–3010.

Han, S., Cui, J. & Li, P. (2009). Post-processing for frequency-domain blind source separation in hearing aids, *Proc. Int. Conf. on Information, Communications and Signal Processing, 2009. ICICS 2009*, pp. 1 –5.

Haykin, S. (2000). *Unsupervised Adaptive Filtering*, Vol. 1: Blind Source Separation, John Wiley and Sons.

IEEE Subcommittee (1969). IEEE recommended practice for speech quality measurements, *IEEE Trans. Audio Electroacoust.* pp. 225–246.

ITU-R (2003). Recommendation BS.1534-1: Method for the subjective assessment of intermediate quality levels of coding systems.

Jan, T., Wang, W. & Wang, D. (2011). A multistage approach to blind separation of convolutive speech mixtures, *Speech Communication* 53(4): 524 – 539.

Jeub, M., Schafer, M. & Vary, P. (2009). A binaural room impulse response database for the evaluation of dereverberation algorithms, *Proc. Int. Conf. Digital Signal Process.*, pp. 1 –5.

Kocinski, J. (2008). Speech intelligibility improvement using convolutive blind source separation assisted by denoising algorithms, *Speech Commun.* 50(1): 29 – 37.

Loizou, P. & Kim, G. (2011). Reasons why current speech-enhancement algorithms do not improve speech intelligibility and suggested solutions, *IEEE Transactions on Audio, Speech, and Language Processing* 19(1): 47 –56.

Marin-Hurtado, J. I., Parikh, D. N. & Anderson, D. V. (2011). Binaural noise-reduction method based on blind source separation and perceptual post processing, *Proc. Interspeech 2011*, Vol. 1, Florence, Italy, pp. 217–220.

Marin-Hurtado, J. I., Parkih, D. N. & Anderson, D. V. (2012). Perceptually inspired noise-reduction method for binaural hearing aids, *IEEE Transactions on Audio, Speech and Language Processing* 20(4): 1372–1382.

Moore, B. C. J. (2007). Binaural sharing of audio signals: Prospective benefits and limitations, *The Hearing Journal* 60(11): 46–48.

Mori, Y., Takatani, T., Saruwatari, H., Shikano, K., Hiekata, T. & Morita, T. (2007). High-presence hearing-aid system using dsp-based real-time blind source separation module, *Proc. IEEE Int. Conf. Acoustics, Speech and Signal Process., 2007. ICASSP 2007*, Vol. 4, pp. IV–609 –IV–612.

Noohi, T. & Kahaei, M. (2010). Residual cross-talk suppression for convolutive blind source separation, *Proc. Int. Conf. Comp. Eng. Technology (ICCET)*, Vol. 1, pp. V1–543 –V1–547.

Parikh, D., Ikram, M. & Anderson, D. (2010). Implementation of blind source separation and a post-processing algorithm for noise suppression in cell-phone applications, *Proc. IEEE Int. Conf. Acoust., Speech Signal Process., ICASSP*, pp. 1634 –1637.

Parikh, D. N. & Anderson, D. V. (2011). Blind source separation with perceptual post processing, *Proc. IEEE 2011 DSP/SPE Workshop*.

Park, K. S., Park, J., Son, K. & Kim, H. T. (2006). Postprocessing with wiener filtering technique for reducing residual crosstalk in blind source separation, *Signal Processing Letters, IEEE* 13(12): 749 –751.

Reindl, K., Zheng, Y. & Kellermann, W. (2010). Speech enhancement for binaural hearing aids based on blind source separation, *Proc. Int. Symp. Commun. Control Signal Process. (ISCCSP)*, pp. 1 –6.

Ris, C. & Dupont, S. (2001). Assessing local noise level estimation methods: Application to noise robust ASR, *Speech Commun.* 34(1-2): 141 – 158.

RWTH Aachen University (2010). Aachen impulse response (AIR) database - version 1.2. http://www.ind.rwth-aachen.de/AIR.

Smith, P., Davis, A., Day, J., Unwin, S., Day, G. & Chalupper, J. (2008). Real-world preferences for linked bilateral processing, *The Hearing Journal* 61(7): 33–38.

Sockalingam, R., Holmberg, M., Eneroth, K. & Shulte, M. (2009). Binaural hearing aid communication shown to improve sound quality and localization, *The Hearing Journal* 62(10): 46–47.

Takafuji, R., Mori, Y., Saruwatari, H. & Shikano, K. (2008). Binaural hearing-aid system using simo-model-based ica and directivity-dependency-reduced binary masking, *Proc. 9th Int. Conf. Signal Process., ICSP 2008*, pp. 320 –323.

Takahashi, Y., Takatani, T., Osako, K., Saruwatari, H. & Shikano, K. (2009). Blind spatial subtraction array for speech enhancement in noisy environment, *IEEE Transactions on Audio, Speech and Language Processing* 17(4): 650 –664.

Takatani, T., Ukai, S., Nishikawa, T., Saruwatari, H. & Shikano, K. (2005). Evaluation of SIMO separation methods for blind decomposition of binaural mixed signals, *Proc. International Workshop on Acoustic Echo and Noise Control (IWAENC)*, pp. 233–236.

Van den Bogaert, T., Klasen, T. J., Moonen, M., Van Deun, L. & Wouters, J. (2006). Horizontal localization with bilateral hearing aids: Without is better than with, *J. Acoust. Soc. Amer.* 119(1): 515–526.

Wehr, S., Puder, H. & Kellermann, W. (2008). Blind source separation and binaural reproduction with hearing aids: An overview, *Proc. ITG Conf. Voice Communication (SprachKommunikation)* pp. 1 –4.

Wehr, S., Zourub, M., Aichner, R. & Kellermann, W. (2006). Post-processing for BSS algorithms to recover spatial cues, *Proc. Int. Workshop Acoust. Echo Noise Control (IWAENC)*.

12

ICA-Based Fetal Monitoring

Rubén Martín-Clemente and José Luis Camargo-Olivares
University of Seville
Spain

1. Introduction

Independent Component Analysis (ICA) has numerous applications in biomedical data processing (James & Hesse, 2005; Nait-Ali, 2009; Tanskanen & Viik, 2012). For example, in the last decade lots of contributions have been made in the field of EEG/MEG[1] analysis (artifact detection and removal, analysis of event-related brain responses, ... see e.g. Zarzoso (2009), and the references therein, for a more detailed information). More recently, several researchers have oriented their efforts towards developing ICA-based approaches to the interpretation of the electrocardiogram (ECG) and the information that can be derived from it (Castells et al., 2007). For example, Vetter *et al* (Vetter et al., 2000) have shown the great potential of ICA in the analysis of the control of the heart by the autonomic nervous system. Arrhythmia detection and, in particular, atrial fibrillation, also constitute possible applications, and several successful examples can be found in the literature (Llinares & Igual, 2009; Rieta et al., 2004).

A particularly appealing problem in tococardiography is that of applying ICA-based methods to the extraction of the fetal ECG (fECG) from maternal cutaneous potential measurements. The present clinic standard procedure for recording the fECG consists in inserting a small electrode into the mother's vagina, through the cervix, and directly under the skin of the fetus's scalp (Symonds et al., 2001). The major shortcoming of this technique is its invasiveness. The placement of the fetal scalp electrode presents certain risks to fetal safety, and cases where the mother is infected have been reported as well. Last but not least, this procedure is not suitable for use during all stages of pregnancy, and can only be performed under limited clinical circumstances: e.g. measuring the fECG with a scalp electrode is only possible during labor, as requires a dilated cervix and the rupture of the amniotic membranes. Due to these and other inconveniences, the scalp electrode is almost exclusively reserved for high risk births.

There exists, by contrast, an increasing interest in non-invasive fECG recording techniques (Hasan et al., 2009). These techniques should enable monitoring in stages of pregnancy earlier than labor, i.e. when the membranes protecting the fetus are not broken (*antepartum*), as well as being comfortable to women, while avoiding the risks of infection or trauma to the fetal scalp. A method for non-invasive fECG monitoring measures the fECG by means of surface electrodes placed on the mother's abdomen. It turns out that the electrical signals recorded

[1] EEG and MEG are abbreviations for, respectively, electro-encephalography and magneto-encephalography.

by the electrodes are a mixture of several electrophysiological signals and noise. Examples of the former are the maternal electrocardiogram (mECG), the electrohysterogram (EHG, the electrical activity of the uterus) and the fECG. The EHG usually lies below 3 Hz and can be reduced significantly by the use of a simple high-pass filter (Devedeux et al., 1993). The main source of noise is the power line signal (50 – 60 Hz, depending on your country), that can be eliminated by a notch filter. The limiting factor in non-invasive fetal electrocardiography is the low amplitude of the fetal electrocardiogram compared to the mECG[2]. As there is a considerable overlap between the frequency bands of the mECG and the fECG (Abboud & Sadeh, 1989), the mECG cannot be suppressed by a simple linear filter. A variety of different approaches have been proposed to extract the fECG from abdominal recordings (Hasan et al., 2009). In this Chapter, we describe and illustrate the specific application of ICA to this exciting problem. Potential readers are assumed very familiar with ICA —if not, they are directed to the classical textbooks Cichocki & Amari (2002); Comon & Jutten (2010); Hyvärinen et al. (2001). The Chapter is organized as follows: in Section 2, we introduce some basic concepts of fetal electrocardiography. Sections 3 and 4 discuss a simple mathematical model of the fECG and its implications for ICA. In Section 5, we review some ICA-based approaches for the fECG extraction problem. Rather than surveying superficially several methods, we shall concentrate on some of the more conceptually appealing concepts. Section 6 introduces a recent and powerful approach, namely to use the mECG as reference for the ICA algorithms. Experiments, using real data, are presented in Section 7. Finally, Section 8 is devoted to the Conclusions.

2. Basic background in cardiac physiology

The heart consist of four chambers: the right and left atrium and the right and left ventricle.

- In the **adult**, the atria are collecting chambers that receive the blood from the body and lungs, whereas the ventricles act as pumping chambers that send out the blood to the body tissues and lungs. Blood circulates as follows (Guyton & Hall, 1996):
  1. Oxygen-depleted blood flows into the right atrium from the body, via the vena cava.
  2. From the right atrium the blood passes into the right ventricle.
  3. The right ventricle pumps the blood, through the pulmonary arteries, into the lungs, where carbon dioxide is exchanged for oxygen.
  4. The oxygenated blood returns to the heart, via the pulmonary vein, into the left atrium.
  5. From the left atrium the blood passes into the left ventricle.
  6. The left ventricle pumps the oxygenated blood into all parts of the body through the aorta artery, and the cycle begins again.
- In the **fetus**, things are slightly different. The fetus receives the oxygen across the placenta and, as a consequence, does not use its lungs until birth. To prevent the blood to be pumped to the lungs, the pulmonary artery is connected to the aorta by a blood vessel called the arterial duct (*ductus arteriousus*). Thus, after the right ventricle contraction, most blood flows through the duct to the aorta. The fetal heart also has an opening between the right and left atria called the *foramen ovale*. The foramen ovale allows oxygenated blood

[2] Whereas the mECG shows an amplitude of up to 10 mV, the fECG often does not reach more that 1 μV.

to flow from the right atrium to the left atrium, where it gets pumped around the body, again avoiding the lungs. Both the ductus arteriousus and the foramen ovale disappear after birth over the course of a few days or weeks (Abuhamad & Chaoui, 2009).

2.1 The fECG

The electrocardiogram (ECG) reflects the electrical activity of the heart as seen from the body surface. The heart generates electrical currents that radiate on all directions and result in electrical potentials. The potential difference between a pair of electrodes placed in predefined points of the surface of the body (*cutaneous recordings*), visualized as a function of time, is what we call the ECG.

The fetal electrocardiogram (fECG), like that of the adult, consists of a P wave, a QRS complex and a T wave, separated by the PR and ST intervals (see Fig. 1) (Symonds et al., 2001). These waves represent the summation of the electrical potentials within the heart. Contraction (depolarization) of both atria begins at about the middle of the P wave and continues during the PR segment. The QRS complex precedes ventricular contraction: pumping of blood normally begins at the end of the QRS complex and continues to the end of the T wave. Finally, the T wave corresponds to the electrical activity produced when the ventricles are recharging for the next contraction (repolarizing)[3]. Note that the repolarization of the atria is too weak to be detected on the ECG. The fECG cannot be usually detected between 28 and 32 weeks (sometimes 34 weeks) of gestation due to the isolating effect of the *vernix caseosa*, a sebum that protects the skin of the fetus (Oostendorp et al., 1989a).

The fECG provides useful information about the health and the condition of the fetus (Pardi et al., 1986): for example, the duration of the ST-segment is important in the diagnosis of fetal hypoxia (i.e. a continued lack of oxygen), and it has been also shown that both the QT interval and T-wave changes are predictive of fetal acidemia (Jenkins et al., 2005). The human heart begins beating at around 21 days after conception with frequency about 65 beats per minute (bpm). This frequency increases during the gestation up to 110 – 160 bpm before delivery. When it is not within this range, it may be indicative of serious potential health issues: e.g. if the fetal heart rate (FHR) is below 110 bpm for 10 minutes or longer (bradycardia), it is considered a *late* sign of hypoxia (there is a depression of the heart activity caused by the lack of oxygen), and a fetal emergency (Freeman & Garite, 2003). On the contrary, an FHR that exceeds 160 bpm (tachycardia) may be an *early* sign of hypoxia (other conditions that increase the FHR include fetal infection, maternal dehydration, medication, et cetera) (Afriat & Kopel, 2008).

3. Mathematical model of the ECG

In the **adult**, the cardiac surface potentials can be approximately considered as originated from a current dipole located in the heart (Symonds et al., 2001). Assuming that the body is a volume conductor, homogeneous and infinite, the potential due to a dipole of moment $\mathbf{p}(t)$ at a point on the skin specified by the position vector $\mathbf{r}$ is given by (Keener & Sneyd, 2009):

[3] Both atria contract and pump the blood together. Both ventricles also contract together. But the atria contract before the ventricles. Nevertheless, all the four chambers relax (stop pushing in) together.

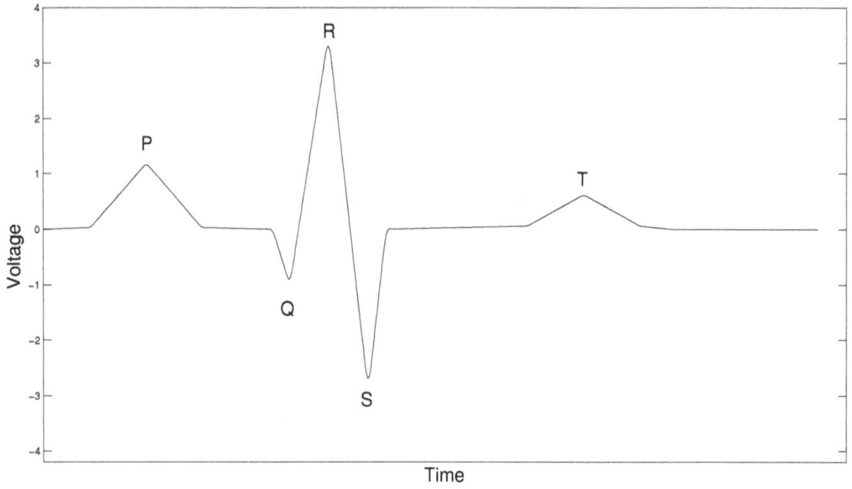

Fig. 1. Sketch of a typical single ECG recording. Note that the actual size and polarity of each wave depend on the location of the recording electrodes (Chan, 2008; Keener & Sneyd, 2009).

$$v(t) = \mathbf{p}(t)\,\frac{\mathbf{r}}{4\pi\sigma\,|\mathbf{r}|^3} \tag{1}$$

where σ is the permittivity of the medium. Let $\mathbf{e}_1, \mathbf{e}_2, \mathbf{e}_3$ be orthonormal basis vectors in the real three-dimensional space and let $\{s_1(t), s_2(t), s_3(t)\}$ be the coordinates of $\mathbf{p}(t)$ in this basis, i.e.,

$$\mathbf{p}(t) = \sum_{i=1}^{3} s_i(t)\,\mathbf{e}_i.$$

The body surface potential at $\mathbf{r}$ can be then written as linear combination of the signals $s_i(t)$:

$$v(t) = \sum_{i=1}^{3} a_i\, s_i(t), \tag{2}$$

where:

$$a_i \stackrel{def}{=} \mathbf{e}_i\,\frac{\mathbf{r}}{4\pi\sigma\,|\mathbf{r}|^3}.$$

It is quite noteworthy that $\mathbf{p}(t)$ is allowed to change in orientation and strength as a function of time. For the reference of the reader, the tip of the vector traces out a loop in the space that is called *vectorcardiogram* (VCG) (Symonds et al., 2001). Different models for $\mathbf{p}(t)$ can be found in the literature. For example, Sameni, Clifford, Jutten & Shamsollahi (2007), based on McSharry et al. (2003)[4], have proposed the following differential equations for the dipole vector:

[4] In essence, McSharry et al. (2003) describe each wave of the ECG (P, Q, R, S and T) by a Gaussian function whose amplitude, width and temporal location have to be determined.

$$
\begin{aligned}
\dot{\vartheta} &= \omega, \\
\dot{s}_1 &= -\sum_i \frac{\alpha_i^1 \omega}{(b_i^1)^2} \Delta\vartheta_i^1 \exp\left[-\frac{(\Delta\vartheta_i^1)^2}{2(b_i^1)^2}\right], \\
\dot{s}_2 &= -\sum_i \frac{\alpha_i^2 \omega}{(b_i^2)^2} \Delta\vartheta_i^1 \exp\left[-\frac{(\Delta\vartheta_i^2)^2}{2(b_i^2)^2}\right], \\
\dot{s}_3 &= -\sum_i \frac{\alpha_i^3 \omega}{(b_i^3)^2} \Delta\vartheta_i^1 \exp\left[-\frac{(\Delta\vartheta_i^3)^2}{2(b_i^3)^2}\right],
\end{aligned} \tag{3}
$$

where $\Delta\vartheta_i^1 = (\vartheta - \vartheta_i^1) \bmod(2\pi)$, $\Delta\vartheta_i^2 = (\vartheta - \vartheta_i^2) \bmod(2\pi)$, $\Delta\vartheta_i^3 = (\vartheta - \vartheta_i^3) \bmod(2\pi)$ and $\omega = 2\pi f$, where f is the beat-to-beat rate. Note also that the equation $\dot{\vartheta} = \omega$ generates periodic signals with the frequency of the heart rate. The problem of estimating the parameters $\alpha_i^k, b_i^k, \vartheta_i^k$ of the model is complicated and has been addressed, e.g., in (Clifford et al., 2005; Sameni, Shamsollahi, Jutten & Clifford, 2007).

3.1 The fECG case

Is the previous dipole-based model capable of describing the potential distribution created by the fetal heart at the maternal abdomen? It depends. At early pregnancies, from week 20 until 28 of gestation, the amplitude of the fECG increases and the model seems to be appropriate and fits the observations well (Oostendorp et al., 1989a). Late in pregnancy, however, great care is needed: we have already mentioned that the fECG is in general impossible to measure between 28th to 32th week of gestation due to the isolating effect of the *vernix caseosa,* the fatty layer that protects the skin of the fetus (Wakai et al., 2000). After the 32th week, the fECG is detected again, but the apparent fetal vectorcardiogram (fVCG), as calculated from the recorded surface potentials, describes almost a straight line (Oldenburg & Macklin, 1977). Hence it no longer corresponds to the activity of the fetal heart vector in an intelligible way. It has been hypothesized that, as the fetus grows, several holes appear in the vernix[5] and current can escape through them (Peters et al., 2005). It turns out that the potential is split up into a contribution of the current dipole and a contribution of the volume currents induced in the vernix. Experiments confirm that, after the 32th week, the fECG recorded at the mother's abdomen can be still described by a model of the type (2), i.e.,

$$
v(t) = \sum_{i=1}^{n} a_i \, s_i(t), \tag{4}
$$

but the fetal source signals $s_i(t)$ cannot be longer interpretable as coordinates of a single current dipole. Rather, we should think of eqn. (4) more as a *latent variable model.* Note that, by definition, the latent variables $s_i(t)$ correspond to abstract or hypothetical concepts. Experiments also show that the number of summands n may be different (usually less) from three (Oostendorp et al., 1989b; Sameni, Clifford, Jutten & Shamsollahi, 2007). See also (Lewis, 2003) and the references therein.

[5] The most important hole is probably at the mouth. A second relevant hole can be expected at the base of the umbilical cord.

4. ICA model

Thus, in view of the previous Section, ECGs seem to satisfy some of the conditions for classical ICA:

- The body surface potentials are a linear mixture of several source signals.
- Time delays in signal transmission are negligible.
- It is feasible to have more observations than sources[6].

Let $v_1(t), \ldots, v_p(t)$ be *zero-mean* signals recorded from electrodes placed on the mother's body, where $t \in \mathbb{Z}$ is the discrete time. Each measurement signal $v_i(t)$ is modelled as a linear combination of r ($r \leq 6$) bioelectric source signals, that have similar definitions to the ones in eqns. (2)–(4), plus noise:

$$\begin{aligned} v_1(t) &= a_{11}\, s_1(t) + \cdots + a_{1r}\, s_r(t) + n_1(t) \\ &\;\;\vdots \\ v_p(t) &= a_{p1}\, s_1(t) + \cdots + a_{pr}\, s_r(t) + n_p(t) \end{aligned} \tag{5}$$

The noise represents the signal degradation due, for example, to baseline wander, mains interference, uterine contractions, and so on. Eqn. (5) can be rewritten in matrix form as:

$$\mathbf{v}(t) = \mathbf{A}\, \mathbf{s}(t) + \mathbf{n}(t) \tag{6}$$

where $\mathbf{v}(t)$ be the vector whose ith component is $v_i(t)$ and so on. Eqn. (5) represents the superposition of the body surface potentials due to the fetus and the maternal cardiac dipole. Note that $\mathbf{s}(t)$ can be partitioned into a block of maternal signals and a block of fetal signals, and there exists a corresponding partitioning for $\mathbf{A}$:

$$\mathbf{s}(t) = \begin{bmatrix} \mathbf{s}_M(t) \\ \mathbf{s}_F(t) \end{bmatrix}, \quad \mathbf{A} = [\mathbf{A}_M\, \mathbf{A}_F] \tag{7}$$

Thus:

$$\mathbf{v}(t) = \mathbf{A}_M\, \mathbf{s}_M(t) + \mathbf{A}_F\, \mathbf{s}_F(t) + \mathbf{n}(t) \tag{8}$$

The fetal electrocardiogram contributions to the measurement signals can then be obtained by:

$$\mathbf{v}_F(t) = \mathbf{A}_F\, \mathbf{s}_F(t) \tag{9}$$

Observe that (9) allows the estimation of the fetal electrocardiogram contributions to *all* leads. Similarly, the mothers' own ECG is given by:

$$\mathbf{v}_M(t) = \mathbf{A}_M\, \mathbf{s}_M(t) \tag{10}$$

Note that $\mathbf{v}_M(t)$ belongs to the column space of $\mathbf{A}_M$, which is usually renamed as the *mECG subspace*. Similarly, the column space of $\mathbf{A}_F$ will be denoted as the *fECG subspace*. Recalling again the discussion in the previous Section, the mECG space can be assumed a three dimensional vector space. However, the dimension of the fECG space is not necessarily equal to three (three is its maximum value) and is subject to changes during the pregnancy (Sameni, Clifford, Jutten & Shamsollahi, 2007).

[6] According to the model, there are, at most, six cardiac bioelectric sources.

The main assumption of the ICA model, the independence between sources, leads to some confusion. Even though this assumption is usually adopted (De Lathauwer et al., 2000a), there is no evidence to support it. The source signals can be actually partitioned into groups (a maternal group and a fetal group); components from different groups are statistically independent (i.e., there is a lack of dependency between coincident source activations), but components of the same group may be dependent. This is the concept of Multidimensional Independent Component Analysis (MICA), which is closely related to Independent Subspace Analysis (ISA) (Comon & Jutten, 2010; Hyvärinen et al., 2001). The idea of MICA, originally proposed in Cardoso (1998), has been further developed in Theis (2004) and Vollgraf & Obermayer (2001) among others. MICA actually proceeds in two steps (Cardoso, 1998): firstly, it runs a traditional ICA algorithm[7] and, then, it is determined which outputs of the algorithm are really independent and which should be grouped together (though the latter may not be trivial). Once we have recovered the source signals, we may use them to estimate (9), i.e. the fECG part of the composite signal, depending on the desired application.

4.1 Electrode placement

The number of electrodes and the positions at which these should be placed is not standardized. The situation is complex due to the fact that the fetal heart position with respect to the maternal abdomen varies with time and cannot be easily determined. Nevertheless, for late pregnancies, it has been observed that the fECG morphology is almost independent of electrode position (Lewis, 2003). A large number of electrodes (more than 30) arranged in a wide belt around the mother's abdomen, also containing some electrodes at the back, has been used in laboratory experiments (Cicinelli et al., 1994; Oostendorp, 1989; Vrins et al., 2004).

5. Algorithms

To the best of our knowledge, De Lathauwer *et al* were the first investigators to show that the application of MICA (Cardoso, 1998), or ISA (Comon & Jutten, 2010; Hyvärinen et al., 2001), to $v_1(t),\ldots,v_N(t)$ succeeds in the extraction of the fECG (De Lathauwer et al., 1995; 2000a). This observation has been subsequently confirmed by many other researchers (Clifford et al., 2011; Sameni et al., 2006; Zarzoso & Nandi, 2001), even in twin and triplet pregnancies (Keralapura et al., 2011; Taylor et al., 2003; 2005). In the literature, we have found numerous general-purpose ICA algorithms that solve the fECG extraction problem. For example, they include the *contrast maximization* (CoM2) method (Comon, 1994), JADE (Cardoso & Souloumiac, 1993), INFOMAX (Bell & Sejnowski, 1995), FastICA (Hyvärinen, 1999), Barros' method (Barros & Cichocki, 2001; Li & Yi, 2008), SOBI (Belouchrani et al., 1997), Pearson-ICA (Karvanen et al., 2000) or MERMAID (Marossero et al., 2003). ICA has been also used in combination with Wavelet transforms (Azzerboni et al., 2005; Vigneron et al., 2003), singular value decompositions (Gao et al., 2003) and neural networks (Yu & Chou, 2008), to cite some few examples. For a review of non-ICA based approaches, see, e.g. (Hasan et al., 2009). Naturally we cannot cover all the existing methods. Instead of surveying superficially several of them, we shall concentrate on some conceptually appealing aspects, some of which are not generally found in the literature.

[7] Some ICA algorithms output signals that are solutions to the MICA problem in the two-step approach described above.

5.1 Subspace analysis (whitening)

Whitening is the classical pre-processing for ICA and it is surely well-known to most of the readers of this book (otherwise, see, e.g., Comon & Jutten (2010)). For this reason, we offer an alternative viewpoint and present here whitening as a classical technique of subspace analysis. The idea is to use whitening to estimate the mECG subspace (or, more precisely, to estimate its orthogonal complement[8]). Then, the mECG can be easily projected out of the dataset. This approach (and its variants) has been conveniently addressed in several papers (see e.g. Callaerts et al. (1990); De Lathauwer et al. (2000b); Kanjilal et al. (1997)). We shall try to focus on the most relevant aspects: consider that we are given q samples $\mathbf{v}(1), \ldots, \mathbf{v}(q)$ of the vector signal $\mathbf{v}(t)$. In order to get rid of the maternal electrocardiogram interference, the eigenvalue decomposition of the data covariance matrix:

$$\mathbf{R}_v = \frac{1}{q} \sum_{t=1}^{q} \mathbf{v}(t)\, \mathbf{v}^T(t)$$

is first computed. Since $\mathbf{R}_v$ is always symmetric and nonnegative definite, it can be factorized into $\mathbf{R}_v = \mathbf{Q}\,\mathbf{D}\,\mathbf{Q}^T$, where:

$$\mathbf{D} = \text{diag}\left(\lambda_1 \geq \lambda_2 \geq \cdots \geq \lambda_p\right)$$

is the $p \times p$ diagonal matrix whose elements are the eigenvalues of $\mathbf{R}_v$ and $\mathbf{Q}$ is the matrix containing the corresponding eigenvectors. If the maternal electrocardiogram is strong enough, it has been shown that the M largest eigenvalues in $\mathbf{D}$ are associated with it. Furthermore, it holds that the eigenvalues have usually the following typical relationship:

$$\lambda_1 \geq \lambda_2 \geq \cdots \geq \lambda_n > \lambda_{n+1} \approx \cdots \approx \lambda_p.$$

This means that the last $(p - n)$ minor eigenvalues correspond to the noise. Matrices $\mathbf{D}$ and $\mathbf{Q}$ can be then partitioned into three groups:

$$\mathbf{D} = \begin{pmatrix} \mathbf{D}_1 & \mathbf{0} & \mathbf{0} \\ \mathbf{0} & \mathbf{D}_2 & \mathbf{0} \\ \mathbf{0} & \mathbf{0} & \mathbf{D}_3 \end{pmatrix} \qquad \mathbf{Q} = \begin{pmatrix} \mathbf{Q}_1 & \mathbf{Q}_2 & \mathbf{Q}_3 \end{pmatrix}$$

where $\mathbf{D}_1$ contains those M largest eigenvalues, and the columns of $\mathbf{Q}_1$ are the corresponding eigenvectors; $\mathbf{D}_2 = \text{diag}\left(\lambda_{M+1} \cdots \lambda_n\right)$ and $\mathbf{Q}_2$ contains the associated eigenvectors, *et cetera*. The maternal electrocardiogram can be then eliminated by projecting the data onto the subspace spanned by $\mathbf{Q}_2$. Specifically, this can be written as:

$$\mathbf{z}(t) = \mathbf{Q}_2^T\, \mathbf{v}(t) \tag{11}$$

where $\mathbf{z}(t)$ is the $(p - M) \times 1$ vector that, in theory, contains *no* maternal contribution, making the identification of the fetal electrocardiogram a feasible task, even by simple inspection[9]. Of course, the determination of M is an important problem. Seminal works considered $M = 3$; however, it has been recently argued that from $M = 4$ to $M = 6$ may be required in some cases. In practice, experiments suggest finding M empirically from the gap between the eigenvalues

[8] The orthogonal complement is the set of all vectors that are orthogonal to the vectors in the mECG space.

[9] In fact, under further hypotheses, it holds that $\mathbf{s}_F(t) = \mathbf{D}_2^{-1/2}\, \mathbf{Q}_2^T\, \mathbf{v}(t)$ and $\mathbf{A}_F = \mathbf{Q}_2\, \mathbf{D}_2^{1/2}$.

of the data covariance matrix. The complete procedure can be accomplished in real time with low computational cost. In any case, the performance of the whitening-based approaches is strongly dependent on the position of the electrodes (Callaerts et al., 1990), which usually becomes a matter of trial and error.

5.2 πCA

The combination of the measured signals $v_i(t)$ to enhance the periodic structure of the fECG also seems to be a promising idea. The algorithm should combine in power (constructive interference) the fetal components and cancel each other out (destructive interference). The best-known approach is to seek for the linear combination $y(t) = \sum_i w_i\, v_i(t) = \mathbf{w}^T\,\mathbf{v}(t)$ that minimizes the following measure of periodicity:

$$\varepsilon(\mathbf{w},\tau) = \frac{\sum_t |y(t+\tau) - y(t)|^2}{\sum_t y^2(t)}, \tag{12}$$

where the time-lag τ is the period of interest (in theory, but not always in practice, $\tau =$ the fetal period —see below). This approach has been named *Periodic Component Analysis* (πCA), and was first used for representing periodic structure in speech (Saul & Allen, 2001). The application of πCA to the fECG extraction problem can be traced back to the paper (Sameni et al., 2008). The minimization of (12) can be easily accomplished in a linear algebra framework. Expanding the right-hand side of (12) gives:

$$\begin{aligned}\varepsilon(\mathbf{w},\tau) &= \frac{\sum_t y^2(t+\tau) + y^2(t) - 2\,y(t+\tau)\,y(t)}{\sum_t y^2(t)}\\ &= \frac{\sum_t \mathbf{w}^T\,\mathbf{v}(t+\tau)\,\mathbf{v}^T(t+\tau)\,\mathbf{w} + \sum_t \mathbf{w}^T\,\mathbf{v}(t)\,\mathbf{v}^T(t)\,\mathbf{w} - 2\sum_t \mathbf{w}^T\,\mathbf{v}(t+\tau)\,\mathbf{v}^T(t)\,\mathbf{w}}{\sum_t \mathbf{w}^T\,\mathbf{v}(t)\,\mathbf{v}^T(t)\,\mathbf{w}}\\ &= 2\left[1 - \frac{\mathbf{w}^T\,\mathbf{C}_v(\tau)\,\mathbf{w}}{\mathbf{w}^T\,\mathbf{C}_v(0)\,\mathbf{w}}\right],\end{aligned} \tag{13}$$

where $\mathbf{C}_v(\tau)$ is the sample covariance matrix defined by:

$$\mathbf{C}_v(\tau) = \frac{1}{q}\sum_t \mathbf{v}(t+\tau)\,\mathbf{v}(t)^T. \tag{14}$$

Now consider the whitened data:

$$\mathbf{z}(t) = \mathbf{D}_1^{-1/2}\,\mathbf{U}_1^T\,\mathbf{v}(t) \tag{15}$$

where $\mathbf{D}_1$ and $\mathbf{U}_1$ are the respective eigenvalue and eigenvector matrices of $\mathbf{C}_v(0)$, i.e.,

$$\mathbf{C}_v(0) = \mathbf{U}_1\,\mathbf{D}_1\,\mathbf{U}_1^T$$

Then we have:

$$\begin{aligned}\mathbf{C}_z(0) &= \mathbf{D}_1^{-1/2}\,\mathbf{U}_1^T\,\mathbf{C}_v(0)\,\mathbf{U}_1\,\mathbf{D}_1^{-1/2} = \mathbf{I}\\ \mathbf{C}_z(\tau) &= \mathbf{D}_1^{-1/2}\,\mathbf{U}_1^T\,\mathbf{C}_v(\tau)\,\mathbf{U}_1\,\mathbf{D}_1^{-1/2}\end{aligned} \tag{16}$$

where $\mathbf{C}_z(\tau) = \frac{1}{q}\sum_t \mathbf{z}(t+\tau)\,\mathbf{z}(t)^T$. Let us define:

$$\underline{\mathbf{w}} = \mathbf{D}_1^{1/2}\,\mathbf{U}_1^T\,\mathbf{w} \tag{17}$$

With this new vector, (13) can be rewritten as follows:

$$\varepsilon(\mathbf{w},\tau) = 2\left[1 - \frac{\underline{\mathbf{w}}^T\,\mathbf{C}_z(\tau)\,\underline{\mathbf{w}}}{\underline{\mathbf{w}}^T\,\mathbf{C}_z(0)\,\underline{\mathbf{w}}}\right] = 2\left[1 - \frac{\underline{\mathbf{w}}^T\,\mathbf{C}_z(\tau)\,\underline{\mathbf{w}}}{\underline{\mathbf{w}}^T\,\underline{\mathbf{w}}}\right] \tag{18}$$

Then:

Proposition 1. *By the Rayleigh-Ritz theorem of linear algebra (Bai et al., 2000), the vector weight* $\underline{\mathbf{w}}$ *minimizing* (18) *is given by the eigenvector of the matrix* $\mathbf{C}_z(\tau)$ *with the largest eigenvalue.*

Denoting this eigenvector by $\underline{\mathbf{w}}_{max}$, πCA then outputs:

$$\begin{aligned} y(t) &= \underline{\mathbf{w}}_{max}^T\,\mathbf{z}(t) \\ &= \underline{\mathbf{w}}_{max}^T\,\mathbf{D}_1^{-1/2}\,\mathbf{U}_1^T\,\mathbf{v}(t) \\ &= \mathbf{w}_{max}^T\,\mathbf{v}(t) \end{aligned} \tag{19}$$

with $\mathbf{w}_{max}^T \stackrel{def}{=} \underline{\mathbf{w}}_{max}^T\,\mathbf{D}_1^{-1/2}\,\mathbf{U}_1^T$.

It is interesting to note that πCA is actually a particularization of the well-known *Algorithm for Multiple Unknown Signals Extraction* (AMUSE) (Tong et al., 1991): by assuming $\mathbf{D}_2$ as the full diagonal eigenvalue matrix of $\mathbf{C}_z(\tau)$, with eigenvalues sorted in descending order, and $\mathbf{U}_2$ being the corresponding eigenvector matrix[10], one can write the eigendecomposition:

$$\mathbf{C}_z(\tau) = \mathbf{U}_2\,\mathbf{D}_2\,\mathbf{U}_2^T.$$

Then, $\mathbf{y}(t) = \mathbf{U}_2^T\,\mathbf{z}(t)$ verifies:

$$\begin{aligned} \mathbf{C}_y(0) &= \mathbf{U}_2^T\,\mathbf{C}_z(0)\,\mathbf{U}_2 = \mathbf{U}_2^T\,\mathbf{U}_2 = \mathbf{I} \\ \mathbf{C}_y(\tau) &= \mathbf{U}_2^T\,\mathbf{C}_z(\tau)\,\mathbf{U}_2 = \mathbf{D}_2 \end{aligned} \tag{20}$$

with $\mathbf{C}_y(\tau) = \frac{1}{q}\sum_t \mathbf{y}(t+\tau)\,\mathbf{y}(t)^T$. Taking eqn. (16) into eqn. (20), we get that both matrices $\mathbf{C}_v(0)$ and $\mathbf{C}_v(\tau)$ are simultaneously diagonalized by matrix $\mathbf{Q} = \mathbf{U}_2^T\,\mathbf{D}_1^{-1/2}\,\mathbf{U}_1^T$:

$$\begin{aligned} \mathbf{Q}\,\mathbf{C}_v(0)\,\mathbf{Q}^T &= \mathbf{I} \\ \mathbf{Q}\,\mathbf{C}_v(\tau)\,\mathbf{Q}^T &= \mathbf{D}_2 \end{aligned} \tag{21}$$

As can be easily verified, this implies that:

$$\begin{aligned} \mathbf{C}_v(\tau)\,\mathbf{Q}^T &= \mathbf{C}_v(0)\,\mathbf{Q}^T\,\mathbf{D}_2 \\ \mathbf{C}_v^{-1}(0)\,\mathbf{C}_v(\tau)\,\mathbf{Q}^T &= \mathbf{Q}^T\,\mathbf{D}_2 \end{aligned} \tag{22}$$

i.e. $\mathbf{D}_2$ and $\mathbf{Q}^T$ are the eigenvalues and eigenvectors, respectively, of the matrix $\mathbf{C}_v^{-1}(0)\,\mathbf{C}_v(\tau)$. Then, $\mathbf{Q}$ can be identified by the simultaneous diagonalization of $\mathbf{C}_v(0)$ and $\mathbf{C}_v(\tau)$. This is the basic idea behind AMUSE.

[10] $\underline{\mathbf{w}}_{max}$ is the first column of $\mathbf{U}_2$ and so on.

Proposition 2. *Let $\mathbf{v}(t) = \mathbf{A}\,\mathbf{s}(t)$, where $\mathbf{A}$ is of full column rank, and the sources $s_i(t)$ are zero-mean WSS processes uncorrelated with each other. Let us choose any time delay τ for which $\mathbf{C}_v^{-1}(0)\,\mathbf{C}_v(\tau)$ has non-zero distinct eigenvalues. Then $\mathbf{y}(t) = \mathbf{Q}\,\mathbf{v}(t)$ is an estimate of the source signals except for the usual scaling and ordering indetermination.*

This proposition readily follows from (Tong et al., 1991). For virtually any time-lag $\tau \neq 0$, AMUSE is able to output the (fetal and maternal) source signals. In addition, the requirement of the sources being mutually uncorrelated is much weaker than the classical ICA condition of mutual independence[11]. The transformation $\mathbf{y}(t) = \mathbf{Q}\,\mathbf{v}(t) = [y_1(t), \ldots, y_p(t)]^T$ can be also interpreted as follows: $y_1(t)$ is the most periodic component *with respect to the period of interest* τ, $y_p(t)$ is the least periodic, and the intermediate components are ranked in descending order of periodicity (Sameni et al., 2008). Of course, this does not mean that, in particular, $y_1(t)$ is (even approximately) periodic with period τ, neither does it mean that $y_1(t)$ has physical meaning. Both questions depend on the specific choice of τ.

5.2.1 Period calculation

No specific strategy for selecting τ was provided in (Tong et al., 1991). A natural approach in our context is to set τ to the value of the fetal heart beat period (which, for simplicity, is assumed to be an integer multiple of the sampling interval). However, such an approach is difficult to implement in practice, since the fetal heart beat period has to be estimated on-line, which requires the prior extraction of the fetal R peaks.

As an alternative, (Sameni et al., 2008) reports good results when τ is chosen as the maternal ECG period. In this way, the most periodic components span the mECG subspace. In addition, see (20), the periodic components $y_1(t), \ldots, y_p(t)$ happen to be uncorrelated with each other. Hence, the space spanned by the less periodic components is orthogonal to the mECG subspace. It follows that this method is similar in spirit to the whitening-based approaches described in the previous Section.

A more challenging problem arises from the fact that the heart beat period is actually *time-dependent*. Hence, the period has to be updated on a beat-to-beat basis (see Sameni et al. (2008) for a possible solution).

5.2.2 Extensions

AMUSE (and, subsequently, πCA) suffers from the limitation that the choice of τ is very critical. To overcome this drawback, one powerful approach is to perform the simultaneous diagonalization of more covariance matrices than just two, as is the case with AMUSE. For example, SOBI (Belouchrani et al., 1997) seeks the matrix $\mathbf{Q}$ as the joint diagonalizer of a set of covariance matrix $\mathbf{C}_v(\tau_i)$ for a preselected set of time-lags $\{\tau_1, \tau_2, \tau_3, \ldots\}$. Some steps to investigate the optimal choice of $\{\tau_1, \tau_2, \tau_3, \ldots\}$ in context of the fECG extraction problem have been done by (Tsalaile et al., 2009).

5.3 HOS-based approach

It is well-known that, implicitly or explicitly, most ICA methods actually rely on higher-order statistics (HOS) (Comon & Jutten, 2010). Let us briefly review one of the simplest approaches:

[11] Correlation can be always removed by an orthogonal transformation, i.e., a change of basis in 3D space.

the maximization of the kurtosis (Hyvärinen et al., 2001). Let $\mathbf{z}(1), \ldots, \mathbf{z}(q)$ be the whitened data. Given an arbitrary vector $\mathbf{w}$, it follows from the central limit theorem that

$$y(t) = \mathbf{w}^T \mathbf{z}(t) \tag{23}$$

is more Gaussian when it is a sum of the fECG and the interferences than when it is equal to only one of them[12]. In consequence, to find $\mathbf{w}$ in such a way that the distribution of $y(t)$ is as far as possible from Gaussian seems to be a sound idea. This general approach to the problem of 'unmixing' mixed signals very common in ICA and is usually referred to as *maximisation of non-Gaussianity* (Hyvärinen et al., 2001). The simplest measure of non-Gaussianity is the kurtosis, defined by:

$$\kappa_y = \frac{1}{q} \sum_{t=1}^{q} y^4(t) - \frac{3}{q} \sum_{t=1}^{q} y^2(t) \tag{24}$$

We maximise the kurtosis of $y(t)$ under the unit-power constraint

$$\frac{1}{q} \sum_{t=1}^{q} y^2(t) = 1 \tag{25}$$

which avoids the solution $y(t) \to \infty$. It is easily shown that this is equivalent to constraining the norm of $\mathbf{w}$ to be the unity. Traditional ICA algorithms, such as FastICA (Hyvärinen, 1999), maximize the kurtosis using standard procedures. As an alternative, we review here the FFD (Fast Fetal Detection) method (Martín-Clemente et al., 2011) which, paradoxically, does not require to compute HOS. Consider first the following theorem, whose proof is straightforward:

Theorem 1. *Let $\{x(t), t = 1, \ldots, q\}$ be the samples of a generic discrete-time signal. The kurtosis of $x(t)$, defined by*

$$\kappa_x = \frac{1}{q} \sum_{t=1}^{q} x^4(t) - \frac{3}{q} \sum_{t=1}^{q} x^2(t)$$

is maximized under the unit-power constraint $\frac{1}{q} \sum_{t=1}^{q} x^2(t) = 1$ by signals of the form

$$x_*(t) = \pm\sqrt{q}\, e_k(t)$$

where $e_k(t)$ is a discrete-time signal that equals one at $t = k$ and is zero elsewhere.

To explore the vicinity of the maximum $\sqrt{q}\, e_k(t)$, where $k \in \{1, \ldots, q\}$, we perform a first order Taylor expansion of the kurtosis around this point (see Martín-Clemente et al. (2011) for the details):

$$\kappa_y \approx q - 3 - 2 \sum_{t=1}^{q} \left(y(t) - \sqrt{q}\, e_k(t)\right)^2 \tag{26}$$

Hence κ_y is maximized when

$$\sum_{t=1}^{q} \left(y(t) - \sqrt{q}\, e_k(t)\right)^2 \tag{27}$$

is minimum: i.e., the optimum $y(t)$ is the signal that is as close as possible to $\sqrt{q}\, e_k(t)$. To determine the best value for the time index k, note that the accuracy of (26) increases as (27)

[12] The fECG can be assumed independent from the others as it has a different physical origin.

decreases. Consequently, we minimize (27) among all possible values of k. Taking into account that $y(t) = \mathbf{w}^T \mathbf{z}(t)$, a bit of algebra shows that the minimum is obtained simply by setting

$$\mathbf{w}_* = \frac{\mathbf{z}(K)}{\|\mathbf{z}(K)\|}, \text{ where } K = \underset{k}{\operatorname{argmax}} \|\mathbf{z}(k)\| \tag{28}$$

Consider the following additional interpretation: by construction, $y(t)$ is the signal that is as close as possible to the impulse signal $\sqrt{q}\, e_K(t)$. If $\mathbf{z}(t)$ is periodic, one can prove easily that $y(t)$ is also the best approximation to an impulse train having the same period and centered upon $t = K$. The ECG resembles an impulse train, but the interferences degrade the measurements. The algorithm restores this property and, as result, restores the signal itself. The method may be then considered as a particular application of the class of waveform-preserving methods for recovering ECG signals.

Finally, to extract sequentially more signals, we can use the procedure described in Chapter 4 of (Cichocki & Amari, 2002). Basically, we remove $y(t)$ from the mixture by $\mathbf{z}'(t) = \mathbf{z}(t) - \mathbf{w}\, y(t)$. Then whitening is applied again to reduce the dimensionality in one unit. The algorithm is repeated until all the desired signals are recovered.

6. The mECG as reference

Incorporating prior information into ICA may reduce the computational cost while improving the performance of the algorithms. The use of a reference signal has been proposed in Adib & Aboutajdine (2005), using an approach similar to that in Martín-Clemente et al. (2004). To get such a reference, ICA is often applied to data sets that include mECG signals taken at the mother's thoracic region. In this Section, we describe the variant proposed in Camargo-Olivares et al. (2011). The architecture of the proposed system is shown in Figure 2 and each block is described separately next:

1. Pre-processing block: it aims to remove the baseline wander, the electromyografic (EMG) noise, and the power line interference from each signal $v_i(t)$. This is usual in most contemporary ECG processing systems.
2. MECG recording: in most previous approaches, the mECG is measured at the chest of the mother. By contrast, in this paper we propose recovering the mECG directly from the mother's abdomen. We face a problem of recovering a signal (the mECG) corrupted by 'noise' (the fECG and others) at, fortunately, a very high signal-to-noise ratio. A state-of-the-art solution is that proposed in Sameni, Shamsollahi, Jutten & Clifford (2007). This filter actually *generates* a *synthetic* mECG whose morphology and parameters (R-R interval and so on) are calculated from the filter input. The proposed procedure is hence as follows: 1.) Filter each signal taken at the mother's abdomen by the filter described in Sameni, Shamsollahi, Jutten & Clifford (2007). 2.) Perform a linear mapping of the filter outputs to a lower dimensional space using whitening to reduce the number of mECG signals under consideration.
3. ICA block: the inputs to ICA are the pre-processed abdominal maternal signals and the mECG estimates (outputs of block 2).
4. Post-processing block: (optional) the fECG is filtered again with the filter described in Sameni, Shamsollahi, Jutten & Clifford (2007) to improve the final signal to noise ratio.

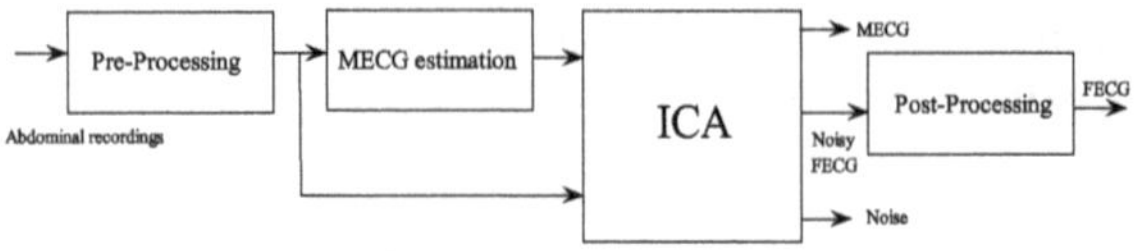

Fig. 2. Block diagram of the proposed system.

7. Examples

7.1 First example

Eight real cutaneous potential recordings of a pregnant woman were obtained from the Database for the Identification of Systems (DaISy)[13]. The data, see Fig. 3, consist of eight channels of ECG signals: the first five channels correspond to electrodes placed on the woman's abdominal region. The last three signals correspond to electrodes located on the mother's region. For many years, these recordings have been extensively used as the standard test data of fECG extraction algorithms (e.g., see Zarzoso & Nandi (2001)). Even thought the

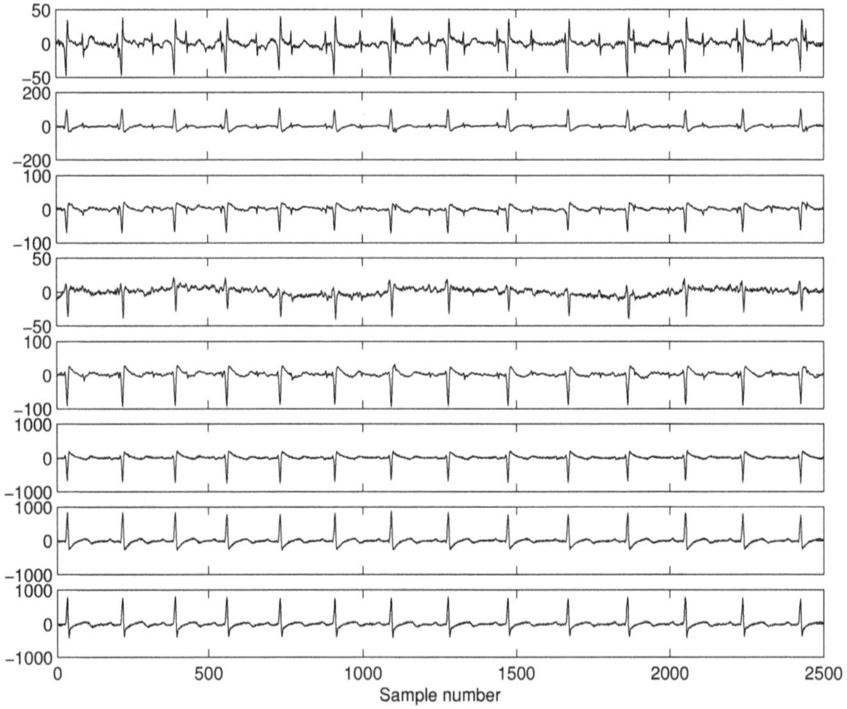

Fig. 3. Cutaneous electrode recordings from a pregnant woman.

[13] ftp://ftp.esat.kuleuven.be/pub/SISTA/data/biomedical/

fECG is much weaker than the mECG, it is slightly visible in the abdominal recordings. We applied the following ICA algorithms to these data: JADE (Cardoso & Souloumiac, 1993), FastICA (Hyvärinen, 1999), FFD (Martín-Clemente et al., 2011), SOBI (Belouchrani et al., 1997) and πCA (Sameni et al., 2008). Apart from whitening, no other pre-processing is used.

All the algorithms succeeded in estimating two fetal source signals. Fig. 4 shows the first one of them, as recovered by each algorithm. All methods produced very similar results. Note that the R wave is perfectly visible, allowing the easy calculation of the fetal heart rate.

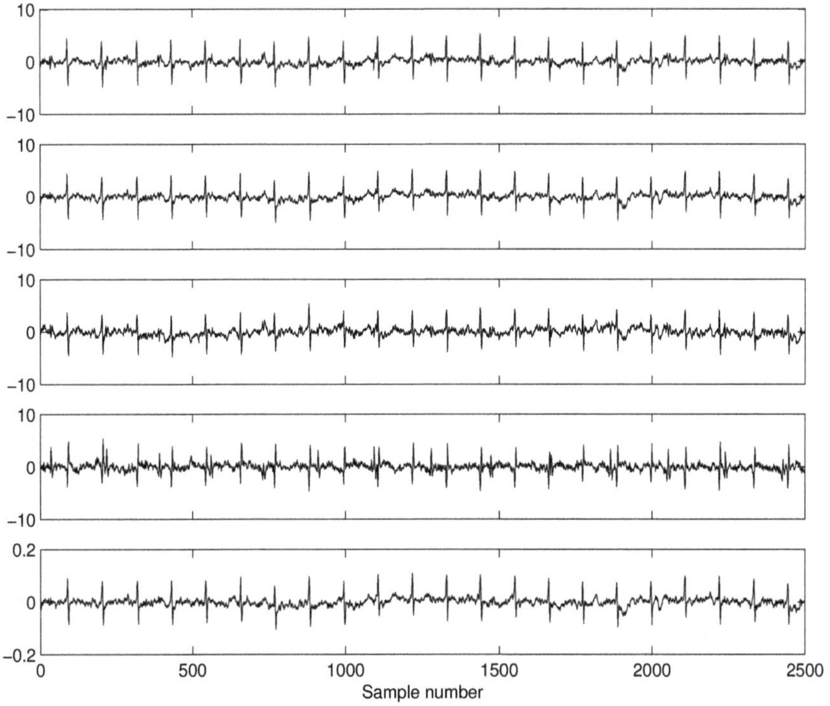

Fig. 4. Fetal source signals detected from the recordings of Fig. 3 by, from the top to the bottom, JADE, FastICA, FFD , SOBI and πCA.

7.2 Second example

The methods are now tested using experimental data from the Non-invasive Fetal Electrocardiogram database[14]. This public database contains a series of 55 multichannel thoracic and abdominal non-invasive recordings, taken from a single pregnant woman between 21 and 40 weeks of pregnancy. The ones used in this experiment correspond to the 21 week of gestation and are shown in Fig. 5. The first two signals from the top correspond to electrodes located on the mother's thoracic region, and the last three signals correspond to

[14] http://physionet.org/pn3/nifecgdb/

electrodes located on the woman's abdomen. The recordings have been pre-processed: the baseline was eliminated using a low-pass filter with cutoff frequency equal to 0.7 Hz, and the powerline interference was attenuated using a notch filter.

Fig. 6 shows the source signals estimated by the same ICA algorithms used in the previous example (JADE, FastICA, FFD, SOBI and πCA). Only the maternal source signals can be recognized. We must conclude that, even though ICA is generally reliable, it sometimes fail.

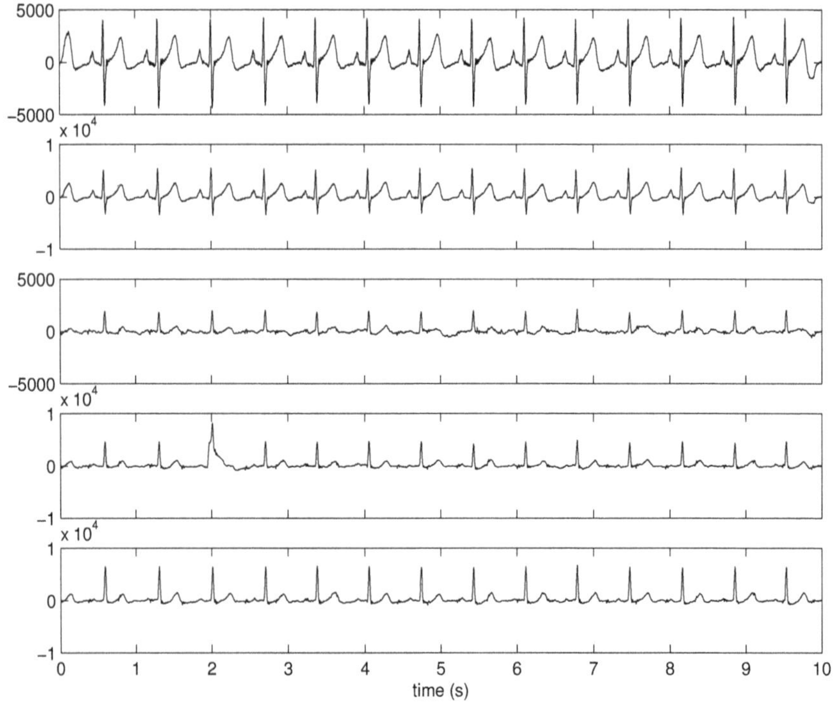

Fig. 5. Cutaneous electrode recordings from a pregnant woman in the 21 week of gestation.

7.3 Third example

We now repeat the previous (failed) experiment using the mECG as reference for the FFD methodd, as explained in Section 6. FFD has been chosen as representative of the ICA methods, but results are similar when any other of the algorithms is used. The estimated source signals are depicted in Fig. 7. Unlike in the previous experiment, the fECG is visible in the third plot from the top, and the fetal heart rate can be estimated even though the signal-to-noise ratio is low. Further denoising may necessary using other techniques –see, e.g., Sameni, Shamsollahi, Jutten & Clifford (2007); Vigneron et al. (2003)– but this is beyond the scope of the present Chapter.

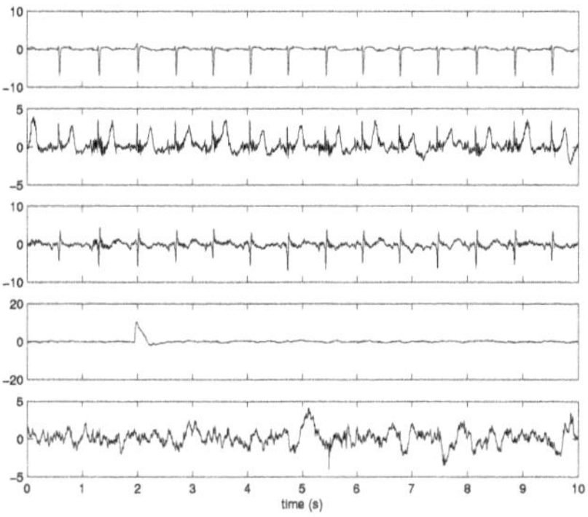

(a) JADE

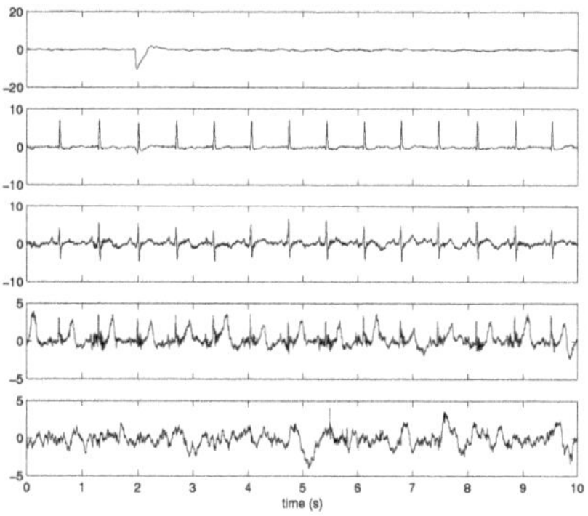

(b) FastICA

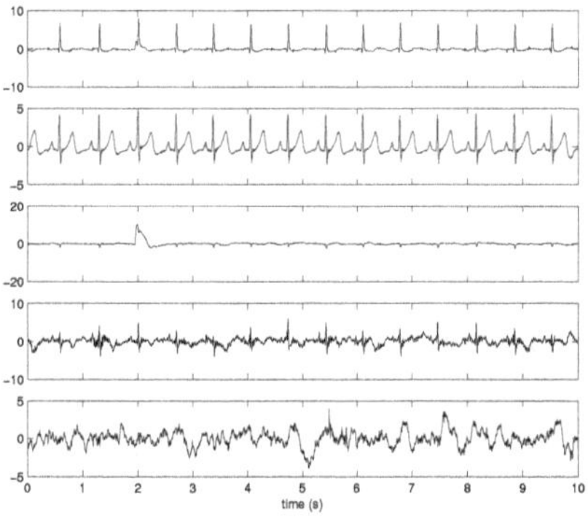

(c) FFD

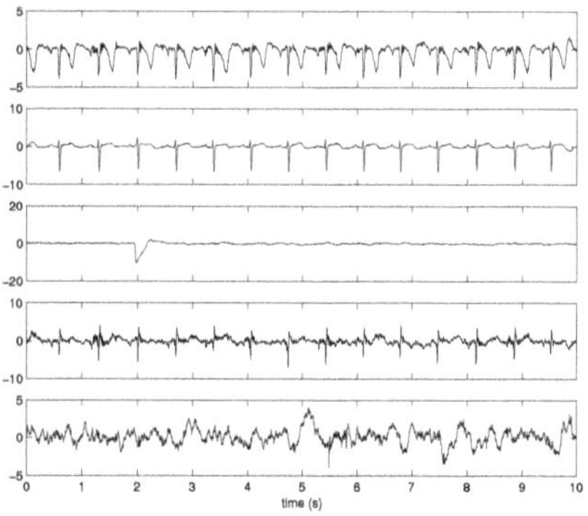

(d) SOBI

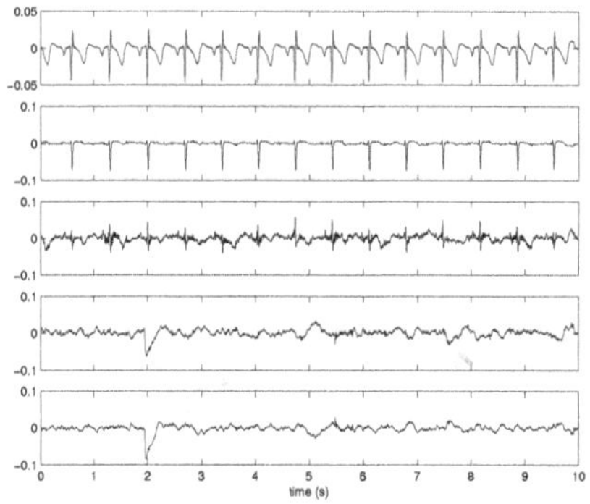

(e) πCA

Fig. 6. Source signals estimated by the different algorithms from the recordings of Fig. 5.

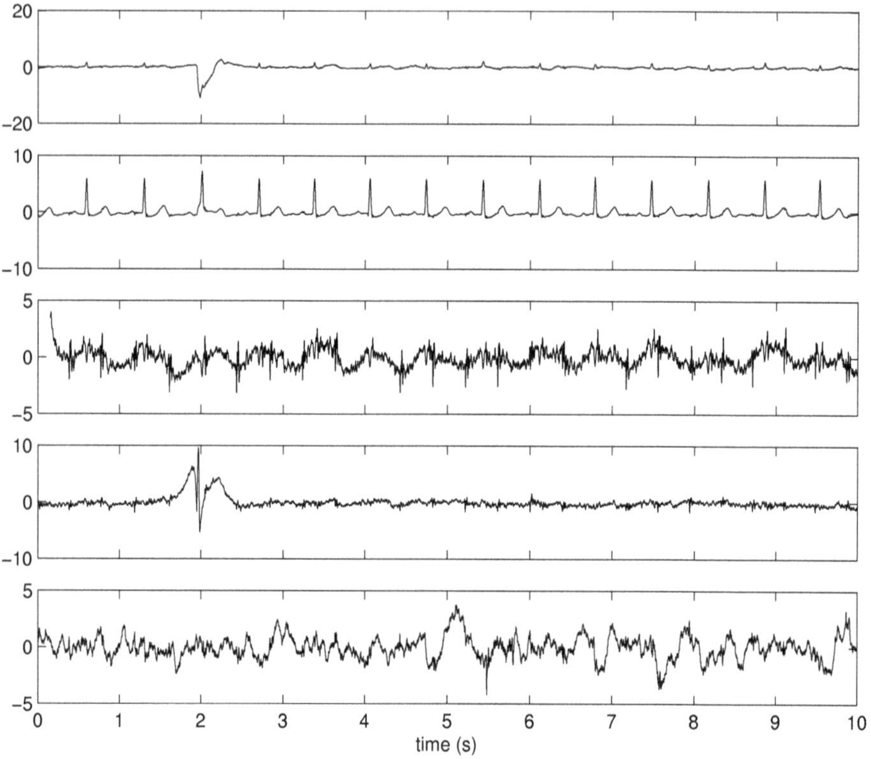

Fig. 7. Source signals detected from the recordings of Fig. 5 by using the mECG as reference.

8. Conclusions

This Chapter has presented a review of the state-of-the-art in the use of ICA for the fECG detection problem. A significant improvement in technical support for fetal monitoring has been obtained in the last decades. Compared to alternative techniques (e.g., filtering, average beat subtraction ...), ICA has proven to be a powerful and leading-edge approach. The most remarkable feature of higher-order ICA methods is that they do not seem to be very sensitive to the location of the electrodes. However, it should be pointed out that, even though promising results have been obtained (the fetal heart rate can be almost routinely determined), there is at present a total lack of accuracy in the detection of the smallest waves (P, Q, S and T) of the fECG. Though it is true that in current clinical practice the physician only considers the fetal cardiac rate, further research is therefore needed to improve accuracy of wave detection. The use of prior information (e.g., reference signals, or the knowledge about the fECG waveform) may be the strategy to achieve this goal. The physical interpretation of the estimated source signals also seems to be an exciting field for future work, and the independence of the sources need to be elucidated.

9. Acknowledgement

This work was supported by a grant from the "Junta de Andalucía" (Spain) with reference P07-TIC-02865.

10. References

Abboud, S. & Sadeh, D. (1989). Spectral analysis of the fetal electrocardiogram, *Computers in biology and medicine* 19(6): 409–415.

Abuhamad, A. & Chaoui, R. (2009). *A Practical Guide to Fetal Echocardiography: Normal and Abnormal Hearts*, Lippincott Williams & Wilkins.

Adib, A. & Aboutajdine, D. (2005). Reference-based blind source separation using a deflation approach, *Signal Processing* 85: 1943 – 1949.

Afriat, C. & Kopel, E. (2008). *Electronic Fetal Monitoring*, 2nd. edn, Lippincott Williams & Wilkins.

Azzerboni, B., La Foresta, F., Mammone, N. & Morabito, F. (2005). A new approach based on wavelet-ica algorithms for fetal electrocardiogram extraction, *Proceeding of European simposium on Artificial Neural Networks* .

Bai, Z., Demmel, J., Dongarra, J., Ruhe, A. & van der Vorst, H. (eds) (2000). *Templates for the Solution of Algebraic Eigenvalue Problems: A Practical Guide.*, SIAM.

Barros, A. & Cichocki, A. (2001). Extraction of specific signals with temporal structure, *Neural Computation* 13(9): 1995–2003.

Bell, A. J. & Sejnowski, T. (1995). An information-maximization approach to blind separation and blind deconvolution, *Neural Computation* 7: 1129–1159.

Belouchrani, A., Abed-Meraim, K., Cardoso, J. & Moulines, E. (1997). A blind source separation technique using second-order statistics, *Signal Processing, IEEE Transactions on* 45(2): 434–444.

Callaerts, D., Moor, B. D., Vandewalle, J. & Sansen, W. (1990). Comparison of svd methods to extract the foetal ecg from cutaneous electrode signals, *Medical & Biolog. Eng. & Computing* 28: 217–224.

Camargo-Olivares, J., Martín-Clemente, R., Hornillo, S., Elena-Pérez, M. & Román-Martínez, I. (2011). The maternal abdominal ecg as input to mica in the fetal ecg extraction problem., *IEEE Signal Processing Letters* 18(3): 161 –164.

Cardoso, J. F. (1998). Multidimensional independent component analysis, *Proc. ICASSP'98*, Seattle, pp. 1941–1944.

Cardoso, J. F. & Souloumiac, A. (1993). Blind beamforming for non gaussian signals, *IEE Proceedings - F* 140(6): 363 – 370.

Castells, F., Cebrián, A. & Millet, J. (2007). The role of independent component analysis in the signal processing of ecg recordings, *Biomedizinische Technik* 52(1): 18 – 24.

Chan, A. (2008). *Biomedical device technology: principles and design*, Charles C. Thomas.

Cichocki, A. & Amari, S.-I. (2002). *Adaptive blind signal and image processing: learning algorithms and applications*, Wiley.

Cicinelli, E., Bortone, A., Carbonara, I., Incampo, G., Bochicchio, M., Ventura, G., Montanaro, S. & Aloisio, G. (1994). Improved equipment for abdominal fetal electrocardiogram recording: description and clinical evaluation, *International journal of bio-medical computing* 35(3): 193 – 205.

Clifford, G., Sameni, R., Ward, J., Robinson, J. & Wolfberg, A. J. (2011). Clinically accurate fetal ecg parameters acquired from maternal abdominal sensors, *American Journal of Obstetrics and Gynecology* 205 (1): 47.e1–47.e5.

Clifford, G., Shoeb, A., McSharry, P. & Janz, B. (2005). Model-based filtering, compression and classification of the ecg, *Proceedings of Bioelectromagnetism and 5th International Symposium on Noninvasive Functional Source Imaging within the Human Brain and Heart (BEM & NFSI)*.

Comon, P. (1994). Independent component analysis, a new concept?, *Signal Processing (Special Issue Higher-Order Statistics)* 36 (3): 287–314.

Comon, P. & Jutten, C. (eds) (2010). *Handbook of blind source separation, independent componet analysis and applications*, Elsevier.

De Lathauwer, L., Callaerts, D., De Moor, B. D. & Vandewalle, J. (1995). Fetal electrocardiogram extraction by source subspace separation, *Proc. IEEE SP/ATHOS Workshop on HOS*, Girona, Spain, pp. 134–138.

De Lathauwer, L., De Moor, B. D. & Vandewalle, J. (2000a). Fetal electrocardiogram extraction by blind source subspace separation, *IEEE Transactions on Biomedical Engineering* 47 (5): 567–572.

De Lathauwer, L., De Moor, B. & Vandewalle, J. (2000b). Svd-based methodologies for fetal electrocardiogram extraction, *Proceedings of the 2000 IEEE Int. Conf. on Acoustics, Speech, and Signal Processing (ICASSP'00)*, Vol. 6, pp. 3771 – 3774.

Devedeux, D., Marque, C., Mansour, S., Germain, G., Duchêne, J. et al. (1993). Uterine electromyography: a critical review., *American journal of obstetrics and gynecology* 169(6): 1636 –1652.

Freeman, R. & Garite, T. (2003). *Fetal Heart Rate Monitoring*, Lippincott Williams & Wilkins.

Gao, P., Chang, E. & Wyse, L. (2003). Blind separation of fetal ecg from single mixture using svd and ica, *Information, Communications and Signal Processing, 2003 and the Fourth Pacific Rim Conference on Multimedia. Proceedings of the 2003 Joint Conference of the Fourth International Conference on*, Vol. 3, IEEE, pp. 1418–1422.

Guyton, A. & Hall, J. (1996). *Texbook of Medical Physiology*, W. B. Saunders Company.

Hasan, M., Reaz, M., Ibrahimy, M., Hussain, M. & Uddin, J. (2009). Detection and processing techniques of fecg signal for fetal monitoring, *Biological procedures online* 11(1): 263 – 295.

Hyvärinen, A. (1999). Fast and robust fixed-point algorithms for independent component analysis, *IEEE Transactions on Neural Networks* 10 (3): 626–634.

Hyvärinen, A., Karhunen, J. & Oja, E. (2001). *Independent Component Analysis*, John Wiley & Sons.

James, C. & Hesse, C. (2005). Independent component analysis for biomedical signals, *Physiological measurement* 26: R15.

Jenkins, H. M. L., Symonds, E. M., Kirk, D. L. & Smith, P. R. (2005). Can fetal electrocardiography improve the prediction of intraparturn fetal acidosis?, *BJOG: An International Journal of Obstetrics & Gynaecology* 93(1): 6–12.

Kanjilal, P., Palit, S. & Saha, G. (1997). Fetal ecg extraction from single-channel maternal ecg using singular value decomposition, *IEEE Tr. on Biomedical Engineering* 44 (1): 51 – 59.

Karvanen, J., Eriksson, J. & Koivunen, V. (2000). Pearson system based method for blind separation, *Proceedings of Second International Workshop on Independent Component Analysis and Blind Signal Separation (ICA2000), Helsinki, Finland*, pp. 585–590.

Keener, J. & Sneyd, J. (2009). *Mathematical Physiology*, Vol. 2, Springer.

Keralapura, M., Pourfathi, M. & Sirkeci-Mergen, B. (2011). Impact of contrast functions in fast-ica on twin ecg separation, *IAENG International Journal of Computer Science* 38:1.

Lewis, M. (2003). Review of electromagnetic source investigations of the fetal heart., *Medical Engineering and Physics* 25 (10): 801 – 810.

Li, Y. & Yi, Z. (2008). An algorithm for extracting fetal electrocardiogram, *Neurocomputing* 71(7 –9): 1538 – 1542.

Llinares, R. & Igual, J. (2009). Application of constrained independent component analysis algorithms in electrocardiogram arrhythmias, *Artificial Intelligence in Medicine* 47: 121–133.

Marossero, D., Erdogmus, D., Euliano, N., Principe, J., Hild, K. et al. (2003). Independent components analysis for fetal electrocardiogram extraction: a case for the data efficient mermaid algorithm, *Neural Networks for Signal Processing, 2003. NNSP'03. 2003 IEEE 13th Workshop on*, IEEE, pp. 399–408.

Martín-Clemente, R., Acha, J. & Puntonet, C. G. (2004). Eigendecomposition of self-tuned cumulant-matrices for blind source separation, *Signal Processing* 84(7): 1201 – 1211.

Martín-Clemente, R., Camargo-Olivares, J., Hornillo-Mellado, S., Elena, M. & Roman, I. (2011). Fast technique for noninvasive fetal ecg extraction, *IEEE Transactions on Biomedical Engineering* 58(2): 227 – 230.

McSharry, P., G., C., Tarassenko, L. & Smith, L. (2003). A dynamical model for generating synthetic electrocardiogram signals, *IEEE Tr. on Biomedical Engineering* 50 (3): 289–294.

Nait-Ali, A. (2009). *Advanced Biosignal Processing*, Springer Verlag.

Oldenburg, J. & Macklin, M. (1977). Changes in the conduction of the fetal electrocardiogram to the maternal abdominal surface during gestation., *American journal of obstetrics and gynecology* 129(4): 425 – 433.

Oostendorp, T. F. (1989). Lead systems for the abdominal fetal electrocardiogram, *Clinical Physics and Physiological Measurements* 10(21): 21 – 26.

Oostendorp, T. F., van Oosterom, A. & Jongsma, H. W. (1989a). The effect of changes in the conductive medium on the fetal ecg throughout gestation, *Clinical Physics and Physiological Measurements* 10 (B): 11 – 20.

Oostendorp, T. F., van Oosterom, A. & Jongsma, H. W. (1989b). The fetal ecg throughout the second half of gestation, *Clinical Physics and Physiological Measurements* 10 (2): 147 – 160.

Pardi, G., Ferrazzi, E., Cetin, I., Rampello, S., Baselli, G., Cerutti, S., Civardi, S. et al. (1986). The clinical relevance of the abdominal fetal electrocardiogram., *Journal of perinatal medicine* 14(6): 371 – 377.

Peters, M., Stinstra, J., Uzunbajakau, S. & Srinivasan, N. (2005). *Advances in electromagnetic fields in living systems*, Springer Verlag, chapter Fetal Magnetocardiography, pp. 1 – 40.

Rieta, J., Castells, F., Sánchez, C., Zarzoso, V. & Millet, J. (2004). Atrial activity extraction for atrial fibrillation analysis using blind source separation, *IEEE Transactions on Biomedical Engineering* 51(7): 1176 – 1186.

Sameni, R., Clifford, G., Jutten, C. & Shamsollahi, M. (2007). Multichannel ecg and noise modeling: Application to maternal and fetal ecg signals, *EURASIP Journal on Advances in Signal Processing* 2007.

Sameni, R., Jutten, C. & M. Shamsollahi, M. (2006). What ica provides for ecg processing:application to non-invasive fetal ecg extraction, *Proc. 2006 IEEE International Symposium on Signal Processing and Information Technology*.

Sameni, R., Jutten, C. & Shamsollahi, M. (2008). Multichannel electrocardiogram decomposition using periodic component analysis, *IEEE Transactions on Biomedical Engineering* 55 (8): 1935 – 1940.

Sameni, R., Shamsollahi, M., Jutten, C. & Clifford, G. (2007). A nonlinear bayesian filtering framework for ecg denoising, *IEEE Tr. on Biomedical Engineering* 54(12): 2172 – 2185.

Saul, L. K. & Allen, J. B. (2001). *Advances in Neural Information Processing Systems 13*, MIT Press: Cambridge, MA, chapter Periodic component analysis: an eigenvalue method for representing periodic structure in speech., pp. 807 – 813.

Symonds, E. M., Sahota, D. & Chang, A. (2001). *Fetal Electrocardiography*, Imperial College Press.

Tanskanen, J. & Viik, J. (2012). *Advances in Electrocardiograms - Methods and Analysis*, InTech, chapter Independent Component Analysis in ECG Signal Processing, pp. 349 – 372.

Taylor, M., Smith, M., Thomas, M., Green, A., Cheng, F., Oseku-Afful, S., Wee, L., Fisk, N. & Gardiner, H. (2003). Non-invasive fetal electrocardiography in singleton and multiple pregnancies, *BJOG: An International Journal of Obstetrics & Gynaecology* 110(7): 668 – 678.

Taylor, M., Thomas, M., Smith, M., Oseku, S., Fisk, N., Green, A., Paterson, S. & Gardiner, H. (2005). Non-invasive intrapartum fetal ecg: preliminary report., *Br. J. Obstet. Gynaecol.* 112: 1016–1021.

Theis, F. J. (2004). Uniqueness of complex and multidimensional independent component analysis., *Signal Processing* 84 (5): 951–956.

Tong, L., Liu, R., Soon, V. C. & Huang, Y.-F. (1991). Indeterminacy and identifiability of blind identification, *IEEE Transactions on Circuit and Systems* 38 (5): 499 – 509.

Tsalaile, T., Sameni, R., Sanei, S., Jutten, C. & Chambers, J. (2009). Sequential blind source extraction for quasi-periodic signals with time-varying period, *IEEE Tr. on Biomedical Engineering* 56(3): 654 – 655.

Vetter, R., Virag, N., Vesin, J., Celka, P. & Scherrer, U. (2000). Observer of autonomic cardiac outflow based on blind source separation of ecg parameters, *IEEE Transactions on Biomedical Engineering* 47: 589–593.

Vigneron, V., Paraschiv-Ionescu, A., Azancot, A., Sibony, O. & Jutten, C. (2003). Fetal electrocardiogram extraction based on non-stationary ica and wavelet denoising, *Signal Processing and Its Applications, 2003. Proceedings. Seventh International Symposium on*, Vol. 2, pp. 69 – 72.

Vollgraf, R. & Obermayer, K. (2001). Multi-dimensional ica to separate correlated sources., *Proceeding of Neural Information Processing System (NIPS)*, pp. 993–1000.

Vrins, F., Jutten, C. & Verleysen, M. (2004). Sensor array and electrode selection for non-invasive fetal electrocardiogram extraction by independent component analysis, *Independent Component Analysis and Blind Signal Separation* pp. 1017–1024.

Wakai, R., Lengle, J. & Leuthold, A. (2000). Transmission of electric and magnetic foetal cardiac signals in a case of ectopia cordis: the dominant role of the vernix caseosa, *Physics in medicine and biology* 45: 1989 – 1995.

Yu, S. & Chou, K. (2008). Integration of independent component analysis and neural networks for ecg beat classification, *Expert Systems with Applications* 34(4): 2841–2846.

Zarzoso, V. (2009). *Advanced Biosignal Processing*, Springer, chapter Extraction of ECG Characteristics Using Source Separation Techniques: Exploiting Statistical Independence and Beyond.

Zarzoso, V. & Nandi, A. K. (2001). Noninvasive fetal electrocardiogram extraction: blind separation versus adaptive noise cancelation., *IEEE Tr. on Biomedical Engineering* 48 (1): 12–18.

13

ICA Applied to VSD Imaging of Invertebrate Neuronal Networks

Evan S. Hill[1], Angela M. Bruno[1,2], Sunil K. Vasireddi[1] and William N. Frost[1]
[1]*Department of Cell Biology and Anatomy,*
[2]*Interdepartmental Neuroscience Program,*
Rosalind Franklin University of Medicine and Science, North Chicago, IL
USA

1. Introduction

Invertebrate preparations have proven to be valuable models for studies addressing fundamental mechanisms of nervous system function (Clarac and Pearlstein 2007). In general the nervous systems of invertebrates contain fewer neurons than those of vertebrates, with many of them being re-identifiable in the sense that they can be recognized and studied in any individual of the species. The large diameter of many invertebrate neurons makes them amenable for study with intracellular recording techniques, allowing for characterization of synaptic properties and connections, leading to circuit diagrams of neuronal networks. Further, there is often a rather straight-forward connection between neuronal networks and the relatively simple behaviors that they produce. For example, years of experimentation on the nervous systems of leeches, sea-slugs and crabs/lobsters have led to significant advances in the understanding of how small neuronal networks produce a variety of different behaviors (Harris-Warrick and Marder 1991; Hawkins et al. 1993; Katz 1998; Kristan et al. 2005). For the most part, these investigations have been carried out using sharp electrode recordings from about three to four neurons at a time (although see (Briggman and Kristan 2006)). Intracellular recording has been a very productive and fruitful technique for revealing details of neuronal connectivity and for studying synaptic changes caused by modulators or by simple forms of learning. However, since even simple behaviors are produced by the activity of populations of dozens to hundreds of neurons, the limited view offered by recording from only four neurons at a time makes it an inadequate technique for understanding larger-scale, network level phenomena that underlie behavior.

In order to understand how populations of neurons produce behaviors, methods are needed to simultaneously monitor the spiking activity of large numbers (dozens to hundreds) of individual neurons. Voltage-sensitive dye (VSD) imaging is a technique for accomplishing precisely this. However, after a promising start showing the immense power and potential of VSD imaging for understanding invertebrate neuronal networks (London et al. 1987; Wu et al. 1994a; Wu et al. 1994b; Zecevic et al. 1989), the technique has not been widely adopted by the field. This is possibly due to the difficulties inherent to the technique - the optical signals of interest are extremely small and are often mixed, redundant and noisy. These factors make it difficult to track the activity of individual neurons from recording trial to trial based solely on the raw optical data. Previous researchers used a spike-template matching technique to uncover single neuron spiking traces from VSD imaging data, but this method was very time consuming and involved substantial human judgment (Cohen et al. 1989). Automated, accurate and fast methods are thus needed to reliably and quickly extract single neuron spiking activity from such complex data sets.

In this chapter, we demonstrate the utility and accuracy of Infomax ICA for extracting single neuron spiking activity (i.e. spike-sorting) from VSD imaging data of populations of neurons located in the central ganglia of two invertebrate preparations, *Tritonia diomedea* and *Aplysia californica*, that are models for topics such as learning, modulation, pattern generation and pre-pulse inhibition (Brown et al. 2001; Cleary et al. 1998; Frost et al. 1998; Frost et al. 2006; Frost et al. 2003; Getting 1981; Katz and Frost 1995; Katz et al. 1994; Lennard et al. 1980). We also demonstrate certain features of the optical data sets that strongly influence the ability of ICA to return maximal numbers of components that represent the spiking activity of individual neurons (neuronal independent components or nICs).

2. Methods

2.1 Preparation

Tritonia diomedea central ganglia consisting of the bilaterally symmetric cerebral, pleural and pedal ganglia, and *Aplysia californica* central ganglia consisting of the cerebral, pleural and pedal ganglia, were dissected out and pinned onto the bottom of a Sylgard (Dow Corning) lined Petri dish containing Instant Ocean artificial seawater (Aquarium Systems). The thick connective tissue covering the ganglia and nerves was removed with fine forceps and scissors (for intracellular recording experiments the thin protective sheath covering the neurons was also removed). The preparation was then transferred and pinned to the Sylgard-lined coverslip bottom of the recording chamber used for optical recording (PC-H perfusion chamber, Siskiyou). In many experiments, to increase the number of neurons in focus, the ganglion to be imaged was flattened somewhat by pressing a cover slip fragment down upon it that was held in place with small blobs of Vaseline placed on the recording chamber floor.

2.2 Optical recording

Imaging was performed with an Olympus BX51WI microscope equipped with either 10x 0.6NA or 20x 0.95NA water immersion objectives. Preparation temperature was maintained at 10 - 11°C for *Tritonia* and 16 - 17°C for *Aplysia*, using Instant Ocean passed through a feedback-controlled in-line Peltier cooling system (Model SC-20, Warner Instruments). Temperature was monitored with a BAT-12 thermometer fitted with an IT-18 microprobe (Physitemp, Inc) positioned near the ganglion being imaged. For staining, the room was darkened and the perfusion saline was switched to saline containing the fast voltage sensitive absorbance dye RH-155 (Anaspec). Staining was carried out in one of two ways: either 5 min of 0.3 mg/ml or 1.5 hr of 0.03 mg/ml RH-155 in saline. Preparations were then perfused with 0.03 mg/ml RH-155 or dye-free saline throughout the experiment. Trans-illumination was provided with light from a 100W tungsten halogen lamphouse that was first passed through an electronic shutter (Model VS35 Vincent Associates), a 725/25 bandpass filter (Chroma Technology), and a 0.9 NA flip top achromat Nikon condenser on its way to the preparation. 100% of the light from the objective was directed either to an Optronics Microfire digital camera used for focusing and to obtain an image of the preparation to superimpose with the imaging data, or to the parfocal focusing surface of a 464-element photodiode array (NeuroPDA-III, RedShirtImaging) sampled at 1600 Hz.

2.3 Sharp electrode recordings

Intracellular recordings were obtained with 15-30 MΩ electrodes filled with 3M KCl or 3M K-acetate connected to a Dagan IX2-700 dual intracellular amplifier. The resulting signals were digitized at 2 KHz with a BioPac MP 150 data acquisition system.

2.4 Data analysis

Optical data were bandpass filtered in the Neuroplex software (5 Hz high pass and 100 Hz low pass Butterworth filters; RedShirtImaging), and then processed with ICA in MATLAB to yield single neuron action potential traces (independent components), see (Hill et al. 2010) for details. ICA run on 60 s of optical data typically takes about 5 minutes on a computer equipped with an Intel I7 processor. Statisical analyses were performed in Sigmaplot.

3. Results

We bath applied the VSD RH-155 to the central ganglia of *Tritonia diomedea* and *Aplysia californica*, and used a 464-element photodiode array to image the action potential activity of populations of neurons on the surface of various ganglia during two rhythmic motor programs: escape swimming in *T. diomedea*, and escape crawling in *A. californica*. Here we show examples of ICA's ability to extract single neuron activity from mixed, redundant and noisy raw optical data in both *T. diomedea* **(Fig. 1)** and *A. californica* **(Fig. 2)** and we demonstrate the ability of ICA to return maps of the neurons' locations in the ganglia (based on the inverse weight matrix).

3.1 Validation of the accuracy of ICA

We previously demonstrated the accuracy of ICA spike-sorting by performing simultaneous intracellular and optical recordings in both *Tritonia* and *Aplysia* (Hill et al. 2010): in 34 out of 34 cases, one of the independent components returned by ICA matched up perfectly spike-for-spike with the intracellularly recorded data. **Figure 3** shows an intracellular recording from a *Tritonia* pedal ganglion neuron while simultaneously recording the optical activity of many neurons in the same ganglion. The activity of the intracellularly recorded neuron was detected by many diodes (**Fig. 3Ai**). ICA returned 50 nICs, one of which matched up perfectly spike-for-spike with the intracellularly recorded neuron **(Fig. 3Aii, iii and 3B)**. Plotting the data points for the intracellular and matching component traces against each other reveals a strong positive correlation (**Fig. 3C**; R^2 value = 0.631). Another of the nICs burst in the same phase of the swim motor program as did the intracellularly recorded neuron (**Fig. 3D**), however plotting the values of the intracellular and non-matching component against each other revealed no correlation whatsoever (**Fig. 3E**). **Figure 4** shows an intracellular recording from an *Aplysia* buccal ganglion neuron while simultaneously recording the optical activity of many neurons in the same ganglion. The activity of the intracellularly recorded neuron was detected by many diodes (**Fig. 4Ai**). ICA returned 10 nICs, one of which matched up perfectly spike-for-spike with the intracellularly recorded neuron **(Fig. 4Aii, iii and 4B)**. Plotting the data points for the intracellular and matching component traces against each other revealed a strong positive correlation (**Fig. 4C**; R^2 value = 0.629).

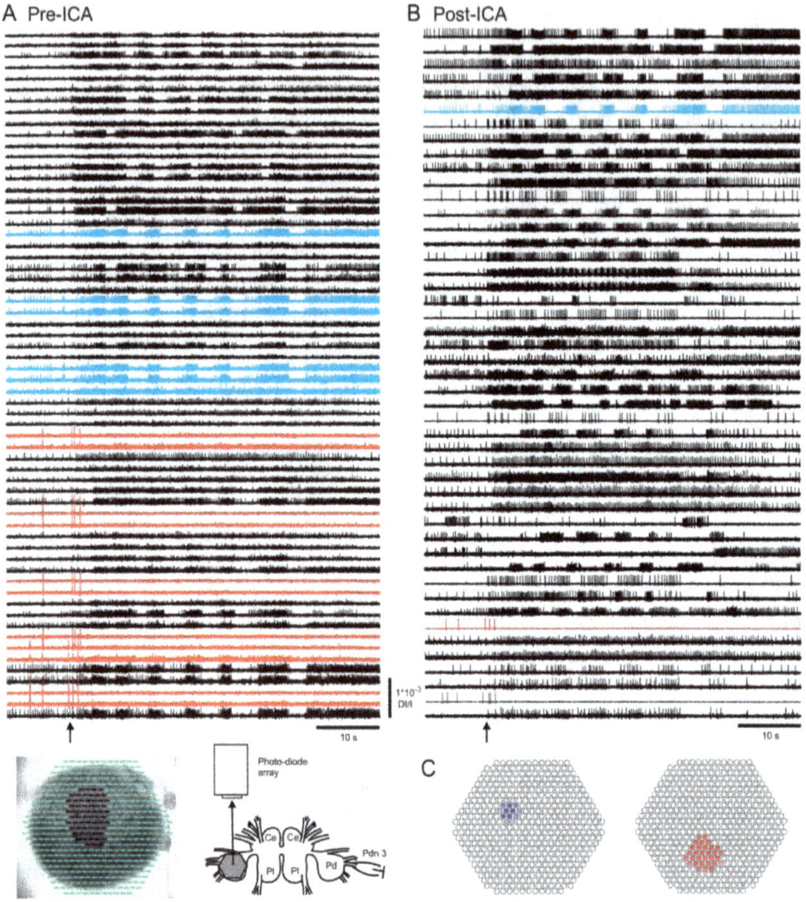

Fig. 1. ICA returns single neuron traces from noisy, redundant, and mixed raw optical data of a *Tritonia diomedea* swim motor program.
A The subset of diodes shown in black superimposed over an image of the pedal ganglion (inset) detected the spiking activity of many pedal ganglion neurons. The optical signals are redundant in the sense that many diodes detect the activity of the same neurons, and mixed in the sense that many diodes detect the activity of more than one neuron. Note that diode traces shown in blue redundantly detected the activity of the same neuron, as did the diode traces shown in red. Experimental set-up shown below the optical traces. Ce = cerebral, Pl = pleural, Pd = pedal ganglion, and pdn 3 = pedal nerve 3. **B** 123 of the 464 independent components returned by ICA represented the activity of single neurons (47 shown here). The redundancy of the optical data was eliminated by ICA - note that the blue and red independent components represent the activity of the two neurons that were detected by multiple diodes in **A**. **C** The maps returned by ICA show the ganglion locations of the neurons whose spiking activity is shown in blue and red in **A** and **B**. Arrows - stimulus to pedal nerve 3 to elicit swim motor program (10 V, 10 Hz, 2 s).

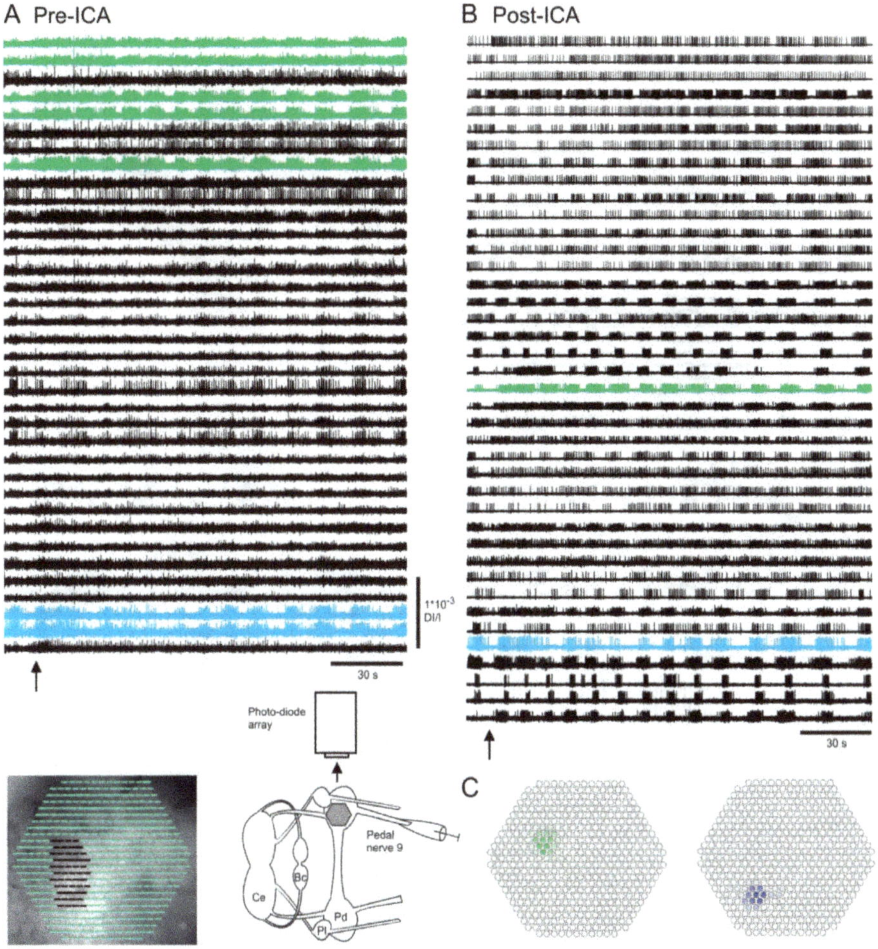

Fig. 2. ICA extracts the activity of individual pedal ganglion neurons from noisy, redundant, and mixed raw optical data of an *Aplysia californica* locomotion motor program.
A The neural activity of many pedal ganglion neurons was detected by a subset of the 464 diodes shown in black (inset, diode array superimposed over an image of the pedal ganglion). Note that diode traces shown in green detected the activity of the same neuron, as did the diode traces shown in blue. Experimental set-up showing the region of the pedal ganglion imaged shown in the inset. Bc = buccal, Ce = cerebral, Pl = pleural, and Pd = pedal ganglion. **B** 95 of the 464 independent components returned by ICA represented the activity of single neurons (45 shown here). Note that the redundancies shown in **A** were eliminated by ICA. **C** The maps returned by ICA show the ganglion locations of the neurons whose spiking activity is shown in green and blue in **A** and **B**. Arrows - stimulus to pedal nerve 9 to elicit the locomotion motor program (10 V, 1 Hz, 155 s).

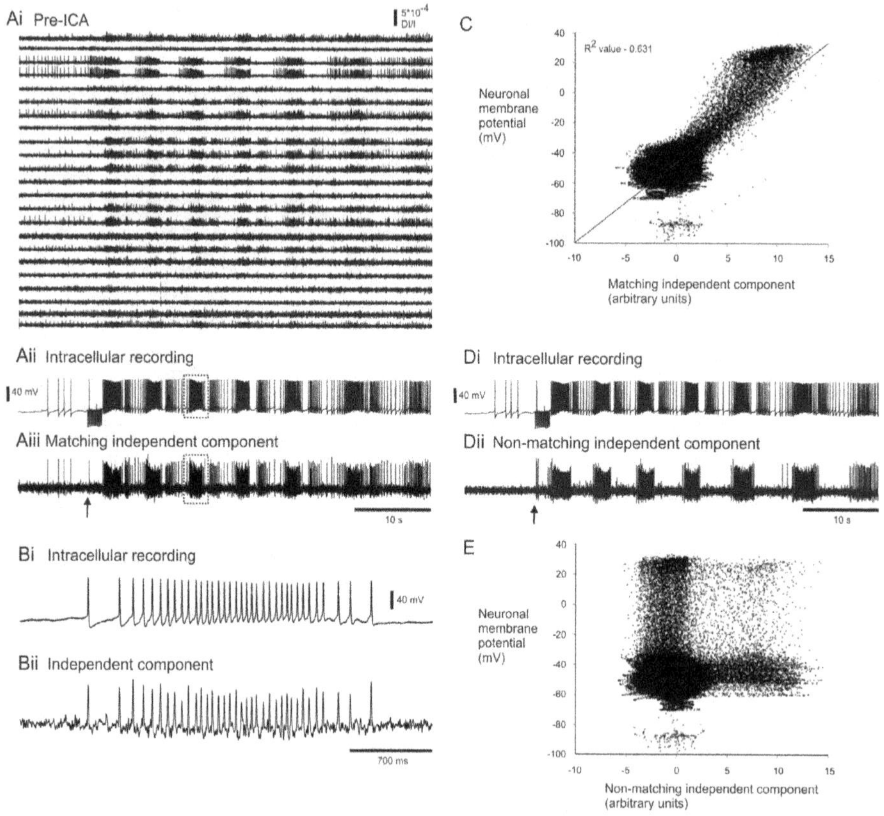

Fig. 3. Validation of the accuracy of ICA in *Tritonia diomedea*.
A Simultaneous intracellular recording from a pedal ganglion neuron and VSD imaging from the same ganglion of *T. diomedea* during a swim motor program. **Ai** Many of the diode traces contained the spiking of activity of the intracellularly recorded neuron. **Aii, Aiii** After performing ICA on the entire data set (464 filtered optical traces), one of the independent components returned by ICA matched up exactly spike-for-spike with the intracellular trace. **B** Expanded view of the traces shown in **Aii** and **Aiii** (dashed boxes). **C** Plotting the values of the intracellular recording trace versus the matching component revealed a positive correlation between the data points. **D** Another of the components returned by ICA appeared similar to the intracellular recording trace in that it burst in the same phase of the motor program. **E** Plotting the values of the intracellular recording trace versus the non-matching component showed no correlation at all between the data points. Arrows - stimulus to pedal nerve 3 (10v, 10 Hz, 2 s).

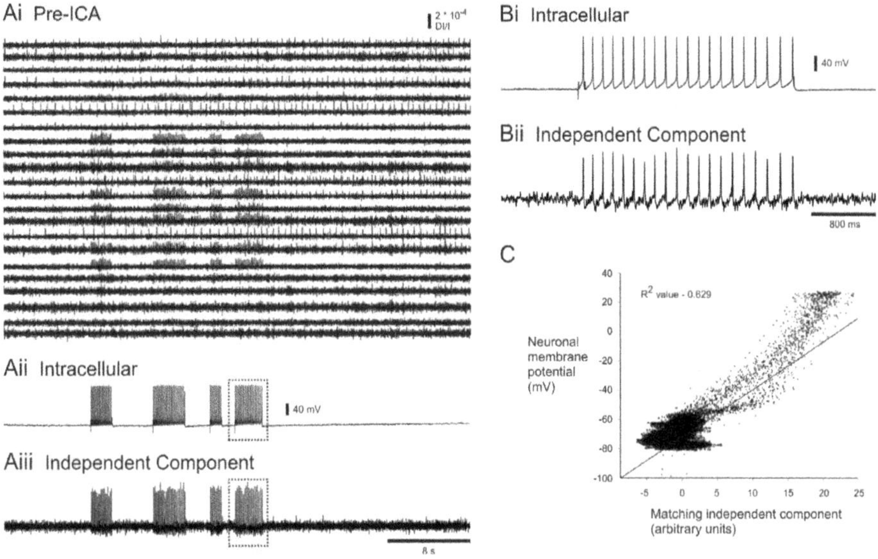

Fig. 4. Validation of the accuracy of ICA in *Aplysia californica*.
A Simultaneous intracellular recording from a buccal ganglion neuron and VSD imaging from the same ganglion of *A. californica*. Current pulses were injected into the impaled neuron to make it to fire trains of action potentials. **Ai** Many of the diode traces contained the spiking of activity of the intracellularly recorded neuron. **Aii, Aiii** After running ICA on the entire data set, one of the independent components returned by ICA matched up exactly spike-for-spike with the intracellular trace. **B** Expanded view of the traces shown in **Aii** and **Aiii** (dashed boxes). **C** Plotting the values of the intracellular recording trace versus the matching component revealed a positive correlation between the data points.

3.2 Certain features of the data sets influence the number of nICs returned by ICA

Next we discuss our findings that certain features of the optical data sets strongly influence the number of nICs returned by ICA. First, we found that simply increasing the number of data points in the optical recording greatly increases the number of nICs returned by ICA (**Fig. 5A**). This could be due to the fact that with longer files, ICA is simply given more information, and is thus better able to determine which components are independent of each other. Increasing the file length only continues to increase the number of nICs returned by ICA up to a certain point though, usually around 45 s (for our data sets). Increasing file length also greatly decreases the variability of the number of nICs returned by ICA **(Fig. 5A)**. We have also found that including spontaneous spiking activity (at least 10 s) preceding the *Tritonia* escape swim motor program greatly increases the number of nICs returned by ICA (**Fig. 5B, C**). For example, for seven preparations we found that when ICA was performed on 40 s of optical data including 10 s of spontaneous firing data preceding the swim motor program it returned a mean of 68.8 nICs, while ICA returned a mean of only 44.5 nICs when it was run on 40 s optical files that didn't include any spontaneous firing preceding the motor program (**Fig. 5C**). Presumably neurons are more independent of each other when they are firing spontaneously than when they are bursting in near synchrony during the swim motor program. Thus, this period of greater independence is important for ICA to return maximal numbers of nICs.

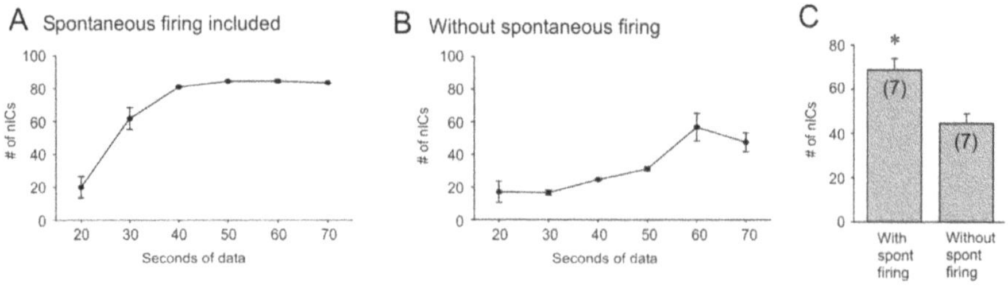

Fig. 5. Increasing file length and including spontaneous firing data improve the performance of ICA.
A Increasing file length leads to an increase in the number of nICs returned by ICA. The points on the graph show the average number of nICs returned by ICA for each data length. Note that with the shorter data lengths there is a fairly large variance in the number of nICs returned by ICA run on the exact same data set. This variance decreases greatly with increasing file length. After a certain point (~ 45 s) having more data points didn't increase the number of nICs returned by ICA. **B** Including spontaneous spiking data preceding the rhythmic swim motor program data is important for ICA to return a maximal number of nICs. The points on the graph show the average number of nICs returned by ICA for each data length. Without the spontaneous firing data (10 s) included before the swim motor program, ICA returned fewer nICs even with longer files (same data set as in **A**). **C** In seven preparations, including 10 s of spontaneous firing data preceding the swim motor program significantly increased the average number of nICs returned by ICA for optical files of equal length (With spont firing = 10 s spontaneous firing + 30 s motor program, Without spont firing = 40 s motor program; paired t-test, * = $p < 0.05$).

3.3 Pre-filtering the data also strongly influences the number of nICs returned by ICA

Finally, we have found that filtering the optical data prior to running ICA to remove high frequency (100 Hz LP) and low frequency noise (5 Hz HP) consistently increases the number of nICs returned by ICA, and increases the signal-to-noise ratio of the nICs. **Figure 6** shows an example of the effect of pre-filtering on the number of nICs returned by ICA. Without pre-filtering, ICA returned 34 nICs (**Fig. 6A**) whereas after pre-filtering it returned 73 nICs (**Fig. 6B**). Removing the low frequency noise in particular should make the components more independent of each other.

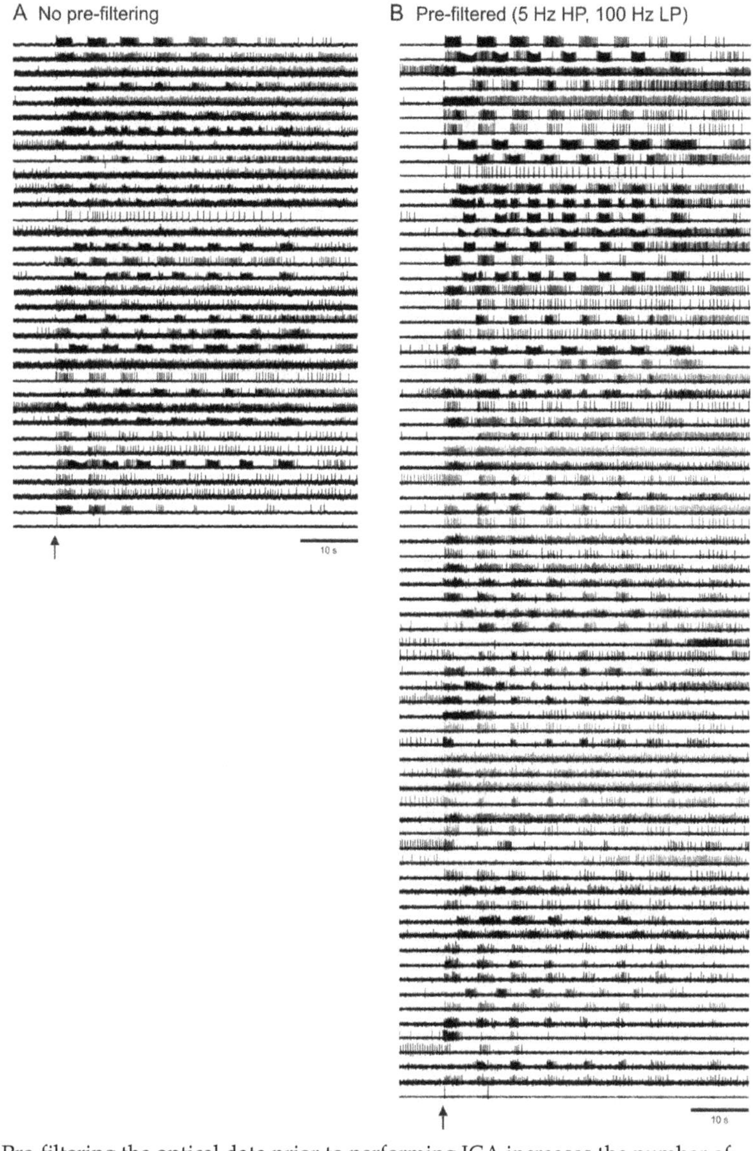

Fig. 6. Pre-filtering the optical data prior to performing ICA increases the number of nICs returned.

A ICA run on unfiltered optical data returned 34 nICs. **B** After pre-filtering the same data set (5 Hz HP, 100 Hz LP), ICA returned 73 nICs. Additionally pre-filtering the optical data increased the signal-to-noise ratio of the nICs. Arrows - stimulus to pedal nerve 3 (10 V, 10 Hz, 2 s).

4. Conclusions

ICA is an ideal technique for quickly and accurately extracting single neuron spiking activity from noisy, mixed and redundant VSD imaging data. The lack of a technique such as ICA has possibly hindered widespread use of fast VSD imaging as a tool to study invertebrate neuronal networks. Here we have shown examples of ICA's ability to extract single neuron spiking activity from optical files of escape motor programs in *Tritonia* and *Aplysia*. Additionally, we have demonstrated the accuracy of ICA with simultaneous intracellular and optical recording in the two molluscan species. We have also demonstrated that features of the optical data sets such as file length and the inclusion of spontaneous firing data are important for ICA to return a maximal number of components that represent the activity of individual neurons. Finally, we have shown that pre-filtering the optical data to remove high and low frequency noise is also beneficial for ICA to return maximal numbers of components that represent the activity of individual neurons.

5. Future directions

The combination of fast VSD imaging and ICA may lead to a resurgence in the use of fast VSDs for deciphering how invertebrate neuronal networks operate and are modified with experience. While the combination of VSD imaging and ICA makes it possible to monitor the activity of well over one hundred neurons during various motor programs, methods are now needed to reveal how sub-groups or ensembles of neurons within these datasets behave during the production of various behaviors. Fortunately, many researchers have developed methods that do precisely this. For example, a recently developed spike-train correlation analysis method called the Functional Clustering Algorithm (Feldt et al. 2009) reveals clusters of neurons that fire spikes together in a statistically significant manner. Using the FCA, we have examined how functional clusters of neurons in the *Tritonia* pedal ganglion change with the transition from escape swimming to post-swim crawling (Hill et al. 2011). Finally, since ICA can be performed rapidly, during the course of an experiment, it will be possible to identify neurons of interest and then to impale those neurons with sharp electrodes and determine their roles in network function.

6. Acknowledgments

The authors thank Caroline Moore-Kochlacs and Terry Sejnowski for helping us implement ICA and Jean Wang for technical assistance with some of the optical recordings.

7. References

Briggman KL, and Kristan WB, Jr. Imaging dedicated and multifunctional neural circuits generating distinct behaviors. *J Neurosci* 26: 10925-10933, 2006.

Brown GD, Yamada S, and Sejnowski TJ. Independent component analysis at the neural cocktail party. *Trends in neurosciences* 24: 54-63, 2001.

Clarac F, and Pearlstein E. Invertebrate preparations and their contribution to neurobiology in the second half of the 20th century. *Brain research reviews* 54: 113-161, 2007.

Cleary LJ, Lee WL, and Byrne JH. Cellular correlates of long-term sensitization in Aplysia. *J Neurosci* 18: 5988-5998, 1998.

Cohen L, Hopp HP, Wu JY, Xiao C, and London J. Optical measurement of action potential activity in invertebrate ganglia. *Annu Rev Physiol* 51: 527-541, 1989.

Feldt S, Waddell J, Hetrick VL, Berke JD, and Zochowski M. Functional clustering algorithm for the analysis of dynamic network data. *Physical review* 79: 056104, 2009.

Frost WN, Brandon CL, and Mongeluzi DL. Sensitization of the Tritonia escape swim. *Neurobiology of learning and memory* 69: 126-135, 1998.

Frost WN, Brandon CL, and Van Zyl C. Long-term habituation in the marine mollusc Tritonia diomedea. *The Biological bulletin* 210: 230-237, 2006.

Frost WN, Tian LM, Hoppe TA, Mongeluzi DL, and Wang J. A cellular mechanism for prepulse inhibition. *Neuron* 40: 991-1001, 2003.

Getting PA. Mechanisms of pattern generation underlying swimming in Tritonia. I. Neuronal network formed by monosynaptic connections. *Journal of neurophysiology* 46: 65-79, 1981.

Harris-Warrick RM, and Marder E. Modulation of neural networks for behavior. *Annual review of neuroscience* 14: 39-57, 1991.

Hawkins RD, Kandel ER, and Siegelbaum SA. Learning to modulate transmitter release: themes and variations in synaptic plasticity. *Annual review of neuroscience* 16: 625-665, 1993.

Hill ES, Moore-Kochlacs C, Vasireddi SK, Sejnowski TJ, and Frost WN. Validation of independent component analysis for rapid spike sorting of optical recording data. *Journal of neurophysiology* 104: 3721-3731, 2010.

Hill ES, Vasireddi S, Wang J, Maruyama D, Zochowski M, and Frost WN. A method for monitoring the temporal structure of neuronal networks. In: *Society for Neuroscience Annual Meeting*. Washington, D.C.: 2011.

Katz PS. Neuromodulation intrinsic to the central pattern generator for escape swimming in Tritonia. *Annals of the New York Academy of Sciences* 860: 181-188, 1998.

Katz PS, and Frost WN. Intrinsic neuromodulation in the Tritonia swim CPG: the serotonergic dorsal swim interneurons act presynaptically to enhance transmitter release from interneuron C2. *J Neurosci* 15: 6035-6045, 1995.

Katz PS, Getting PA, and Frost WN. Dynamic neuromodulation of synaptic strength intrinsic to a central pattern generator circuit. *Nature* 367: 729-731, 1994.

Kristan WB, Jr., Calabrese RL, and Friesen WO. Neuronal control of leech behavior. *Progress in neurobiology* 76: 279-327, 2005.

Lennard PR, Getting PA, and Hume RI. Central pattern generator mediating swimming in Tritonia. II. Initiation, maintenance, and termination. *Journal of neurophysiology* 44: 165-173, 1980.

London JA, Zecevic D, and Cohen LB. Simultaneous optical recording of activity from many neurons during feeding in Navanax. *J Neurosci* 7: 649-661, 1987.

Wu JY, Cohen LB, and Falk CX. Neuronal activity during different behaviors in Aplysia: a distributed organization? *Science* 263: 820-823, 1994a.

Wu JY, Tsau Y, Hopp HP, Cohen LB, Tang AC, and Falk CX. Consistency in nervous systems: trial-to-trial and animal-to-animal variations in the responses to repeated applications of a sensory stimulus in Aplysia. *J Neurosci* 14: 1366-1384, 1994b.

Zecevic D, Wu JY, Cohen LB, London JA, Hopp HP, and Falk CX. Hundreds of neurons in the Aplysia abdominal ganglion are active during the gill-withdrawal reflex. *J Neurosci* 9: 3681-3689, 1989.

Section 4

ICA: Time-Frequency Analysis

14

A Study of Methods for Initialization and Permutation Alignment for Time-Frequency Domain Blind Source Separation

Auxiliadora Sarmiento, Iván Durán, Pablo Aguilera and Sergio Cruces
Department of Signal Theory and Communications, University of Seville, Seville Spain

1. Introduction

The problem of the blind signal separation (BSS) consists of estimating the latent component signals in a linear mixture, referred to as the sources, starting from several observed signals, without relying on any specific knowledge of the sources. In particular, when the sources are audible, this problem is known as to the cocktail-party problem, making reference to the ability of the human ear to isolate the conversation of our interest among several conversations immersed in a noisy environment with many people talking at the same time.

The complexity of the blind separation problem greatly depends on the mixture model, the number of sources and sensors that better adjust to reality and the presence of noise in the mixture. The simplest case regards the linear and instantaneous mixture, that is, when the sources are mixed affected only by some scaling. However, in a real room recording the situation becomes more difficult, since the source signals do not only follow the direct path from the source to the sensor, but there are also other paths coming from the reflections in the walls. Hence, the problem becomes convolutive rather than instantaneous, and the mixture process is then modelled by means of a convolution of the sources by some acoustic mixing filters. In the present chapter we assume that the channel between sources and microphones is time-invariant, the number of sources equals the number of sensors, that is, the $N \times N$ case, and there is no additive noise. In such recording environments the separation is very complex, especially in highly reverberant conditions where the mixing filters can be very long (greater than 250 ms) and can contains strong peaks corresponding to the echoes.

Several component analysis techniques solve the instantaneous and determined case in the time domain. One of the most popular is the Independent Component Analysis (ICA), which is a method to recover statistically independent sources by using implicitly or explicitly high-order statistics Comon (1994). Some of those techniques have been extended to solve the convolutive case in the time domain. However, its use for the separation of real speech recordings is limited because of the high length of the acoustical mixing filters (of the order of hundreds of milliseconds). Since it is required the adjustment of too many parameters, those method present convergence problems and a high computational cost. An extended strategy, referred to as fd-ICA in the literature, consists of formulating the problem in the time-frequency domain instead of the time domain (Smaragdis, 1998). The main reason is that the convolutive mixture can be approximated by a set of instantaneous mixtures, one for each frequency bin, that can be solved independently by applying several separation algorithms. However, this simplification introduces some additional problems referred to as scaling and permutation problems, since the obtained solutions in each frequency exhibit an arbitrary complex scaling and order. The scaling ambiguity introduces a filtering effect in the estimated sources, that can be removed by introducing some constraint on the separation filters. Nevertheless, the permutation problem leads to non-consistent time-frequency representations of the estimated signals, and needs to be addressed to successfully recover the original sources. During the last years, several algorithms have been proposed in order to solve this problem, although nowadays there is no satisfactory solution, specially for highly reverberant environments. Furthermore, the problem increases in complexity with the number of sources in the mixture, and the developed algorithms usually deal with the case of two source signals and two observations only.

Here, we will focus our attention on the initialization of the separation algorithms and the permutation problem. The initialization procedure is very important since some separation algorithms are very sensitive to initial conditions, and often there are some frequency bins in which the separation algorithm fails to converge. Furthermore, a suitable initialization can achieve a spectacular reduction of permutation misalignments, since it favours the preservation of the order of the separated components in wide frequency blocks, which facilitate the permutation correction. A new permutation algorithm is also proposed based on the spectral coherence property of the speech signal. For that, we will derive a contrast function which maximization achieves the solution to the permutation. For the assessment of the developed algorithms, an exhaustive study has been performed in both synthetic measurements and real recordings, for various mixture environments.

2. Model of BSS of convolutive mixtures in the time-frequency domain

It is well known that any acoustic signal acquired from microphones in a real recording environment suffers from reflections on the walls and surfaces inside the room. In this sense, the recorded signals can be accurately modelled as a convolutive mixture, where the mixing filter is usually considered a high-order FIR filter. The standard convolutive mixing model of N sources, $s_j(n), j = 1, \cdots, N$, in a noiseless situation can be written as

$$x_i(n) = \sum_{j=1}^{N} \sum_{k=-\infty}^{\infty} h_{ij}(k) s_j(n-k), \quad i = 1, \cdots, N \ , \tag{1}$$

where $x_i(n), i = 1, \ldots, N$ are the N sensor signals, and $h_{ij}(n)$ is the impulse response from source j^{th} to microphone i^{th}. In order to blindly recover the original speech signals (sources) one can apply a matrix of demixing filters to the observations $x_i(n)$ that yields an estimate of each of the sources

$$y_i(n) = \sum_{j=1}^{N} \sum_{k=0}^{Q-1} b_{ij}(k) x_j(n-k), \quad i = 1, \cdots, N, \tag{2}$$

where the coefficients $b_{ij}(k)$ denote the impulse response of demixing system filter of Q taps.

The transformation of time-domain signals to the time-frequency domain is usually performed by the short-time Fourier transform (STFT). The main advantage of using the time-frequency domain is that convolutive mixture in Equation (1), can be approximated in the time-frequency domain by a set of of instantaneous mixtures of complex values, one for each frequency bin, that is an easier problem for which many algorithms have been developed.

Let $X_i(f,t)$ and $S_i(f,t)$ be, respectively, the STFT of $x_i(n)$ and $s_i(n)$, and $H_{ij}(f)$ be the frequency response of the channel $h_{ij}(n)$. From Equation (1) we obtain

$$X_i(f,t) \simeq \sum_{j=1}^{N} H_{ij}(f) S_j(f,t), \quad i = 1, \cdots, N \ , \tag{3}$$

which can be rewritten, in matrix notation, as $\mathbf{X}(f,t) \simeq \mathbf{H}(f)\mathbf{S}(f,t)$, where the observation and source vectors for each time-frequency point are $\mathbf{X}(f,t) = [X_1(f,t), \ldots, X_N(f,t)]^T$ and $\mathbf{S}(f,t) = [S_1(f,t), \ldots, S_N(f,t)]^T$, respectively, and $\mathbf{H}(f)$ is the frequency response of the mixing filter whose elements are $H_{ij}(f) = [\mathbf{H}(f)]_{ij} \ \forall i,j$. The superscript T represents the matrix transpose operator. From now on, we will assume that the mixing matrices $\mathbf{H}(f)$ are full rank. In practice, the approximation (3) is considered valid when the length of the DFT is significantly greater than the length of the mixing filters Parra & Spence (2000). For instance, in fd-ICA context for speech separation it is common that the DFT is twice as long as the length of the mixing filters Araki et al. (2003).

Each of the separation matrices $\mathbf{B}(f)$ can be estimated independently with a suitable algorithm for instantaneous mixtures of complex values, which is computationally very efficient. The vector of outputs or estimated sources $\mathbf{Y}(f,t) = [Y_1(f,t), \ldots, Y_N(f,t)]^T$ is thus given by applying $\mathbf{B}(f)$ to the observations in each frequency bin,

$$\mathbf{Y}(f,t) = \mathbf{B}(f)\mathbf{X}(f,t). \tag{4}$$

Nevertheless, the simplification (3) has some disadvantages that need to be solved to successfully recover the sources. As each instantaneous separation problem is solved independently, the recovered signals will have an arbitrary permutation and scaling in each frequency bin. Those ambiguities are inherent to the problem of the blind source separation. In consequence, $\mathbf{Y}(f,t)$ is usually modelled as

$$\mathbf{Y}(f,t) \approx \mathbf{\Pi}(f)\mathbf{D}(f)\mathbf{S}(f,t) \ , \tag{5}$$

where $\mathbf{\Pi}(\mathbf{f})$ is a permutation matrix and $\mathbf{D}(f)$ is an arbitrary nonsingular diagonal matrix of complex scalars, representing respectively the permutation and scaling ambiguities.

The scaling ambiguity is not a serious problem. In fact, it causes an overall filtering of the sources. However, the correction of the permutation is essential. Even when perfect separation is achieved in all frequency bins, the transformation of the recovered signals into the time domain will be erroneous if the order of the extracted components are not the same in all frequency bins. Therefore, it is necessary to determine the permutation matrix $\mathbf{P}_*(f)$ in each frequency bin in such way that the order of the outputs remains constant over all the frequencies,

$$\mathbf{Y}(f_k,t) \leftarrow \mathbf{P}_*(f_k)\mathbf{Y}(f_k,t). \tag{6}$$

Once the separated components are well aligned, the sources can be finally recovered by converting the time-frequency representations $Y_j(f,t)$ back to the time domain. It is also possible to estimate the sources by first transforming the separation matrices $\mathbf{B}(f)$ to the time domain, correcting previously the ambiguities, and then by applying the Equation (2).

3. The separation stage

The most widely used methods for solving the instantaneous separation problems in the standard fd-ICA approach relies on the statistical independence among different sources and on the notion of contrast function. The statistical independence of the sources is a plausible assumption in real-room recordings, since each speakers acts independently of the others. On the other hand, the notion of contrast function defines a correspondence between the distribution of the estimated sources and the real line which is only maximized when the sources are mutually independent Comon (1994).

In the fd-ICA context, it is important to note that the separation algorithm must be capable of handling complex data, given that the separation problem is formulated in the time-frequency domain. Nowadays, most of the ICA methods that work with complex data often use a preliminary whitening step that leads to $\mathbf{Z} \equiv \mathbf{Z}(f)$ the spatially whitened observations. This preprocessing simplifies the problem and, in some cases, it is also used because it improves the convergence of the algorithm. The whitening procedure consists of a linearly transform of the observed variables to zero mean and unit variance, that can be accomplished by e.g. Principal Component Analysis (Comon, 1994). One of the most widely used algorithm is FastICA Hyvärinen & Oja (1997), which exploits the property of the non-Gaussianity of the sources. The extension to complex data was formulated in Bingham & Hyvärinen (2000). The solution is obtained by finding the extrema of the following contrast function

$$\Psi_{BH}(\mathbf{u}) = E\left[G\left(\left|\mathbf{u}^{\mathrm{H}}\mathbf{Z}\right|^{2}\right)\right] \quad \text{s. t. } E\left[\left|\mathbf{u}^{\mathrm{H}}\mathbf{Z}\right|^{2}\right] = 1\,, \tag{7}$$

where E represents expectation, $\mathbf{u}$ is the extraction vector (a row of the separating matrix $\mathbf{U}^{\mathbf{H}}$), while G is a smooth even function whose expectation measures the departure (in a given sense) from the Gaussian distribution. Some usual choices for function G can be found in Bingham & Hyvärinen (2000).

The optimization of the contrast function (7) is performed by the Newton's method, resulting the following update rule for the fixed-point algorithm for one unit

$$\begin{aligned}
\mathbf{u}^{(i)} &= E\left[\mathbf{Z}\left(u^{(i-1)^{\mathbf{H}}}\mathbf{Z}\right)^{*} g\left(|\mathbf{u}^{\mathrm{H}}\mathbf{Z}|^{2}\right)\right] - E\left[g\left(|\mathbf{u}^{\mathrm{H}}\mathbf{Z}|^{2}\right) + |\mathbf{u}^{\mathrm{H}}\mathbf{Z}|^{2} g'\left(|\mathbf{u}^{\mathrm{H}}\mathbf{Z}|^{2}\right)\right]\mathbf{u}^{(i-1)} \\
\mathbf{u}^{(i)} &\leftarrow \frac{\mathbf{u}^{(i)}}{\left\|\mathbf{u}^{(i)}\right\|}\,,
\end{aligned} \tag{8}$$

where i is the iteration index, and $g(\cdot)$ and $g'(\cdot)$ denote the first and the second derivatives of $G(\cdot)$, respectively. This method can be combined with a deflation procedure to retrieve all the original components. An optimized variant of FastICA consists of introducing an adaptive choice of function G. For this purpose, the distributions of the independent components can be modelled by a generalized Gaussian distribution. The resulting algorithm is called efficient FastICA or simply EFICA Koldovský et al. (2006).

The ICA algorithms previously commented ignore the time structure of source signals. However, for the speech signals, nearby samples are highly correlated and when comparing the statistics for distant samples the nonstationary behaviour is revealed. It is possible to exploit any of these features to achieve the separation using only second order statistics (SOS). One important advantage of the SOS based systems is that they are less sensitive to noise and outliers. One popular method of this family of algorithms is the second order blind identification (SOBI) algorithm, proposed in Belouchrani et al. (1997).

Under the assumption of spatial decorrelation of the sources, the correlation matrices of the sources $\mathbf{R}_s(\tau) = E[s(t+\tau)s^*(t)]$ for any nonzero time lag τ are diagonal, where superscript * denotes conjugate operation. If we consider now time-delayed correlation matrices of whitened observations, the next relation for prewhitened sensor signals is satisfied

$$\mathbf{R}_z(\tau) = \mathbf{W}\mathbf{R}_x(\tau)\mathbf{W}^H = \mathbf{U}\mathbf{R}_s(\tau)\mathbf{U}^H \tag{9}$$

where $\mathbf{W}$ is the whitening matrix and $\mathbf{U}$ is the unitary mixing matrix. Since $\mathbf{R}_s(\tau)$ is diagonal, the separation matrix $\mathbf{U}^H$ may be estimated by enforcing an unitary diagonalization of a covariance matrix $\mathbf{R}_z(\tau)$ for some non zero lag. Instead of use only one time lag, SOBI approximated jointly diagonalizes a set of covariance matrices computed for a fixed set of time lags.

An extension of this algorithm that jointly exploits the non-stationary and the temporal structure of the source signals is second-order non-stationary source separation (SEONS) algorithm proposed in Choi & Cichocki (2000). This method estimates a set of covariance matrices at different time-frames. For that, the whitened observations are divided into non-overlapping blocks, where different time-delayed covariance matrices are computed. Then, a joint approximate diagonalization method is applied to this set of matrices to estimate the separation matrix. The application of the SEONS algorithm in the simulations of this chapter considers covariance matrices for $\tau = 0$ and one sample in each block.

3.1 The ThinICA algorithm

The higher order cumulants of the outputs have been one of the first class of contrast functions proposed in the context of blind deconvolution Donoho (1981) and later extensively used in the context of blind source separation Comon (1994); Cruces et al. (2004a). In its simpler form, the contrast function takes the form of a sum of the fourth-order cumulants of the outputs

$$\Psi(\mathbf{U}) = \sum_{i=1}^{N} |Cum\,(Y_i(t), \cdots, Y_i(t))|^2, \tag{10}$$

subject to a unitary constraint on the separation matrix ($\mathbf{U}^H\mathbf{U} = \mathbf{I}$). Indeed, the first implementation of the Fast-ICA algorithm Hyvärinen & Oja (1997) considered the maximization of (10). Nearly at the same time, other authors developed in DeLathauwer et al. (2000) a higher-order power method that consider the separation of the sources with a contrast function based on a least squares fitting of a higher-order cumulant tensor.

The ThinICA algorithm was proposed in Cruces & Cichocki (2003) as a flexible tool to address the simultaneous optimization of several correlation matrices and/or cumulant tensors. These matrices and tensors can be arbitrarily defined, so the algorithm is able to exploit not only the independence of the sources, but also their individual temporal correlations and also their possible large-term non-stationarity Cruces et al. (2004b).

For simplicity, it is assumed that sources are locally stationary and standardized to zero mean and unit variance. In order to determine the statistics that take part in the ThinICA contrast function one should specify the order of the cumulants q and the considered time tuples $\theta = (t_1, \cdots, t_q)$ which are grouped in the set $\Theta = \{\theta_m \in \mathbb{R}^q, m = 1, \cdots, r\}$. The algorithm works with positive weighting scalars w_θ and with q unitary matrix estimates $\mathbf{U}^{[k]}, k = 1, \cdots, q$, of the mixing system $\mathbf{WA}$ and their respective linear estimates $\mathbf{Y}^{[k]}(t) = \mathbf{U}^{[k]H}\mathbf{Z}(t)$, $k = 1, \ldots, q$, of the vector of desired sources. It was shown in Cruces et al. (2004b) that the function

$$\Phi_\Theta(\mathbf{U}^{[1]}, \ldots, \mathbf{U}^{[q]}) = \sum_{i=1}^{N} \sum_{\theta \in \Theta} w_\theta \left| Cum\left(Y_i^{[1]}(t_1), \cdots, Y_i^{[q]}(t_q)\right) \right|^2, \tag{11}$$

is a contrast function whose global maxima are only obtained when all the estimates agree ($\mathbf{U}^{[1]} = \cdots = \mathbf{U}^{[q]}$) and the sources of the mixture are recovered. Moreover, the constrained maximization of the previous contrast function is equivalent to the constrained minimization of the weighted least squares error between a set of q-order cumulant tensors of the observations $\{\mathcal{C}_q^{\mathbf{Z}}(\theta), \forall\theta \in \Theta\}$ and their best approximations that take into account the mutual independence statistical structure of the sources $\{\hat{\mathcal{C}}_q^{\mathbf{Z}}(\mathbf{D}_\theta, \mathbf{U}^{[1]}, \ldots, \mathbf{U}^{[q]}), \forall\theta \in \Theta\}$. If $\mathbf{D}_\theta$ denote diagonal matrices, the maximization of (11) is equivalent to the minimization of

$$\epsilon_\Theta(\mathbf{U}^{[1]}, \ldots, \mathbf{U}^{[q]}) = \sum_{\theta \in \Theta} w_\theta \min_{\mathbf{D}_\theta} \|\mathcal{C}_q^{\mathbf{Z}}(\theta) - \hat{\mathcal{C}}_q^{\mathbf{Z}}(\mathbf{D}_\theta, \mathbf{U}^{[1]}, \ldots, \mathbf{U}^{[q]})\|_F^2 \tag{12}$$

with respect to the unitary matrices $\mathbf{U}^{[k]}, k = 1, \cdots, q$. See Cruces et al. (2004b) for more details on the equivalence between the contrast functions (11) and (12).

The optimization of the ThinICA contrast function can be implemented either hierarchically or simultaneously, with respective implementations are based on the thin-QR and thin-SVD factorizations. A MatLab implementation of this algorithm can be found at the ICAlab toolbox icalab (2012), or obtained from the authors upon request.

The ThinICA contrast function and the algorithm has been also extended in Durán & Cruces (2007) to allow the simultaneous combination of correlation matrices and cumulant tensors of arbitrary orders. In this way, the algorithm is able to simultaneously exploit the information of different statistics of the observations, what makes it suitable for obtaining accurate estimates from a reduced set of observations.

The application of the ThinICA algorithm in the simulations of this chapter tries to exploit the non-stationarity behavior of the speech signals by considering $q = 2$ and the set $\Theta = \{(t_m, t_m) \in \mathbb{R}^q, m = 1, \cdots, r\}$, i.e., it uses the information of several local autocorrelations of the observations in frequency domain in order to estimate the latent sources.

4. Initialization procedure for ICA algorithms

The ICA algorithm used for estimating the optimal separation system in each frequency bin, is often randomly initialized. However, there are several advantages to make a suitable initialization of the algorithm. For instance, if the algorithm is initialized near the optimal solution, one can guarantee a high convergence speed. Also, the permutation ambiguity can be avoided if the mixture have some properties.

One interesting approach to develop an appropriate initialization method is to consider the continuity of the frequency response of the mixing filter $\mathbf{H}(f)$ and its inverse. Under this assumption, it seems reasonable to initialize the separation system $\mathbf{B}_{ini}(f)$ from the value of the optimal separation system at the previous frequency $\mathbf{B}_o(f-1)$. However, we can not directly apply $\mathbf{B}(f) = \mathbf{B}_o(f-1)$ in those separation algorithms that whiten the observations as a preprocessing step.

The whitening is performed by premultiplying the observations with an $N \times N$ matrix $\mathbf{W}(f)$ as $\mathbf{Z}(f,t) = \mathbf{W}(f)\mathbf{X}(f,t)$, where $\mathbf{W}(f)$ is chosen so as to enforce the covariance of $\mathbf{Z}(f,t)$ to be the identity matrix $\mathbf{C}_Z(f,t) = \mathbf{I}_N$. The computation of the whitening matrix can be accomplished by e.g. Principal Component Analysis (Comon, 1994). After that, the new observations $\mathbf{Z}(f,t)$ can be expressed as a new mixture of the sources through a new unitary mixing matrix $\mathbf{U}_o(f) = \mathbf{W}(f)\mathbf{H}(f)$,

$$\mathbf{Z}(f,t) = \mathbf{U}(f)\mathbf{S}(f,t). \tag{13}$$

Given an estimate of the unitary mixing matrix $\mathbf{U}_o(f)$, then it is immediate to see that the separation matrix $\mathbf{U}(f)^{-1} = \mathbf{U}(f)^H$ is also unitary. Therefore, the estimated components or outputs

$$\mathbf{Y}(f,t) = \mathbf{U}(f)^H\mathbf{Z}(f,t) = \mathbf{B}(f)\mathbf{X}(f,t) \tag{14}$$

yields the decomposition of the separation matrix $\mathbf{B}(f)$ as the product of an unitary matrix and the whitening system $\mathbf{B}(f) = \mathbf{U}(f)^H\mathbf{W}(f)$. Due to the variability of the sources spectra, even at contiguous frequencies, the whitening matrices $\mathbf{W}(f)$ and $\mathbf{W}(f-1)$ are different. Consequently, in general we violate the unitary assumption of $\mathbf{U}(f)$ by solving directly for $\mathbf{U}_{ini}(f)^H = \mathbf{B}_o(f-1)\mathbf{W}^{-1}(f)$.

An alternative method to initialize from previous solutions while avoiding the previously described problem, consists of initially preprocessing the observations at frequency f by the separation matrix determined for the previous frequency. This technique, referred on now as classical initialization, first computes the new observations as

$$\mathbf{X}_{new}(f,t) = \mathbf{B}(f-1)\mathbf{X}(f,t), \tag{15}$$

and then, determines the matrix $\mathbf{W}(f)$ which whitens these new observations. Finally, the separation matrices are obtained by any preferred ICA method on those new observations. In brief, this classical initialization method decomposes the overall separation matrix in the following three factors

$$\mathbf{B}(f) = \mathbf{U}(f)^H\mathbf{W}(f)\mathbf{B}(f-1). \tag{16}$$

Instead of this classical initialization, here we aim to exploit the continuity of the frequency response of the separation filter in a different way. We propose to initialize the separation system $\mathbf{B}_{ini}(f)$ from its joint closest value to a set of optimal separation systems already computed at nearby frequencies (Sarmiento et al., 2009; 2010) This leads to the following constrained minimization problem

$$\arg\min_{\mathbf{U}(f)^H} \sum_i \alpha_i \|\mathbf{B}(f-i) - \mathbf{B}(f)\|_F^2 \quad \text{s.t. } \mathbf{U}(f)^H\mathbf{U}(f) = \mathbf{I}_N\ , \tag{17}$$

where $\|\cdot\|_F$ denotes the Frobenius norm and α_i are weights assigned to the separation matrices of nearby frequencies. This problem can be solved by applying Lagrange multipliers, where the corresponding Lagrangian function $\mathcal{L}$ is given by

$$\begin{aligned}\mathcal{L} = \mathrm{Tr}\Big\{\sum_i \alpha_i \Big[\Big(\mathbf{B}(f-i) - \mathbf{U}(f)^H\mathbf{W}(f)\Big)^H \Big(\mathbf{B}(f-i) - \mathbf{U}(f)^H\mathbf{W}(f)\Big)\Big] \\ -\mathbf{\Lambda}\Big(\mathbf{U}(f)^H\mathbf{U}(f) - \mathbf{I}_N\Big)\Big\},\end{aligned} \tag{18}$$

where Λ is the Hermitian matrix of multipliers and $\mathrm{Tr}\{\cdot\}$ denotes the trace of the argument. The minimization of the Lagrangian is obtained solving for $\mathbf{U}(f)$ from the equation

$$\nabla_{\mathbf{U}(f)^*}\mathcal{L} = -\left[\sum_i \alpha_i \mathbf{W}(f)\left(\mathbf{B}(f-i)^H - \mathbf{W}(f)^H\mathbf{U}(f)\right) + \mathbf{U}(f)\mathbf{\Lambda}\right] = \mathbf{0}_N, \tag{19}$$

where $\mathbf{0}_N$ denotes null matrix of dimension $N \times N$. After some manipulations, one obtains the desired solution

$$\mathbf{U}_{ini}(f)^H = \mathbf{Q}_R\mathbf{Q}_L^H. \tag{20}$$

where $\mathbf{Q}_L$ and $\mathbf{Q}_R$ are, respectively, the left and right singular vectors of the following factorization

$$[\mathbf{Q}_L, \mathbf{D}, \mathbf{Q}_R] = \mathrm{svd}\left(\mathbf{W}(f)\sum_i \alpha_i \mathbf{B}(f-i)^H\right). \tag{21}$$

As we will se below, this initialization procedure helps to preserve the ordering of the separated components across the frequencies. However, we can not guarantee that all the frequencies will be correctly aligned. In fact, in the audio context, the mixing filters, and therefore the demixing filters, can contain strong echoes. Thus, in general, the assumption of continuity of the filter frequency response is not valid in all the frequency bins.

Furthermore, it can exist some isolated frequency bins in which the separation problem is ill conditioned, and in consequence, the estimated separation matrices should not correspond to the optimal solution. Despite those aspects, in practice, the initialization procedure can achieve a spectacular reduction of permutation misalignments when it is applied to various ICA separation algorithms.

In order to corroborate this point, we now present various 2×2 separation experiments. In Figure 1, we show the number of transitions in the ordering of the estimated components when we apply both, the classical and the initialization procedures aforementioned to various standard ICA algorithm that whitens the observations to the estimation of the separation matrices. For comparison, we have selected three representative ICA algorithms: ThinICA, SEONS and EFICA. As it can be seen, the initialization overperforms the classical procedure, achieving a drastic reduction in the number of permutations in all the cases. Although it is possible to take into account the separation matrices from several frequencies to estimate the initial separation matrix, in our experience the best performing initialization is achieved when we use only one preceding frequency.

The initialization procedure also preserves the ordering of the separated components in wide frequency blocks. This last property is illustrated in Figure 2, where it is shown the spectrograms of the original and estimated components from a simulation for separating two speech sources from a synthetic convolutive mixture. The estimated components have been obtained by using the ThinICA algorithm initialized with the procedure described above, but without correcting the permutation ambiguity. In this simulation there are only four transitions in the order of the estimated components, where it is easy to see that the components are well aligned in wide frequency blocks. This property is particularly very interesting for our purposes, because it could be used to alleviate the computational burden of the algorithms that solve the permutation problem, although this issue will not be discussed in this chapter.

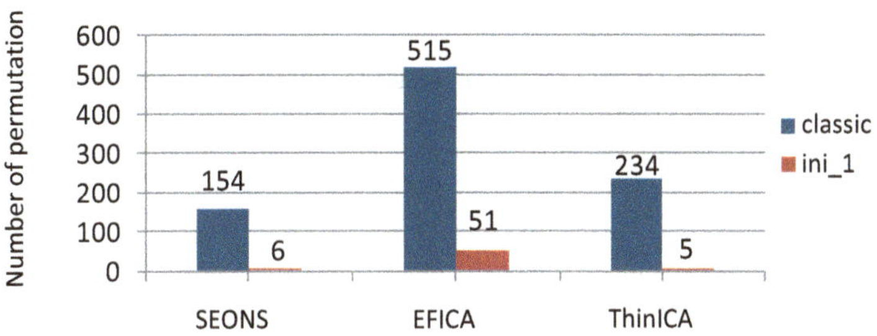

Fig. 1. Number of transitions in the ordering of the estimated components by applying the classical and the initialization procedure to several ICA separation algorithms, explored on synthetic 2×2 convolutive mixtures. Results are the average number over 10 different mixtures.

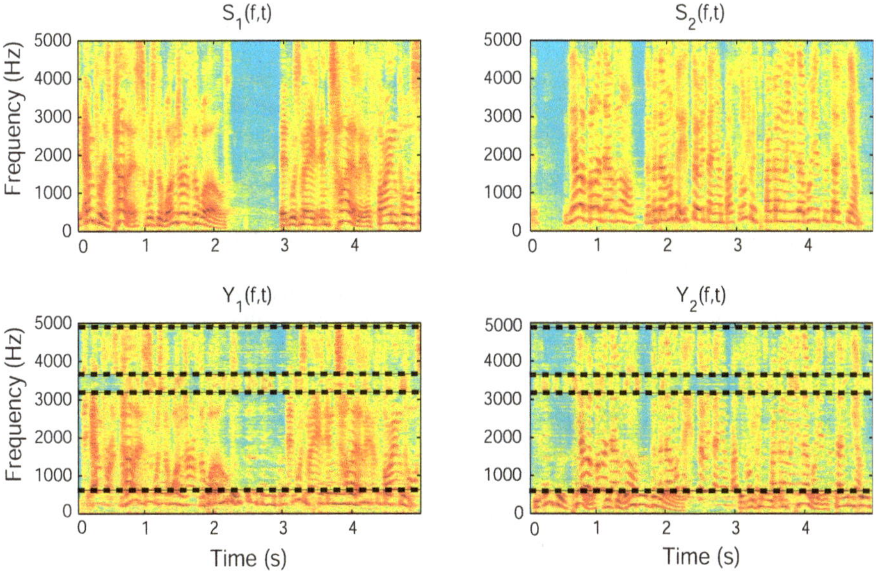

Fig. 2. Spectrograms estimated by the ThinICA algorithm using the initialization procedure. In the first row it is shown the spectrogram of the two original speech sources, whereas in the second row it is shown the estimated spectrograms. There are only four frequencies in which the order is not preserved, indicated by a dotted line.

5. Avoiding the indeterminacies

As we described above, due to the decoupled nature of the solutions across different frequencies, the correspondence between the true sources and their estimates, in general suffers from scaling and ordering ambiguities. Hereinafter, we describe some existing method to try to avoid these ambiguities.

5.1 The scaling ambiguity

The scale ambiguity can be fixed by setting some constraints to the separation filters or by using some *a priori* knowledge of the source signals. One option is to constraint the separating matrices to have unit determinant Smaragdis (1998), whereas another one is to constraint the diagonal elements of the separating matrices to unity Parra & Spence (2000). However, the most extended option is based on the minimal distortion principle introduced in Matsuoka & Nakashima (2001). The goal of this procedure is to obtain the signals as received by the microphones, that is, including the distortion of the mixing system while not adding other distortion effects. The solution consists of multiplying the separation matrices in each frequency bin by the diagonal of the matrix $\mathbf{B}(f)^{-1}$

$$\mathbf{B}(f) \leftarrow \text{diag}\{\mathbf{B}(f)^{-1}\}\mathbf{B}(f). \tag{22}$$

5.2 The permutation ambiguity

Nowadays the permutation ambiguity constitutes the main problem in fd-ICA of acoustic signals, and it is still not satisfactory solved in high reverberating environments or for a large number of sources and observations. In order to tackle the problem, it is necessary to take some aspects into consideration. First, it is important to note that, when there are N sources in the mixtures, there are $N!$ possible permutations in each frequency bins, so the problem becomes difficult as the number of sources increase. In fact, a great number of the existing methods work only on the 2×2 case, and unfortunately, such methods cannot be directly extended to the general $N \times N$ case. On the other hand, in general, we can not guarantee the optimal solution of the instantaneous separation problem in all frequency bins, since the source signals are not homogeneous in their statistical properties along different frequencies. Therefore, there will be some frequencies in which the estimated sources do not correspond to the original sources. This can affect hardly the robustness of the permutation correction algorithms, and often it causes fatal errors in some of the existing methods.

The general structure of the permutation correction algorithms is presented below. The main goal of the permutation correction algorithms consist of estimating a set of permutation correction matrices, one for each frequency bin, $\mathbf{P}_{:} = \left\{\mathbf{P}_{f_1}, \mathbf{P}_{f_2}, \cdots, \mathbf{P}_{f_{n_F}}\right\}, \mathbf{P}_{f_k} \in \mathcal{P}$, where n_F is the number of frequency bins and $\mathcal{P}$ represents all the possible permutation matrices of dimension $N \times N$. Those permutation matrices are applied either to the outputs $\mathbf{Y}(f,t)$ or to the separation filters $\mathbf{B}(f)$ to fix the permutation problem.

If we denote $\mathbf{\Pi}_{:} = \left\{\mathbf{\Pi}_{f_1}, \mathbf{\Pi}_{f_2}, \cdots, \mathbf{\Pi}_{f_{n_F}}\right\}, \mathbf{\Pi}_{f_k} \in \mathcal{P}$ a set of permutation matrices, one for each frequency bin, that describes mathematically the permutation ambiguity, then it is possible to define the set of global permutation matrices $\mathbf{Q}_{:} = \left\{\mathbf{Q}_{f_1}, \mathbf{Q}_{f_2}, \cdots, \mathbf{Q}_{f_{n_F}}\right\}$, whose elements are $\mathbf{Q}_{f_k} = \mathbf{P}_{f_k}\mathbf{\Pi}_{f_k}$.

Then, it is immediate to deduce that the set $\mathbf{P}_{:}$ will be an optimal solution to the permutation problem if the corresponding set of global permutation matrices $\mathbf{Q}_{:}$ satisfy the following condition,

$$\mathbf{Q}_{f_1} = \mathbf{Q}_{f_2} = \cdots = \mathbf{Q}_{f_{n_F}} = \mathbf{Q}, \quad , \forall \mathbf{Q} \in \mathcal{P}, \tag{23}$$

which implies that the permutation problem has $N!$ possible optimal solutions.

5.2.1 Brief review of existing methods

Here, we present the main ideas emerged in last years to solve the permutation problem, putting special emphasis on the drawbacks and limitations of the techniques. Many different approaches have been proposed during last years. Basically, those methods are based on one of the following two assumptions, or in a combination of both (Pedersen et al., 2008): consistency of the spectrum of the recovered signals and consistency of the filter coefficients.

The first set of methods use the consistency of the spectrum of the recovered signals, which relies on the property of amplitude modulation correlation Anemüller & Kollmeier (2000) or simply co-modulation, of speech signals. This property refers to the spectrogram of a speech signal reveals that there is a pattern in the changes in amplitude at different frequency bins. This can be explained since the energy seems to vary in time in a similar way over different frequency bins, up to a gain factor. In fact, when a speaker starts talking, the power of the signal increases in a similar way at all the frequencies, and the same happens when the speaker stops talking, that is, the power decreases in a similar way in all the frequencies. This pattern is in general different for different speakers, at least at some parts of the recording. Therefore, it is possible to propose a permutation correction algorithm based on some evaluation procedure of the similarity between the envelopes of separated signals.

This idea has been used extensively to propose different methods. One option consists of adjusting the permutation between either adjacent or sorted frequency bins in a sequential order. The method was first proposed in Ikeda & Murata (1998); Murata et al. (2001), where the output components are ordered according to the highest correlation between the frequency bin to be order and a global envelope calculated with the already ordered frequency bins. However, the sequential approach has a major drawback, since an error while estimating the correct permutation at a frequency bin can be propagated to the rest of frequencies to be ordered. In Rahbar & Reilly (2005) it is proposed a dyadic hierarchical sorting scheme to prevent this situation.

Rather than solving the permutation problem *a posteriori*, some methods try to avoid it. One option consists of introducing some constraint that penalizes the permuted solutions in the separation problem in each frequency bin. For instance, in Anemüller & Kollmeier (2000) the separation and permutation problems are solved simultaneously, so the computational cost is limited. However, it does not work well in high reverberation environments. Another option proposed in Kim et al. (2006) is based on the concept of Independent Vector Analysis (IVA), which is an extension of ICA from univariate components to multivariate components. This method models the time-frequency representations of speech signals with a multivariate probability density function, and separates the fellow source components together. The contrast proposed is a multidimensional extension of the maximum likelihood (ML) approach. The method performs successfully in most conditions, recovering high quality speech signals. However, the convergence to local minima limits the robustness of the method, since in those cases it does not successfully separate all the components.

The second set of methods, which are based on the spectral coherence of the separation filters, includes methods based on the continuity and smoothness of the frequency response of the separation filters and methods based on the sparsity of the separation filters. The property of continuity and smoothness refers to the fact that the frequency response of the separation filters has not got abrupt transitions. Under this assumption, in Pham et al. (2003) the permutation is solved by checking if the ratio $\mathbf{R}(f, f-1) = \mathbf{B}(f)\mathbf{B}^{-1}(f-1)$ is close to a diagonal matrix, in which case the frequencies f and $f-1$ are well aligned. In a similar way, in Asano et al. (2003) the permutation is corrected by minimizing a distance measure between the filters evaluated at contiguous frequencies. The main weakness of those methods is the propagation of error, since an error in one frequency bin can lead to wrong permutations over the rest of frequency bins to be solved.

The continuity of the separation filters is equivalent to constraint the separation filters to have short support in the time domain. This idea, proposed in Parra & Spence (2000) is based on the observation that the existence of permutations will produce time domain filters with greater lengths. Therefore, if we impose a short length on the separation filters in the separation stage, one can assume that the estimated filters will preserve the same order in all the frequencies. Unfortunately, this method tends to fail in reverberant acoustic environments since the acoustic filters are already quite long. A recent method introduced in Sudhakar & Gribonval (2009) uses the temporal sparsity of the filters to solve the permutation problem, where the sparsity means that the filters have few non-zero coefficients. The main idea is that the permutation errors decreases the sparsity of the reconstructed filters in the time domain, so it is possible to solve the permutation problem by maximizing the sparsity of the time domain demixing filters. This method has a high computational cost, and also only works in absence of the scaling ambiguity, which is not a realistic assumption.

Another family of methods, closed to the beamforming techniques, are based on the different direction of arrival (DOA) of source signals Parra & Alvino (2002). For that, it is necessary the knowledge of the geometry of the sensor array, and the distance between the microphones to be small enough to prevent the problem of spatial aliasing. Those methods assume that the direct path dominates the mixing filter response, and therefore the frequency response of the mixing filter from the i source to the j sensor can be approximately modelled as an anechoic model,

$$H_{ji}(f) = e^{j2\pi f \tau_{ij}}, \quad \tau_{ij} = \frac{d_{ij}\sin\theta_i}{c}, \tag{24}$$

where θ_i is the direction of arrival of the i source, d_{ij} is the distance between the microphones i and j, and c is the propagation speed of sound. Due to the coherence of the separation filter, some authors, as in Kurita et al. (2000); Saruwatari et al. (2003), assume that the quotient of the frequency response of the mixing filters between a given source and whatever two sensors will present a continuous variation with frequency, so this property is exploited to match the order of the components. However, a correct estimation of the DOAs is not always possible and the method tends to fail in high reverberation conditions or when the sources are near.

Finally, some methods combine the direction of arrival estimation with signal inter-frequency dependence to provide robust solutions, as in Sawada et al. (2004). The method, first fix the permutations by using the DOA approach in those frequencies where the confidence of the method is high enough. Then the remaining frequencies are solved by a correlation approach on nearby frequencies, without changing the permutation fixed by the DOA approach. This method has been extended when the geometry of the sensor array is unknown, in Sawada et al. (2005), and when spatial aliasing happens, in Sawada et al. (2006).

6. A coherence contrast for solving the permutation

In this section we present a method for solving the permutation problem in the general $N \times N$ case, based on the amplitude modulation correlation property of speech signals. For that, we will define a global coherence measure of the separated components that constitutes a contrast for solving the permutation problem Sarmiento et al. (2011). Then, the set of permutation matrices to align the separated components are estimated in an iterative way by using a block-coordinate gradient ascent method that maximize the contrast.

First, we transform the profiles of the separated components in a logarithmic scale, since it will exhibit clearly the coherence property of speech signals. Given a source signal $s_i(k)$ and its STFT $S_i(f,t)$, the spectrogram in dB $|S_i|_{dB}(f,t)$ is defined as

$$|S_i|_{dB}(f,t) = 10\log_{10}|S_i(f,t)|^2 . \tag{25}$$

Consider now two source signals $s_i(k)$ and $s_j(k)$. The correlation coefficient between i component at f_k frequency, $|S_i|_{dB}(f_k,t)$, and j component at f_p frequency, $|S_j|_{dB}(f_p,t)$ is given by

$$\rho_{ij}(f_k,f_p) = \rho(|S_i|_{dB}(f_k,t),|S_j|_{dB}(f_p,t)) = \frac{r_{ij}(f_k,f_p) - \mu_i(f_k)\mu_j(f_p)}{\sigma_i(f_k)\sigma_j(f_p)}, \quad \in [-1,1], \tag{26}$$

where the cross correlation, mean and variance of the spectrograms are respectively

$$r_{ij}(f_k,f_p) = E\left[|S_i|_{dB}(f_k,t),|S_j|_{dB}(f_p,t)\right] \tag{27}$$

$$\mu_i(f_k) = E\left[|S_i|_{dB}(f_k,t)\right] \tag{28}$$

$$\sigma_i(f_k)^2 = E\left[|S_i|^2_{dB}(f_k,t)\right] - \mu_i^2(f_k). \tag{29}$$

Although, in general, the speech signals fulfil the co-modulation property , several authors have stated that the direct comparison between the separated components at different frequencies is not always efficient to solve the permutation problem, mainly owing to the fact that the inter-frequency correlation is degraded in certain conditions. In fact, one speech signal will have high correlation coefficients in nearby frequency bins, but this assumption is not always correct if the frequencies are far apart or the correlation is evaluated in certain frequency range, mainly at very low frequencies or very high frequencies (approximately over 5 kHz).To overcome this, we define the mean correlation coefficient $\rho_{ij}\,(f_k)$, as an averaged measure of the correlation coefficients, in other words, a measure of similarity between the i component at frequency f_k and the j component in all the frequencies

$$\rho_{ij}\,(f_k) = \frac{1}{n_F}\sum_{p=1}^{n_F}\rho_{ij}(f_k,f_p), \quad \in [-1,1], \tag{30}$$

where n_F is the number of frequency bins.

Due to the spectral properties of speech signals, it is reasonable to expect that the mean correlation coefficient between one source at any f_k and itself will be greater than if we compare with another different source. Therefore, given the set of sources $\mathbf{s}(k)$, one can deduce that the following property will be satisfied $\forall f_k$

$$\rho_{ii}\,(f_k) > \rho_{ij}(f_k), \qquad \forall i,j = 1,\cdots,N\,, j \neq i. \tag{31}$$

This property is illustrated in Figure 3, where we it is shown the mean correlation coefficients of a set of 3 sources, evaluated in all frequency bins. From Figure 3, we can see that the assumption of Equation (31) is clearly valid in most frequency bins, except at lower frequencies, as it was expected.

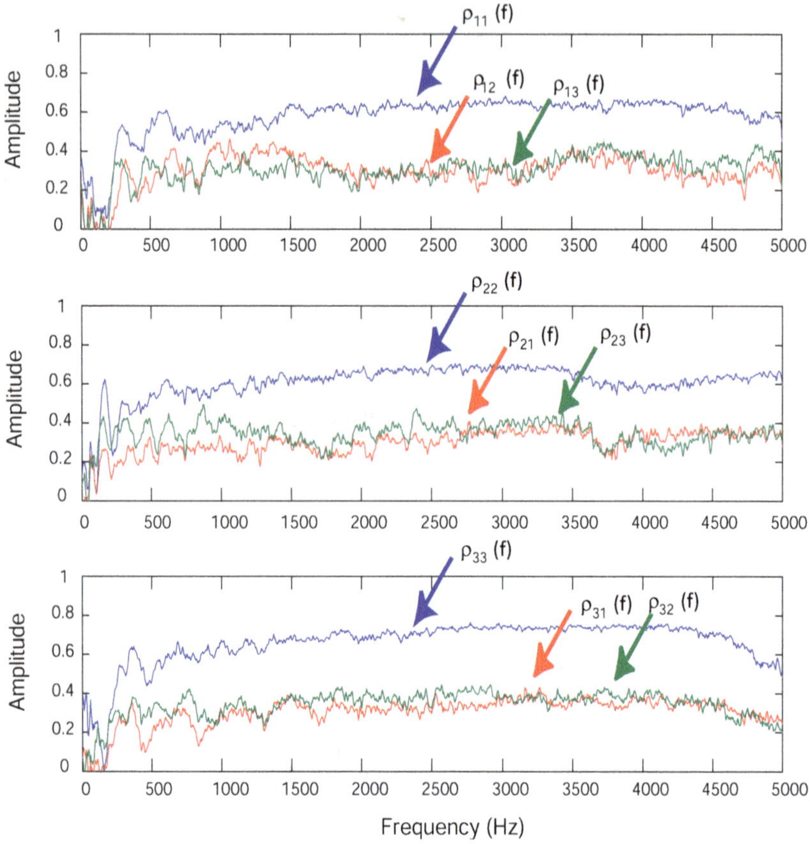

Fig. 3. *Mean correlation coefficients between one speech and itself, denoted as* $\bar{\rho}_{11}$, *and 2 other speech signals, denotes as* $\bar{\rho}_{12}$ *and* $\bar{\rho}_{13}$.

Considering the mean correlation coefficient, we define the global coherence of the source vector $\bar{\rho}$ as the average as in frequency as in components of the mean correlation coefficients, that is

$$\bar{\rho} = \frac{1}{N}\frac{1}{n_F}\sum_{i,f_k}\rho_{ii}(f_k) \qquad \in [-1,1]. \tag{32}$$

6.1 Description of the permutation correction algorithm

Consider the ordering vector $\pi_{f_k} = [\pi_{f_k}(1), \cdots, \pi_{f_k}(N)]$ associated to the existing permutation matrix $\mathbf{\Pi}_{f_k}$, defined in such way that its i element represents the non-nule row index of the i column of $\mathbf{\Pi}_{f_k}$. Therefore, the i component of the estimated source vector $\hat{S}_i(f_k,t)$ at frequency f_k corresponds to the component $\pi_{f_k}(i)$ of the output vector, that is

$$Y_{\pi_{f_k}(i)}(f_k,t) = \hat{S}_i(f_k,t). \tag{33}$$

In order to clarify this point, an example for $N = 3$ sources it is presented,

$$\mathbf{\Pi}_{f_k} = \begin{bmatrix} 0\ 1\ 0 \\ 0\ 0\ 1 \\ 1\ 0\ 0 \end{bmatrix} \rightarrow \pi_{f_k} = [3\ 1\ 2], \tag{34}$$

which means that $\hat{S}_1(f_k) = Y_3(f_k)$, $\hat{S}_2(f_k) = Y_1(f_k)$, y $\hat{S}_3(f_k) = Y_2(f_k)$. The global coherence of the outputs with alignment errors is given then by

$$\bar{\rho}(\mathbf{\Pi}_:) = \frac{1}{N}\frac{1}{n_F}\sum_{i,f_k}\left(\frac{1}{n_F}\sum_{f_p}\rho_{\pi_{f_k}(i),\pi_{f_p}(i)}(f_k, f_p)\right) \qquad \in [-1,1]. \tag{35}$$

From Equation (31) we can deduce that the global coherence of the outputs with alignment errors will be lower than the global coherence of the sources. Hence, it is possible to derive a contrast based on the global coherence which maximization achieves to solve the permutation problem. For that, we define, analogously to Equation (35), a global coherence of the corrected outputs

$$\bar{\rho}(\mathbf{Q}_:) = \frac{1}{N}\frac{1}{n_F}\sum_{i,f_k}\left\{\frac{1}{n_F}\sum_{f_p}\rho_{q_{f_k}(i) q_{f_p}(i)}(f_k, f_p)\right\} \qquad \in [-1,1], \tag{36}$$

where $q_{f_k}(i) = p_{f_k}(\pi_{f_k}(i))$, $i = 1, \ldots, N$ are the elements of the ordering global vector at frequency f_k. This global coherence will be maximum when the global permutation matrices satisfy the condition of Equation (23). Hence, the Equation (36) constitutes a coherence contrast for solving the permutation problem.

In order to calculate the permutation matrices that correct the alignment, it is necessary to solve the next constrained optimization problem

$$\mathbf{P}_: = \left\{\mathbf{P}_{f_1}, \mathbf{P}_{f_2}, \cdots, \mathbf{P}_{f_{n_F}}\right\} = \underset{\mathbf{P}_:}{\arg\max}\ \{\bar{\rho}(\mathbf{Q}_:)\}. \tag{37}$$

However, since it is not possible to find an analytical solution to the optimization problem, it is necessary to estimate the permutation correction matrices in an iterative way. Here we adopt a block coordinated ascent method, since it provides good permutation corrections with an efficient computational cost. In this method, at each iteration, the correction permutation matrices are calculated in a independent manner in all the frequencies as follows,

- Step 1: Calculate the mean correlation coefficients $\rho_{ij}^{(l)}(f_k)$ for all N separated components and for all frequency bins. The superindex (l) denotes the iteration index of the algorithm.

$$\begin{aligned} \rho_{ij}^{(l)}(f_k) &= \frac{1}{n_F}\sum_{p=1}^{n_F}\rho_{ij}^{(l)}(f_k, f_p) \\ &= \frac{1}{n_F}\sum_{p=1}^{n_F}\rho(|Y_i|_{dB}^{(l)}(f_k, t), |Y_j|_{dB}^{(l)}(f_p, t)). \end{aligned} \tag{38}$$

- Step 2: Find at each f_k the permutation matrix $P^{(l)}(f_k)$ that maximizes the sum of the mean correlation coefficients

$$\mathbf{P}_{f_k}^{(l)} = \arg\max_{\mathbf{P}_{f_k} \in \mathcal{P}} \sum_{i=1}^{N} \frac{1}{N} \rho_{p_{f_k}^{(l)}(i),i}(f_k). \tag{39}$$

- Step 3: If $\mathbf{P}_{f_k}^{(l)} \neq \mathbf{I}_N$ for any f_k, reorder the estimated components as

$$\mathbf{Y}^{(l+1)}(f_k, t) = \mathbf{P}^{(l)}(f_k)\mathbf{Y}^{(l)}(f_k, t), \tag{40}$$

set the iteration index $l = l + 1$ and go to step 1. Otherwise, it is considered that the estimated components are well aligned and end the algorithm.

It is important to note that convergence to the optimal solution is not guaranteed. However, in practice, the convergence to local optima that provide highly erroneous solutions is highly improbable.

6.2 Performance in a perfect separation situation

Here we present some experiments that were conducted to illustrate the robustness of the proposed method in perfect separation situation. For that, we artificially applied randomly selected permutation matrices to a set of spectrograms of speech sources $\mathbf{S}(f, t)$ at each frequency bin. The result corresponds with the outputs of a frequency domain blind source separation scheme, when the separation is achieved perfectly in all the frequencies. We used speech sources of 5-seconds long sampled at 10 kHz, randomly chosen from the database of 12 individual male and female sources in sources (2012). The parameter for the calculus of the STFT were FFT length of 2048 points, Hanning windows of 1024 samples and 90 % overlap. Then, we used the correction permutation algorithm in order to recover the original spectrograms. In Table 1, it is presented the average number of unsolved permutations for a set of 30 different simulation for each configuration from $N = 2, \cdots, 8$ speech sources.

	2x2	3x3	4x4	5x5	6x6	7x7
Errors	1.87	2.67	7.87	12.73	14.07	24.13

Table 1. Performance of the permutation correction algorithm in perfect separation situation. Results are the averaged remaining unsolved permutations (errors) when the number of sources are $N = 2, \cdots, 8$ over 30 simulations.

In all the simulations the algorithm correctly order the frequency components, remaining some permuted solutions at lower frequencies as we expected, since the speech sources do not always satisfy the property of spectral coherence. However, those errors do not affect the quality of the recovered speech sources, since they are always located at very low frequencies. In Figure 4, we show the spectrograms of the original, permuted and recovered signals for one simulation of 6 speech sources, where it can be corroborated the robustness of the permutation correction algorithm. Another important feature of the algorithm is its capacity to order the source components by using a reduced number of iterations. For instance, in the previous experiment, the convergence was achieved in only four iterations.

7. Simulations

In this section we are going to test the performance of the initialization procedure and the permutation correction algorithm with both simulated and live recording by means of the

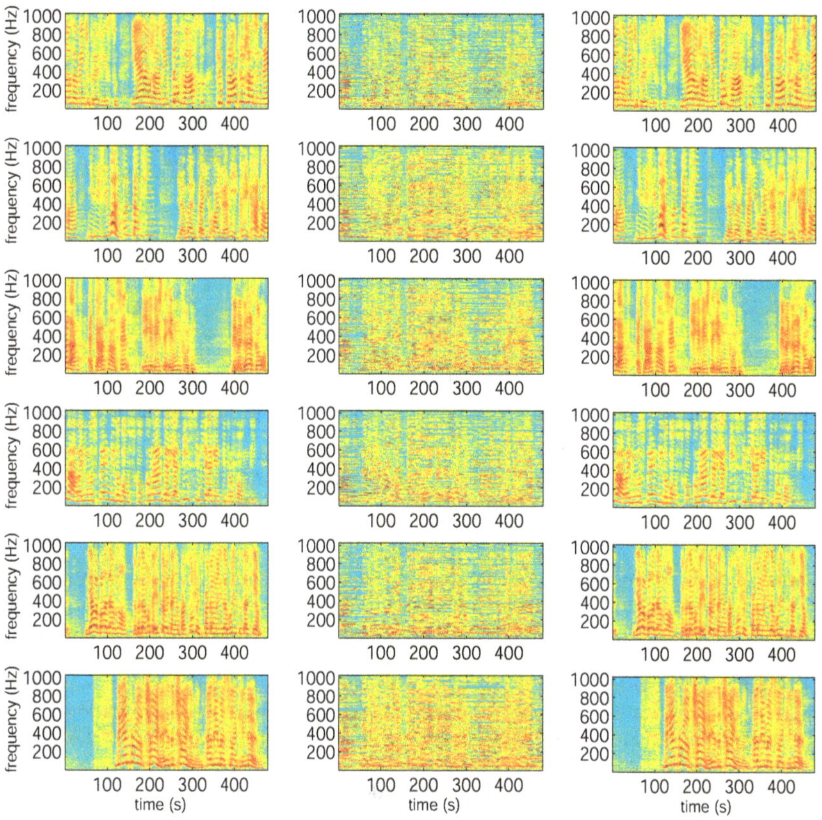

(a) Original sources (b) Permuted sources (c) Recovered sources

Fig. 4. Performance of the proposed correction permutation algorithm in perfect separation situation for $N = 6$ speech sources. For clarity, we have arranged the outputs according to the original sources.

quality of the recovered sources. This quality was measured in terms of both objective and perceptually measures. The objective measures are the Source to Distortion Ratio (SDR), the Source to Interferences Ratio (SIR) and the Source to Artifacts Ratio (SAR) computed by the BSS_EVAL toolbox, Fèvotte et al. (2006), whereas the perceptually measure is the Perceptual Evaluation of Speech Quality (PESQ) with a maximum value of 4.5. The Matlab code for calculating the PESQ index can be found in Loizou (2007).

7.1 Performance for simulated recording

For the synthetic mixtures, we considered the 2×2 and 3×3 mixing system for the configuration of microphones and loudspeakers showed in Figure 5. The corresponding channel impulse responses were determined by the Roomsim toolbox roomsim (2012). The sources were randomly chosen from male and female speakers in a database of 12 individual recordings of 5 s duration and sampled at 10 KHz, available in sources (2012), for the 2×2 case, and in a database of 8 individual recordings from the Stereo Audio Source Separation Evaluation Campaign 2007 Vincent et al. (2007) of 10 second long sampled at 8 KHz for the 3×3 case.

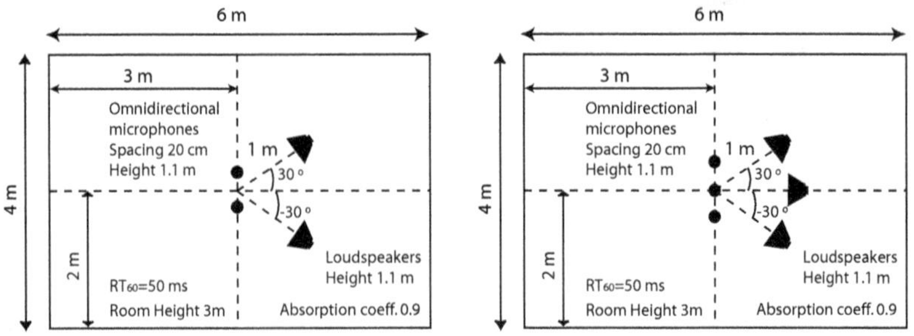

(a) 2×2 simulated room recording (b) 3×3 simulated room recording

Fig. 5. Microphones and loudspeakers positions for the simulated recording rooms.

The separation experiments have been carried out as following. For the computation of the STFT, the parameters chosen were: Hanning windows of length 1024 samples, FFT of 2048 points and 90% overlap. Then, we estimated the separation matrices by initializing the ThinICA algorithm with the two procedures presented in Section 4: the classical initialization and the initialization with $k = 1$ preceding frequency, which will be referred from now on as ThinICA$_{\text{classic}}$ and ThinICA$_{\text{ini1}}$, respectively. After that, we fixed the permutation problem applying the method described in Section 6, and the scaling ambiguity by the Minimal Distortion Principle. Finally, we transformed the separation matrices back to the time domain and filtered the observations to obtain the time domain estimated sources and the quality of those signals was computed with the aforementioned methods. For comparison, we also carried out the same separation experiments by using IVA algorithm. The obtained results are presented In Table 2.

In Figure 6 is depicted an example of original sources, mixtures and demixed sources of one 3×3 separation experiment by using ThinICA$_{\text{ini1}}$ simulation configuration.

Note that, in the simplest case, the 2×2 case, the initialization procedure does not seem to introduce any significant improvement in the quality of the recovered sources respect to the classical procedure. This can be explained when the separation algorithm and the permutation correction algorithm adequately converge in all the experiments. Nevertheless, in the most complex 3×3 case, where the convergence of the separation algorithms can be more difficult to achieve, the initialization procedure over perform the classical procedure. Thus, one can conclude that the initialization procedure can obtain better performances in hard situations, mainly as the increment of the number of sources, or when it is available a reduced number of data. It is important to note that IVA method failed in three of the fifteen simulations. In those experiments IVA recovered only one sources, remaining the other two mixed.

From Table 2 we find another interesting result. The quality of the separated source obtained with fd-ICA methods by means of SIR, SAR and SDR ratios are better than those obtained

		SIR(dB)	SAR(dB)	SDR(dB)	PESQ
2x2	$\text{ThinICA}_{\text{classic}}$	19.87	17.57	15.35	3.05
	$\text{ThinICA}_{\text{ini1}}$	20.34	18.01	15.83	3.05
	IVA	14.93	13.46	10.75	2.94
3x3	$\text{ThinICA}_{\text{classic}}$	19.44	11.33	10.45	2.42
	$\text{ThinICA}_{\text{ini1}}$	22.32	11.88	11.31	2.54
	IVA	12.86	8.87	6.54	2.53
	IVA*	15.54	9.48	8.23	2.68

Table 2. Quality evaluation for 2×2 and 3×3 cases using various separation methods. Results were averaged over 23 mixtures in the 2×2 case, and 15 mixtures in the 3×3 case, except in IVA* case, where results from 3 simulations, in which IVA method failed, have been retracted.

with IVA method. However, the PESQ quality measure obtained in all the cases is similar. This discrepancy can be explained by the conceptual differences between the different quality measures. In general, simulations show that fd-ICA methods obtain better separation ratios than the IVA method, despite hearing the recovered speech reveals that fd-ICA introduce more reverberation in the estimated sources than IVA method. This reverberation degrades the perceived quality of resulting sound, which explains the similar PESQ score of fd-ICA and IVA methods.

7.2 Performance for live recording

In this study, we have reproduce two clean speech sources in a typical office room to obtain a real recording in a noisy environment. A sampling frequency of 10 kHz has been used. The recording setup includes Logitech 5.1 Z-5450 loudspeakers, Shure Lavalier MX_ 180 Serie Microflex microphones and a Digi003Rack recording interface. The source signals were estimated by using both fd-ICA by means of ThinICA algorithm and IVA method. In this case, the STFT were computed using Hanning windows of length 2048 samples, FFT of 4096 points and 90% overlap.

For correctly interpret the results, it is important to note that the mixing conditions on live recordings present significant differences respect to the synthetic mixtures. One of the most important feature is the presence of additive noise coming from different noise sources in the recording environment, such us computers, air conditioning, etc. As a consequence, the estimated components will not correspond to the original sources, since they will have also a component of additive noise. Thus, we have included a component of additive noise in the objective quality measure. This noise component have been estimated in the silence periods of the recording. In Table 3 we show the obtained results. Due to computational limitations in the BSS EVAL toolbox, we present only the SIR and SNR ratios.

As it can be seen in Table 3, the three methods perform well in this situation, although the best performance is achieved by the ThinICA method including the initialization procedure. Moreover, fd-ICA methods present better SIR ratios than IVA method, as in the synthetic mixtures experiments. Also, in Figure 7 is depicted the original sources, real recordings and demixed sources by using $\text{ThinICA}_{\text{ini1}}$ simulation configuration.

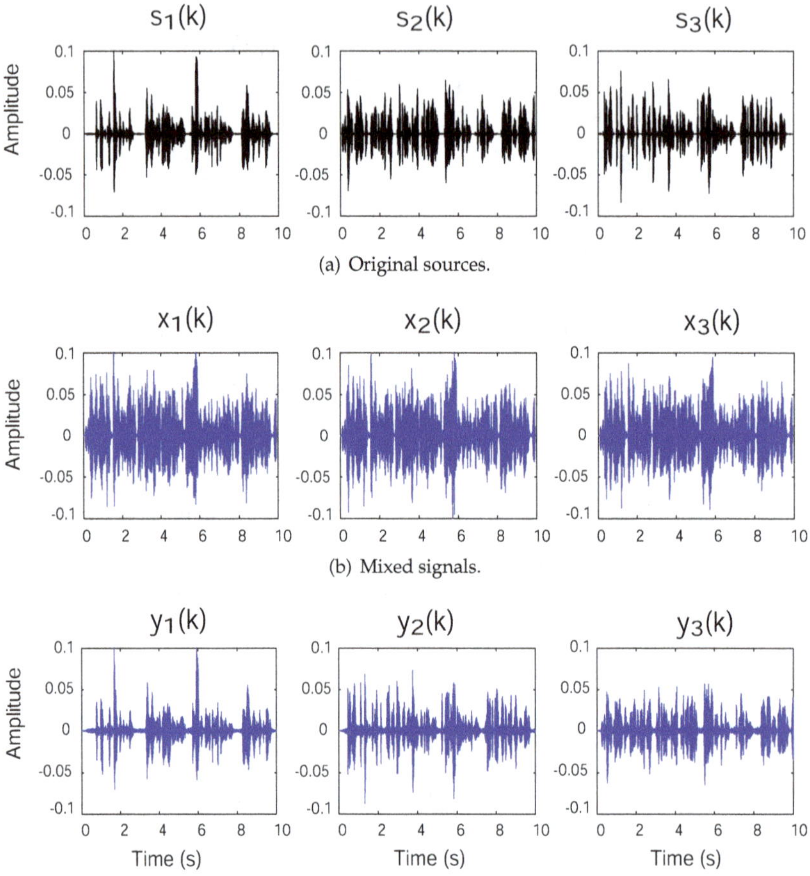

(a) Original sources.

(b) Mixed signals.

(c) Recovered signals.

Fig. 6. Example of 3×3 separation experiment by using ThinICA$_{\text{ini1}}$ simulation configuration.

	SIR(dB)	SNR(dB)	PESQ
ThinICA$_{\text{classic}}$	13.37	14.73	1.96
ThinICA$_{\text{ini1}}$	16.28	18.88	2.07
IVA	8.84	14.27	1.77

Table 3. SIR (dB), SAR (dB), SDR (dB) and PESQ index for 2×2 real recording separation.

To conclude, we have also applied the complete method to a live recording of 3 sources and 3 mixtures provided in SISEC (2012). The quality of the estimated sources was measured in terms of Source to Interferences Ratio (SIR) by E. Vincent, since the original sources are not public. An average SIR of 10.1 dB was obtained.

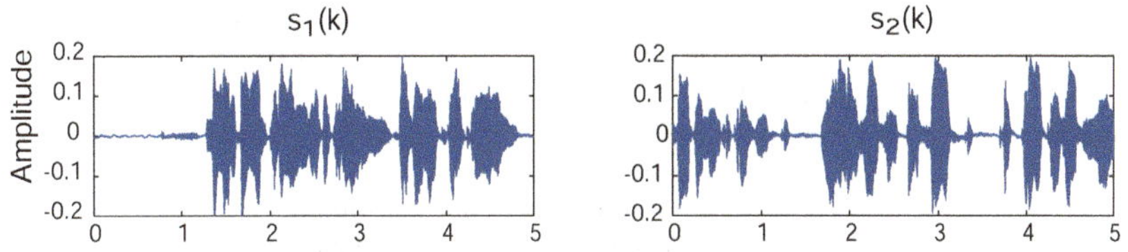

(a) Original sources.

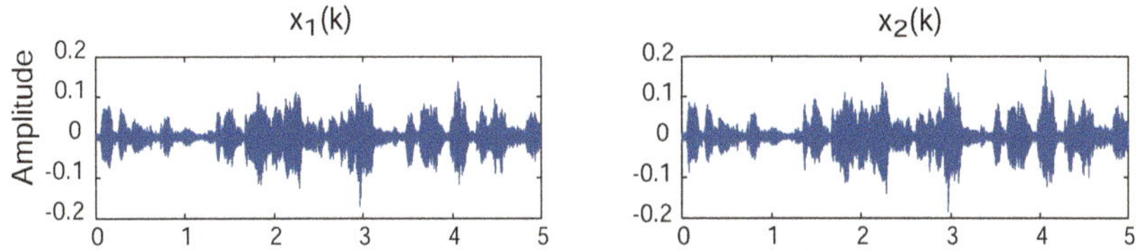

(b) Recordered signals.

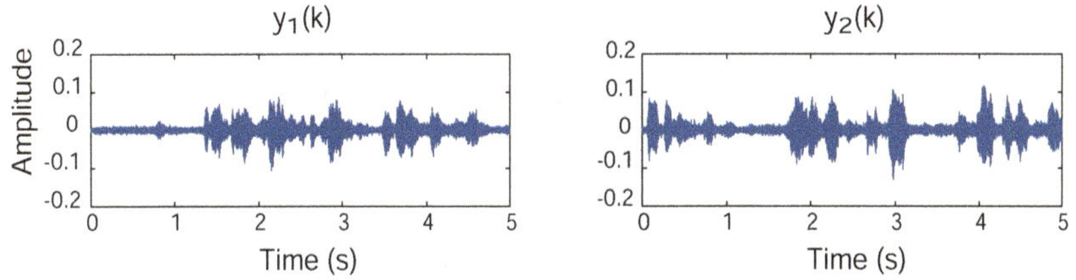

(c) Recovered signals.

Fig. 7. Example of a separation experiment with real recordered signals by using ThinICA$_{\text{ini1}}$ simulation configuration.

8. Conclusions

In this chapter we have considered the problem of the blind separation of speech signals recorded in a real room, when the number of speakers equals the number of simultaneous recordings. We have adopted the time-frequency approach, focusing our attention in the initialization of the separation algorithms and the permutation problem which is ubiquitous to fd-ICA methods. In order to improve the performance of the existing methods we have incorporated an initialization procedure for those ICA algorithms that work in the time-frequency domain and require the whitening of the observations as a preprocessing step. This initialization exploit the local continuity of the demixing filter in the frequency domain, which is a valid property for reverberant filters in a wide range of frequencies. For that, the separation matrix in one frequency bin is initialized from its joint closest value to a set of separation systems already computed at nearby frequencies. Computer simulations show that this initialization, when it is incorporated to the existing ICA algorithms, reduces drastically the number of permutations, preserving the separated components well aligned in wide frequency blocks. Simulations with more than two sources reveal that the proposed initialization also helps to the convergence of the ICA algorithms that solve the separation in each frequency.

The permutation problem becomes a severe problem when the number of sources is large or in high reverberant environments. Nowadays, it is still considered an open problem. For solving the permutation problem, we have present a method, based on the amplitude correlation modulation property of speech signals, that arises the general case of N sources and N observations. We have defined for each frequency bin a measure of coherence based on the amplitude modulation correlation property of speech signals. This measure has been used to formulate a coherence contrast function which maximization allows to successfully arrange the estimated components. An iterative method has been provide for searching the maxima of the contrast. The robustness of the algorithm has been illustrated for artificially permuted sources, which corresponds with a situation of perfect separation. Results show that the algorithm is able to reorder completely the frequency components, except for some very low frequencies that in some cases remained permuted. However, this does not affect to the quality of the recovered sources. Finally, experiments with simulated and live recording in a room with reverberation, for the case where two or three sources are mixed, show that the complete method improves considerably the performance of classical fd-ICA method, as well as IVA method, by means of both objective and perceptually measures.

9. Acknowledgements

This work has been supported by the Ministry of Science and Innovation project of the Government of Spain (Grant TEC2011-23559) and the Andalusian Government (Grant TIC-7869). We thank Emmanuel Vincent collaboration for the evaluation of the results.

10. References

Comon, P., Independent Component Analysis a new concept?, *Signal Processing*, Vol. 36, pp 287-314, 1994.

Smaragdis P. (1998),Blind separation of convolved mixtures in the frequency domain, *Neurocomputing*, Vol. 2, Nov. 1998, pp. 21-34.

Parra, L. & Spence, C., Convolutive blind source separation of non-stationary sources, *IEEE Trans. on Speech and Audio Processing*, May. 2000, pp. 320-327.

Araki, S., Mukai, R., Makino, S., Nishikawa, T. & Saruwatari, H. , The fundamental limitation of frequency domain blind source separation for convolutive mixtures of speech, *IEEE Transactions on Speech and Audio Processing*, Vol. 11, No. 2, Mar. 2003, pp. 109-116.

Hyvärinen, A. & Oja, E., A fast fixed point algorithm for independent component analysis, *Neural Computation*,Vol. 9, 1997, pp. 1483-1492.

Bingham, E. & Hyvärinen, A., A fast fixed point algorithm for independent component analysis of complex valued signals, *International Journal of Neural Systems*, Vol. 10, No. 1, 2000, pp. 1-8.

Koldovský, Z., Tichavský, P. & Oja, E., Efficient variant of algorithm FastICA for independent component analysis attaining the Cramé Rao lower bound, *IEEE Trans. on Neural Networks*, Vol. 17, No. 5, Sep. 2006, pp. 1265-1277.

Belouchrani, A., Abed-Meraim, K., Cardoso, J.F., Moulines, E., A blind source separation technique using second-order statistics, *IEEE Trans. on Signal Processing*, Vol. 45, No. 2, Feb. 1997, pp. 434-444.

Choi, S. & Cichocki, A., Blind separation of nonstationary sources in noisy mixtures, *IEEE Workshop on Neural Networks for Signal Processing (NNSP'2000)*, Sydney, Australia, Dec. 2000, pp. 11-13.

D. Donoho, On Minimun Entropy Deconvolution, *Applied Time Series Analysis II*, D. F. Findley Editor, Academic Press, New York, 1981,pp. 565-608.

Cruces, S., Cichocki, A. & Amari, S., From Blind Signal Extraction to Blind Instantaneous Signal Separation: Criteria, Algortihms and Stability, *IEEE Trans. on Neural Networks*, vol 15(4), July 2004, pp. 859-873.

De Lathauwer, L., De-Moor, B. & Vandewalle, J., On the Best Rank-1 and Rank-(R1;R2;...;RN) Approximation of Higher-order Tensors, *SIAM J. Matrix Anal. Appl.*, vol. 21(4), 2000, pp. 1324-1342.

S. Cruces & A. Cichocki, Combining Blind Source Extraction with Joint Approximate Diagonalization: Thin Algorithms for ICA, *Proceedings of the 4rd International Symposium on Independent Component Analysis and Blind Signal Separation*, Japan, 2003, pp. 463-468.

Cruces, S., Cichocki, A. & De Lathauwer, L., Thin QR and SVD factorizations for simultaneous Blind Signal Extraction, *Proceeding of the European Signal Processing Conference (EUSIPCO'04)*, Viena Austria, 2004, pp. 217-220

Available: http://www.bsp.brain.riken.jp/ICALAB/, Accessed 2012 Feb. 1.

Durán Díaz I. & Cruces, S., A joint optimization criterion for blind DS-CDMA detection, *EURASIP Journal of Applied Signal Processing, Special Issue: Advances in Blind Source Separation*, 2007, pp. 1-11.

Sarmiento, A.; Cruces, S. & Durán, I., Improvement of the initialization of time-frequency algorithms for speech separation,*Proceedings of Int Conf. on Independent Component Analysis and Blind Source Separation (ICA'09)*, 2009, pp. 629-636.

Sarmiento, A., Durán-Díaz & I., Cruces S., Initialization method for speech separation algorithms that work in the time frequency domain, *The Journal of the Acoustical Society of America*, Vol. 127, No. 4, 2010, pp. 121-126.

Matsuoka, K. & Nakashima, S., Minimal distorsion principle for blind source separation,*Proceedings of Int. Conf. on Independent Component Analysis and Blind Source Separation*, 2001, pp. 722-727.

Pedersen, M.S., Larsen, J., Kjems, U. & Parra, L.C., A survey of convolutive blind source separation methods, *Multichannel Speech Processing Handbook*, Eds. Jacob Benesty and Arden Huang, Springer 2007, Chapter 51, pp. 1065-1084.

Anemüller, J. & Kollmeier, B. , Amplitude modulation decorrelation for convolutive blind source separation, *Proceedings of Second International Workshop on Independent Component Analysis and Blind Signal Separation*, Jun. 2000, pp. 215-220.

Ikeda, S. & Murata, N., An approach to blind source separation of speech signals, *Proceedings of International Conference on Artificial Neural Networks (ICANN'98)*, Sep. 1998, Sweden, pp.761-766.

Murata, N., Ikeda,S. & and Ziehe, A., An approach to blind source separation based on temporal structure of speech signals, *Neurocomputing*, Vol. 41, Issue 1-4, Oct. 2001, pp.1-24.

Rahbar, K. & Reilly J.P., A frequency domain method for blind source separation of convolutive audio mixtures, *IEEE Transactions on speech and audio processing*, Vol. 13,No. 5, 2005, pp. 832-844.

Kim, T., Lee, I. & Lee T.W., Independent Vector Analysis: definition and algorithms, *Proceeding of Fortieth Asilomar Conference on Signals, Systems and Computers (ACSSC '06)*, 2006, pp.1393-1396.

Pham, D.T., Serviére, C. & Boumaraf, H. , Blind separation of convolutive audio mixtures using nonstationarity, *Proceedings of Int Conf. on Independent Component Analysis and Blind Source Separation (ICA'03)*, Nara, Japan, Apr. 2003.

Asano, F., Ikeda, S., Ogawa, M., Asoh, H.& Kitawaki, N., Combined approach of array processing and independent component analysis for blind separation of acoustic signals, *IEEE Transactions on Speech and Audio Processing*,Vol.11, No. 3, May. 2003, pp. 204- 215.

Sudhakar, P. & Gribonval, R., A sparsity-based method to solve permutation indeterminacy in frequency-domain convolutive blind source separation, *Proceedings of the 8th International Conference on Independent Component Analysis and Signal Separation (ICA '09)*, pp. 338-345.

Parra, L.C. & Alvino, C.V., Geometric source separation: merging convolutive source separation with geometric beamforming, *IEEE Transactions on Speech and Audio Processing*, Vol.10, No.6, Sep. 2002, pp. 352- 362.

Kurita, S., Saruwatari, H., Kajita, S., Takeda, K. & Itakura, F., Evaluation of blind signal separation method using directivity pattern under reverberant conditions, *Proceedings of IEEE International Conference on Acoustics, Speech, and Signal Processing (ICASSP '00)*, Vol. 5, 2000, pp. 3140-3143.

Saruwatari, H., Kurita, S., Takeda, K., Itakura, F., Nishikawa, T. & Shikano, K., Blind source separation combining independent component analysis and beamforming, *EURASIP Journal on Applied Signal Processing*, Jan. 2003, pp. 1135-1146.

Sawada, H., Mukai, R., Araki, S. & Makino, S., A robust and precise method for solving the permutation problem of frequency-domain blind source separation, *IEEE Transactions on Speech and Audio Processing*, Vol. 12, No. 5, Sept. 2004, pp. 530- 538.

Sawada, H., Mukai, R., Araki, S. & Makino, S., Frequency-domain blind source separation without array geometry information,*Proceedings of Joint Workshop on Hands-Free Speech Communication and Microphone Arrays (HSCMA'05)*, Mar. 2005.

Sawada, H., Araki, S., Mukai, R. & Makino, S., Solving the permutation problem of frequency-domain BSS when spatial aliasing occurs with wide sensor spacing,*Proceedings of the IEEE International Conference on Acoustics, Speech and Signal Processing (ICASSP'06)*, Vol. 5, May. 2006, pp. 77-80.

Sarmiento, A., Durán-Díaz, I., Cruces, S. & Aguilera, P., Generalized method for solving the permutation problem in frequency-domain blind source separation of convolved speech signals, *Proceedings of the 12th Annual Conference of the International Speech Communication Association (INTERSPEECH'11)*, Aug. 2011, pp. 565-568.

Available: http://www.imm.dtu.dk/pubdb/p.php?4400, Accessed 2012 Feb. 1.

Fèvotte, C., Gribonval, R., Vincent, E. ,BSS_ EVAL toolbox user guide, *Tech. Rep. 1706*, IRISA, Rennes, France, 2005, Available: http://www.irisa.fr/metiss/bss_ eval, Accessed 2012 Feb. 1.

Loizou, P.C., *Speech Enhancement. Theory and Practice*, CRC Press, 2007.

Campbell, D., Roomsim Toolbox, Available. http://media.paisley.ac.uk/ campbell/Roomsim/, Accessed 2012 Feb. 1

Vincent,E., Sawada, H., Bofill, P., Makino, S. & Rosca J. P., First Stereo Audio Source Separation Evaluation Campaign: Data, algorithms and results,*Proceedings of the 7th International Conference on Independent Component Analysis and Signal Separation (ICA'07)*, 2007, pp. 552–559.

Available: http://sisec2010.wiki.irisa.fr/tiki-index.php, Accessed 2012 Feb. 1.

15

Advancements in the Time-Frequency Approach to Multichannel Blind Source Separation

Ingrid Jafari[1], Roberto Togneri[1] and Sven Nordholm[2]
[1]*The University of Western Australia*
[2]*Curtin University*
Australia

1. Introduction

The ability of the human cognitive system to distinguish between multiple, simultaneously active sources of sound is a remarkable quality that is often taken for granted. This capability has been studied extensively within the speech processing community and many an endeavor at imitation has been made. However, automatic speech processing systems are yet to perform at a level akin to human proficiency (Lippmann, 1997) and are thus frequently faced with the quintessential "cocktail party problem": the inadequacy in the processing of the target speaker/s when there are multiple speakers in the scene (Cherry, 1953). The implementation of a source separation algorithm can improve the performance of such systems. Source separation is the recovery of the original sources from a set of observations; if no *a priori* information of the original sources and/or mixing process is available, it is termed blind source separation (BSS). Rather than rely on the availability of *a priori* information of the acoustic scene, BSS methods often employ an assumption on the constituent source signals, and/or an exploitation of the spatial diversity obtained through a microphone array. BSS has many important applications in both the audio and biosignal disciplines, including medical imaging and communication systems.

In the last decade, the research field of BSS has evolved significantly to be an important technique in acoustic signal processing (Coviello & Sibul, 2004). The general BSS problem can be summarized as follows. M observations of N sources are related by the equation

$$\mathbf{X} = \mathbf{A}\mathbf{S}\,, \tag{1}$$

where $\mathbf{X}$ is a matrix representing the M observations of the N sources contained in the matrix $\mathbf{S}$, and $\mathbf{A}$ is the unknown $M \times N$ mixing matrix. The aim of BSS is to recover the source matrix $\mathbf{S}$ given simply the observed mixtures $\mathbf{X}$, however rather than directly estimate the source signals, the mixing matrix $\mathbf{A}$ is instead estimated. The number of sensors relative to the number of sources present determines the class of BSS: evendetermined ($M = N$), overdetermined ($M > N$) or underdetermined ($M < N$). The evendetermined system can be solved via a linear transformation of the data; whilst the overdetermined case can be solved by an estimation of the mixing matrix $\mathbf{A}$. However, due to its intrinsic noninvertible nature, the underdetermined BSS problem cannot be resolved via a simple mixing matrix estimation, and the recovery of the original sources from the mixtures is considerably more complex than the other aforementioned BSS instances. As a result of its intricacy, the underdetermined BSS problem is of growing interest in the speech processing field.

Traditional approaches to BSS are often based upon assumptions about statistical properties of the underlying source signals, for example independent component analysis (ICA) (Hyvarinen et al., 2001), which aims to find a linear representation of the sources in the observation mixtures. Not only does this rely on the condition that the constituent source signals are statistically independent, it also requires that no more than one of the independent components (sources) follows a Gaussian distribution. However, due to the fact that techniques of ICA depend on matrix inversion, the number of microphones in the array must be at least equal to, or greater than, the number of sources present (i.e. even- or overdetermined cases exclusively). This poses a significant restraint on its applicability to many practical applications of BSS. Furthermore, whilst statistical assumptions hold well for instantaneous mixtures of signals, in most audio applications the expectation of instantaneous mixing conditions is largely impractical, and the convolutive mixing model is more realistic.

The concept of time-frequency (TF) masking in the context of BSS is an emerging field of research that is receiving an escalating amount of attention due to its ease of applicability to a variety of acoustic environments. The intuitive notion of TF masking in the speech processing discipline originates from analyses on human speech perception and the observation of the phenomenon of masking in human hearing: in particular, the fact that the human mind preferentially processes higher energy components of observed speech whilst compressing the lower components. This notion can be administered within the BSS framework as described below.

In the TF masking approach to BSS, the assumption of sparseness between the speech sources, as initially investigated in (Yilmaz & Rickard, 2004), is typically exploited. There exists several varying definitions for sparseness in the literature; (Georgiev et al., 2005) simply defines it as the existence of "as many zeros as possible", whereas others offer a more quantifiable measure such as kurtosis (Li & Lutman, 2006). Often, a sparse representation of speech mixtures can be acquired through the projection of the signals onto an appropriate basis, such as the Gabor or Fourier basis. In particular, the sparseness of the signals in the short-time Fourier transform (STFT) domain was investigated in (Yilmaz & Rickard, 2004) and subsequently termed W-disjoint orthogonality (W-DO). This significant discovery of W-DO in speech signals motivated the degenerate unmixing estimation technique (DUET) which was proven to successfully recover the original source signals from simply a pair of microphone observations. Using a sparse representation of the observation mixtures, the relative attenuation and phase parameters between the observations are estimated at each TF cell. The parameters estimates are utilized in the construction of a power-weighted histogram; under the assumption of sufficiently ideal mixing conditions, the histogram will inherently contain peaks that denote the true mixing parameters. The final mixing parameters estimates are then used in the calculation of a binary TF mask.

This initiation into the TF masking approach to BSS is oft credited to the authors of this DUET algorithm. Due to its versatility and applicability to a variety of acoustic conditions (under-, even- and overdetermined), the TF masking approach has since evolved as a popular and effective tool in BSS, and the formation of the DUET algorithm has consequently motivated a plethora of demixing techniques.

Among the first extensions to the DUET was the TF ratio of mixtures (TIFROM) algorithm (Abrard & Deville, 2005) which relaxed the condition of W-DO of the source signals, and had a particular focus on underdetermined mixtures for arrays consisting of more than two sensors. However, its performance in reverberant conditions was not established and the observations were restricted to be of the idealized linear and instantaneous case. Subsequent research as in (Melia & Rickard, 2007) extended the DUET to echoic conditions with the DESPRIT (DUET-ESPRIT) algorithm; this made use of the existing ESPRIT (estimation of signal parameters via rotational invariance technique) algorithm (Roy & Kailath, 1989). This ESPRIT algorithm was combined with the principles of DUET, however, in contrast to the DUET, it utilized more than two microphone observations with the sensors arranged in a uniform linear array. However, due to this restriction in the array geometry, the algorithm was naturally subjected to front-back confusions. Furthermore, a linear microphone arrangement poses a constraint upon the spatial diversity obtainable from the microphone observations.

A different avenue of research as in (Araki et al., 2004) composed a two-stage algorithm which combined the sparseness approach in DUET with the established ICA algorithm to yield the SPICA algorithm. The sparseness of the speech signals was firstly exploited in order to estimate and subsequently remove the active speech source at a particular TF point; following this removal, the ICA technique could be applied to the remaining mixtures. Naturally, a restraint upon the number of sources present at any TF point relative to the number of sensors was inevitable due to the ICA stage. Furthermore, the algorithm was only investigated for the stereo case.

The authors of the SPICA expanded their research to nonlinear microphones arrays in (Araki et al., 2005; 2006a;b) with the introduction of the clustering of normalized observation vectors. Whilst remaining similar in spirit to the DUET, the research was inclusive of nonideal conditions such as room reverberation. This eventually culminated in the development of the multiple sensors DUET (MENUET) (Araki et al., 2007). The MENUET is advantageous over the DUET in that it allows more than two sensors in an arbitrary nonlinear arrangement, and is evaluated on underdetermined reverberant mixtures. In this algorithm the mask estimation was also automated through the application of the *k*-means clustering algorithm. Another algorithm which proposes the use of a clustering approach for the mask estimation is presented in (Reju et al., 2010). This study is based upon the concept of complex angles in the complex vector space; however, evaluations were restricted to a linear microphone array.

Despite the advancements of techniques such as MENUET, it is not without its limitations: most significantly, the *k*-means clustering is not very robust in the presence of outliers or interference in the data. This often leads to non-optimal localization and partitioning results, particularly for reverberant mixtures. Furthermore, binary masking, as employed in the MENUET, has been shown to impede on the separation quality with respect to the musical noise distortions. The authors of (Araki et al., 2006a) suggest that fuzzy TF masking approaches bear the potential to reduce the musical noise at the output significantly. In (Kühne et al., 2010) the use of the fuzzy *c*-means clustering for mask estimation was investigated in the TF masking framework of BSS; on the contrary to MENUET, this approaches integrated a fuzzy partitioning in the clustering in order to model the inherent ambiguity surrounding the membership of a TF cell to a cluster. Examples of contributing factors to such ambiguous conditions include the effects of reverberation and additive channel noise at the sensors in the array. However, this investigation, as with many others in the literature, possessed the significant restriction in its limitation to a linear microphone arrangement.

Another clustering approach to TF mask estimation lies with the implementation of Gaussian Mixture Models (GMM). The use of GMMs in conjunction with the Expectation-Maximization (EM) algorithm for the representation of feature distributions has been previously investigated in the sparseness approach to BSS (Araki et al., 2009; Izumi et al., 2007; Mandel et al., 2006). This avenue of research is motivated by the intuitive notion that the individual component densities of the GMM may model some underlying set of hidden parameters in a mixture of sources. Due to the reported success of BSS methods that employ such Gaussian models, the GMM-EM may be considered as a standard algorithm for mask estimation in this framework, and is therefore regarded as a comparative model in this study.

However, each of the TF mask estimation approaches to BSS discussed above are yet to be inclusive of the noisy reverberant BSS scenario. Almost all real-world applications of BSS have the undesired aspect of additive noise at the recording sensors (Cichocki et al., 1996). The influence of additive noise has been described as a very difficult and continually open problem in the BSS framework (Mitianoudis & Davies, 2003). Numerous studies have been proposed to solve this problem: (Li et al., 2006) presents a two-stage denoising/separation algorithm; (Cichocki et al., 1996) implements a FIR filter at each channel to reduce the effects of additive noise; and (Shi et al., 2010) suggests a preprocessing whitening procedure for enhancement. Whilst noise reduction has been achieved with denoising techniques implemented as a pre- or post-processing step, the performance was proven to degrade significantly at lower signal-to-noise ratios (SNR) (Godsill et al., 1997). Furthermore, the aforementioned techniques for the compensation of additive noise have yet to be extended and applied in depth to the TF masking approach to BSS.

Motivated by these shortcomings, this chapter presents an extension of the MENUET algorithm via a novel amalgamation with the FCM as in (Kühne et al., 2010) (see Fig. 1). The applicability of MENUET to underdetermined and arbitrary sensor constellations renders it superior in many scenarios over the investigation in (Kühne et al., 2010); however, its performance is hindered by its non-robust approach to mask estimation. Firstly, this study proposes that the combination of fuzzy clustering with the MENUET algorithm, which will henceforth be denoted as MENUET-FCM, will improve the separation performance in reverberant conditions. Secondly, it is hypothesized that this combination is sufficiently robust to withstand the degrading effects of reverberation and random additive channel noise. For all investigations in this study, the GMM-EM clustering algorithm for mask estimation is implemented with the MENUET (and denoted MENUET-GMM) for comparative purposes. As a side note, it should be observed that all ensuing instances of the term MENUET are in reference to the original MENUET algorithm as in (Araki et al., 2007).

The remainder of the chapter is structured as follows. Section 2 provides a detailed overview of the MENUET and proposed modifications to the algorithm. Section 3 explains the three different clustering algorithms and their utilization for TF mask estimation. Section 4 presents details of the experimental setup and evaluations, and demonstrates the superiority of the proposed MENUET-FCM combination over the baseline MENUET and MENUET-GMM for BSS in realistic acoustic environments. Section 5 provides a general discussion with insight into potential directions for future research. Section 6 concludes the chapter with a brief summary.

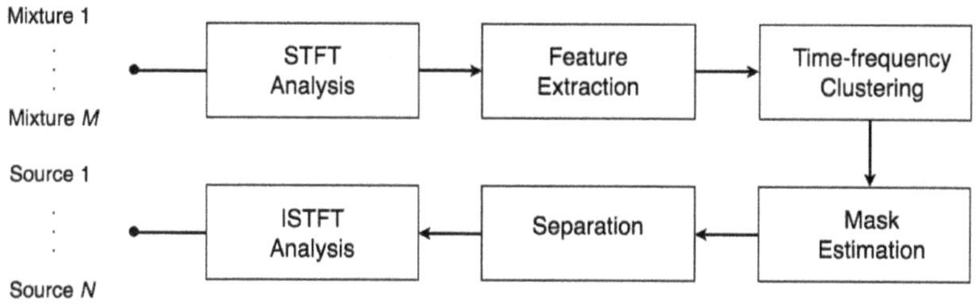

Fig. 1. Basic scheme of proposed time-frequency masking approach for BSS.

2. Source separation with TF masking

This section provides an introduction to the problem statement of underdetermined BSS and insight into the TF masking approach for BSS. The MENUET, MENUET-FCM and MENUET-GMM algorithms are described in greater detail.

2.1 Problem statement

Consider a microphone array made up of M identical sensors in a reverberant enclosure where N sources are present. It is assumed that the observation at the mth sensor can be modeled as a summation of the received images, denoted as $s_{mn}(t)$, of each source $s_n(t)$ by

$$x_m(t) = \sum_{n=1}^{N} s_{mn}(t) \ , \tag{2}$$

where

$$s_{mn}(t) = \sum_p h_{mn}(p) s_n(t-p) + n_m(t) \ , \tag{3}$$

and where t indicates time, $h_{mn}(p)$ represents the room impulse response from the nth source to the mth sensor and $n_m(t)$ denotes the additive noise present at the mth sensor.

The goal of any BSS system is to recover the sets of separated source signal images $\{\hat{s}_{11}(t), \ldots, \hat{s}_{M1}(t)\}, \ldots, \{\hat{s}_{1N}(t), \ldots, \hat{s}_{MN}(t)\}$, where each set denotes the estimated source signal $\hat{s}_n(t)$, and $\hat{s}_{mn}(t)$ denotes the estimate of the nth source image, $s_{mn}(t)$, at the mth sensor. Ideally, the separation is performed without any information about $s_n(t)$, $h_{mn}(p)$ and the true source images $s_{mn}(t)$.

2.2 Feature extraction

The time-domain microphone observations, sampled at frequency f_s, are converted into their corresponding frequency domain time-series $X_m(k,l)$ via the STFT

$$X_m(k,l) = \sum_{\tau=-L/2}^{L/2-1} \mathrm{win}(\tau) x_m(\tau + k\tau_0) e^{-jl\omega_0 \tau} \ , \quad m = 1, \ldots, M, \tag{4}$$

where $k \in \{0, \ldots, K-1\}$ is a time frame index, $l \in \{0, \ldots, L-1\}$ is a frequency bin index, $\mathrm{win}(\tau)$ is an appropriately selected window function and τ_0 and ω_0 are the TF grid resolution parameters. The analysis window is typically chosen such that sufficient information is retained within whilst simultaneously reducing signal discontinuities at the edges. A suitable window is the Hann window

$$\mathrm{win}(\tau) = 0.5 - 0.5\cos(\frac{2\pi\tau}{L}) \,, \quad \tau = 0, \ldots, L-1, \tag{5}$$

where L denotes the frame size.

It is assumed that the length of L is sufficient such that the main portion of the impulse responses h_{mn} is covered. Therefore, the convolutive BSS problem may be approximated as an instantaneous mixture model (Smaragdis, 1998) in the STFT domain

$$X_m(k,l) \approx \sum_{n=1}^{N} H_{mn}(l) S_n(k,l) + N_m(k,l) \,, \quad m = 1, \ldots, M, \tag{6}$$

where (k,l) represents the time and frequency index respectively, $H_{mn}(l)$ is the room impulse response from source n and sensor m. $S_n(k,l)$, $X_m(k,l)$ and $N_m(k,l)$ are the STFT of the mth observation, n^{th} source and additive noise at the mth sensor respectively. The sparseness of the speech signals assumes at most one dominant speech source $S_n(k,l)$ per TF cell (Yilmaz & Rickard, 2004). Therefore, the sum in (6) is reduced to

$$X_m(k,l) \approx H_{mn}(l) S_n(k,l) + N_m(k,l) \,, \quad m = 1, \ldots, M. \tag{7}$$

Whilst this assumption holds true for anechoic mixtures, as the reverberation in the acoustic scene increases it becomes increasingly unreliable due to the effects of multipath propagation and multiple reflections (Kühne et al., 2010; Yilmaz & Rickard, 2004).

In this work the TF mask estimation is realized through the estimation of the TF points where a signal is assumed dominant. To estimate such TF points, a spatial feature vector is calculated from the STFT representations of the M observations. Previous research has identified level ratios and phase differences between the observations as appropriate features in this BSS framework as such features retain information on the magnitude and the argument of the TF points. A comprehensive review is presented in (Araki et al., 2007), with further discussion presented in Section 4.2.1. Should the source signals exhibit sufficient sparseness, the clustering of the level ratios and phase differences will yield geometric information on the source and sensor locations, and thus facilitate effective separation.

The feature vector

$$\boldsymbol{\theta}(k,l) = \left[\theta^L(k,l), \theta^P(k,l)\right]^T, \tag{8}$$

per TF point is estimated as

$$\boldsymbol{\theta}^L(k,l) = \left[\frac{|X_1(k,l)|}{A(k,l)}, \ldots, \frac{|X_M(k,l)|}{A(k,l)}\right], \quad m \neq J, \tag{9}$$

$$\boldsymbol{\theta}^P(k,l) = \left[\frac{1}{\alpha}\arg\left[\frac{X_1(k,l)}{X_J(k,l)}\right], \ldots, \frac{1}{\alpha}\arg\left[\frac{X_M(k,l)}{X_J(k,l)}\right]\right], \quad m \neq J, \tag{10}$$

for $A(k,l) = \sqrt{\sum_{m=1}^{M} |X_m(k,l)|^2}$ and $\alpha = 4\pi c^{-1} d_{max}$, where c is the propagation velocity, d_{max} is the maximum distance between any two sensors in the array and J is the index of the reference sensor. The weighting parameters $A(k,l)$ and α ensure appropriate amplitude and phase normalization of the features respectively. It is widely known that in the presence of reverberation, a greater accuracy in phase ratio measurements can be achieved with greater spatial resolution; however, it should be noted that the value of d_{max} is upper bounded by the spatial aliasing theorem.

The frequency normalization in $A(k,l)$ ensures frequency independence of the phase ratios in order to prevent the frequency permutation problem in the later stages of clustering. It is possible to cluster without such frequency independence, for example (Sawada et al., 2007; 2011); however, the utilization of all the frequency bins in the clustering stage avoids this and also permits data observations of short length (Araki et al., 2007).

Rewriting the feature vector in complex representation yields

$$\theta_j(k,l) = \theta_j^L(k,l) \exp(j\theta_j^P(k,l)) \, , \tag{11}$$

where θ_j^L and θ_j^P are the jth components of (9) and (10) respectively. In this feature vector representation, the phase difference information is captured in the argument term, and the level ratio is normalized by the normalization term $A(k,l)$.

Equivalently (Araki et al., 2007)

$$\bar{\theta}_j(k,l) = |X_j(k,l)| \exp\left[j \frac{\arg[X_j(k,l)/X_J(k,l)]}{\alpha_j f} \right] \, , \tag{12}$$

and

$$\boldsymbol{\theta}(k,l) \leftarrow \frac{\bar{\boldsymbol{\theta}}(k,l)}{\|\boldsymbol{\theta}(k,l)\|} \, , \tag{13}$$

where $\bar{\boldsymbol{\theta}}(k,l) = [\bar{\theta}_1(k,l), \ldots, \bar{\theta}_M(k,l)]^T$. In the final representation of (13), the level and phase information are captured in the amplitude and argument respectively.

Fig. 2(a) and 2(b) depict the histogram of extracted level ratios and phase differences, respectively, in the ideal anechoic environment. The clear peaks in the phase histogram in (b) are distinctively visible and correspond to the sources. However, when the anechoic assumption is violated and reverberation is introduced into the environment, the distinction between peaks is reduced in clarity as is evident in the phase ratio histogram in Fig. 2(c). Furthermore, the degrading effects of additive channel noise can be seen in Fig. 2(d) where the phase ratio completely loses its reliability. It is hypothesized in this study that a sufficiently robust TF mask estimation technique will be competent to withstand the effect of reverberation and/or additive noise in the acoustic environment.

The masking approach to BSS relies on the observation that in an anechoic setting, the extracted features are expected to form N clusters, where each cluster corresponds to a source at a particular location. Since the relaxation of the anechoic assumption reduces the accuracy of the extracted features as mentioned above in Section 2.2, it is imperative that a sufficiently robust TF clustering technique is implemented in order to effectively separate the sources.

The feature vector set $\Theta(k,l) = \{\boldsymbol{\theta}(k,l) \mid \boldsymbol{\theta}(k,l) \in \mathbb{R}^{2(M-1)}, (k,l) \in \Omega\}$ is divided into N clusters, where $\Omega = \{(k,l) : 0 \leq k \leq K-1, 0 \leq l \leq L-1\}$ denotes the set of TF points in the STFT plane. Depending on the selection of clustering algorithm, the clusters are represented by distinct sets of TF points (hard k-means clustering); a set of prototype vectors and membership partition matrix (fuzzy c-means); or a parameter set (GMM-EM approach).

Specifically, the k-means algorithm results in N distinct clusters $C_1, \ldots, C_N$, where each cluster is comprised of the constituent TF cells, and $\sum_{n=1}^{N} |C_n| = |\Theta(k,l)|$ where the operator $|.|$ denotes cardinality. The fuzzy c-means yields the N centroids $\boldsymbol{v}_n$ and a partition matrix $\boldsymbol{U} = \{u_n(k,l) \in \mathbb{R} \mid n \in (1,\ldots,N), (k,l) \in \Omega)\}$, where $u_n(k,l)$ indicates the degree of membership of the TF cell (k,l) to the nth cluster. The GMM-EM clustering results in the parameter set associated with the Gaussian mixture densities $\{\Lambda = \lambda_1, \ldots, \lambda_G\}$ where G is the number of mixture components in the Gaussian densities, and each λ_i vector has a representative mean and covariance matrix. Further details on the three main clustering algorithms used in this study are provided in Section 3.

2.3 Mask estimation and separation

In this work source separation is effectuated by the application of TF masks, which are the direct result of the clustering step.

For the k-means algorithm, a binary mask for the nth source is simply estimated as

$$M_n(k,l) = \begin{cases} 1 & \text{for } \boldsymbol{\theta}(k,l) \in C_n \,, \\ 0 & \text{otherwise.} \end{cases} \tag{14}$$

In the instances of FCM clustering, the membership partition matrix is interpreted as a collection of N fuzzy TF masks, where

$$M_n(k,l) = u_n(k,l) \,. \tag{15}$$

For the GMM-EM algorithm, the mask estimation is based upon the calculation of probabilities from the final optimized parameter set $\Lambda = \{\lambda_1, \ldots, \lambda_n\}$. The parameter set is used to estimate the masks as follows

$$M_n(k,l) \sim \underset{n}{\text{argmax}}\; p(\boldsymbol{\theta}(k,l)|\lambda_n) \,, \tag{16}$$

where λ_n denotes the parameter set pertaining to the nth source, and probabilities $p(\boldsymbol{\theta}(k,l)|\lambda_n)$ are calculated using a simple normal distribution (Section 3.3).

The separated signal image estimates $\{\hat{S}_{11}(k,l), \ldots, \hat{S}_{1M}(k,l)\}, \ldots, \{\hat{S}_{N1}(k,l), \ldots, \hat{S}_{NM}(k,l)\}$ in the frequency domain are then obtained through the application of the mask per source to an individual observation

$$\hat{S}_{mn}(k,l) = M_n(k,l) X_m(k,l) \,, \quad m = 1, \ldots, M. \tag{17}$$

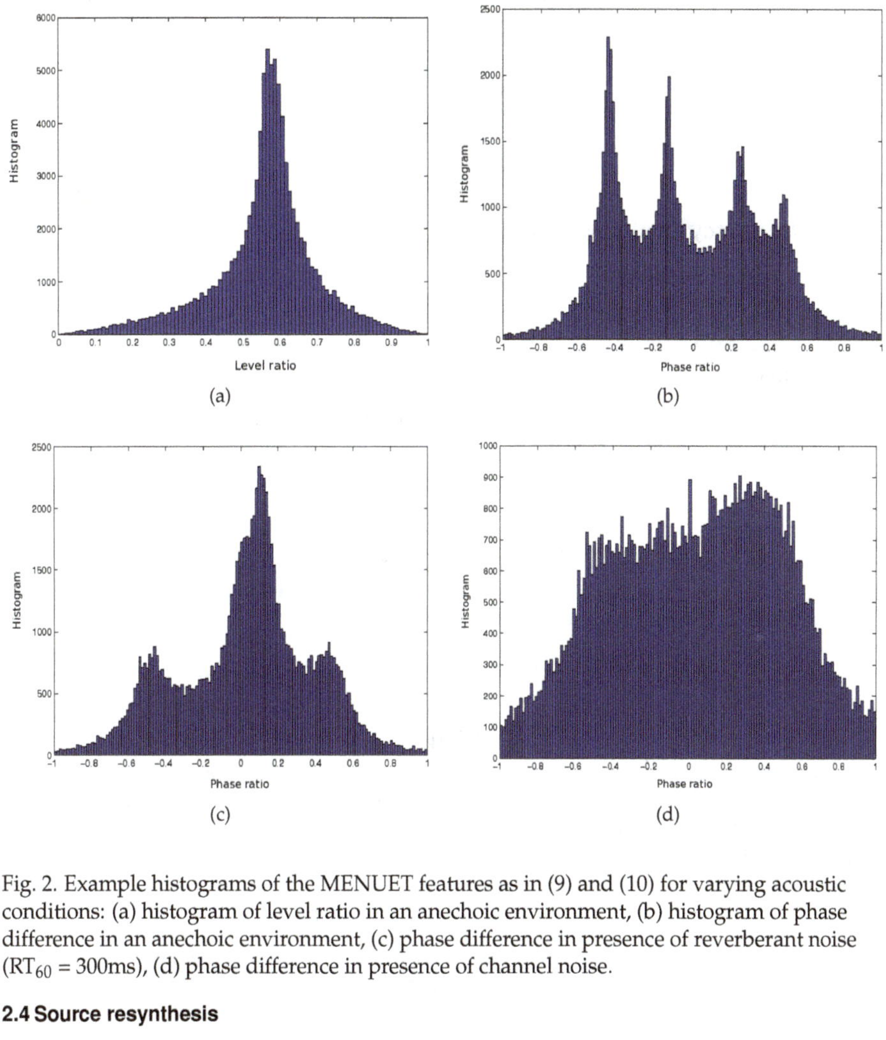

Fig. 2. Example histograms of the MENUET features as in (9) and (10) for varying acoustic conditions: (a) histogram of level ratio in an anechoic environment, (b) histogram of phase difference in an anechoic environment, (c) phase difference in presence of reverberant noise (RT_{60} = 300ms), (d) phase difference in presence of channel noise.

2.4 Source resynthesis

Lastly, the estimated source images are reconstructed to obtain the time-domain separated estimates of the source images $\hat{s}_{mn}(t)$ for $n = 1, \ldots, N$ and $m = 1, \ldots, M$. This is realized with the application of the overlap-and-add method (Rabiner, 1978) onto the separated frequency components $\hat{S}_{mn}(k, l)$. The reconstructed estimate is

$$\hat{s}_{mn}(t) = \frac{1}{C_{\text{win}}} \sum_{k'=0}^{L/\tau_0 - 1} \hat{s}_{mn}^{k+k'}(t) , \tag{18}$$

where $C_{win} = 0.5/\tau_1 0L$ is a Hann window function constant, and individual frequency components of the recovered signal are acquired through an inverse STFT

$$\hat{s}^k_{mn}(t) = \sum_{l=0}^{L-1} \hat{S}_{mn}(k,l) e^{jl\omega_0(t-k\tau_0)} , \tag{19}$$

if $(k\tau_0 \leq t \leq k\tau + L - 1)$, and zero otherwise.

3. Time-frequency clustering algorithms

3.1 Hard *k*-means clustering

Previous methods (Araki et al., 2006b; 2007) employ hard clustering techniques such as the hard *k*-means (HKM) (Duda et al., 2000). In this approach, the feature vectors $\boldsymbol{\theta}(k,l)$ are clustered to form N distinct clusters $C_1, \ldots, C_N$.

The clustering is achieved through the minimization of the objective function

$$J_{kmeans} = \sum_{n=1}^{N} \sum_{\boldsymbol{\theta}(k,l) \in C_n)} \|\boldsymbol{\theta}(k,l) - \boldsymbol{c_n}\|^2 , \tag{20}$$

where the operator $\|.\|$ denotes the Euclidean norm and $\boldsymbol{c_n}$ denotes the cluster centroids. Starting with a random initialization for the set of centroids, this minimization is iteratively realized by the following alternating equations

$$C_n^* = \{\boldsymbol{\theta}(k,l) | n = \underset{n}{\text{argmin}} \, \|\boldsymbol{\theta}(k,l) - \boldsymbol{c_n}\|^2 , \quad \forall n,k,l, \tag{21}$$

$$\boldsymbol{c_n^*} \leftarrow E\{\boldsymbol{\theta}(k,l)\}_{\boldsymbol{\theta}(k,l) \in C_n} , \quad \forall n, \tag{22}$$

until convergence is met, where $E\{.\}_{\boldsymbol{\theta}(k,l) \in C_n}$ denotes the mean operator for the TF points within the cluster C_n, and the (*) operator denotes the optimal value. The resulting N clusters are then utilized in the mask estimation as described in Section 2.3. Due to the algorithm's sensitivity to initialization of the cluster centres, it is recommended to either design initial centroids using an assumption on the sensor and source geometry (Araki et al., 2007), or to utilize the best outcome of a predetermined number of independent runs.

Whilst this binary clustering performed satisfactorily in both simulated and realistic reverberant environments, the authors of (Jafari et al., 2011; Kühne et al., 2010) demonstrate that the application of a soft masking scheme improves the separation performance substantially.

Summary: K-means Algorithm
1 Initialize centroids $\boldsymbol{c}_1, \ldots, \boldsymbol{c}_N$ randomly
2 For $j = 1, 2, \ldots$
3 Update cluster members C_n using (21)
4 Update centroids c_n with calculated clusters C_n according to (22)
5 Repeat until for some j^* the convergence is met
6 Assign $C_1^*, \ldots, C_N^*$ and $\boldsymbol{c}_1^*, \ldots, \boldsymbol{c}_N^*$ to each TF point.

3.2 Fuzzy *c*-means clustering

In the fuzzy c-means clustering, the feature set $\Theta(k,l) = \{\theta(k,l)|\theta(k,l) \in \mathbb{R}^{2(M-1)}, (k,l) \in \Omega\}$ is clustered using the fuzzy c-means algorithm (Bezdek, 1981) into N clusters, where $\Omega = \{(k,l) : 0 \leq k \leq K-1, 0 \leq l \leq L-1\}$ denotes the set of TF points in the STFT plane. Each cluster is represented by a centroid $\boldsymbol{v_n}$ and partition matrix $\boldsymbol{U} = \{u_n(k,l) \in \mathbb{R}|n \in (1,\ldots,N),(k,l) \in \Omega)\}$ which specifies the degree $u_n(k,l)$ to which a feature vector $\theta(k,l)$ belongs to the nth cluster. Clustering is achieved by the minimization of the cost function

$$J_{fcm} = \sum_{n=1}^{N} \sum_{\forall(k,l)} u_n(k,l)^q D_n(k,l) \,, \tag{23}$$

where

$$D_n(k,l) = \|\theta(k,l) - \boldsymbol{v_n}\|^2 \,, \tag{24}$$

is the squared Euclidean distance between the vector $\theta(k,l)$ and the nth cluster centre. The fuzzification parameter $q > 1$ controls the membership softness; a value of q in the range of $q \in (1, 1.5]$ has been shown to result in a fuzzy performance akin to hard (binary) clustering (Kühne et al., 2010). However, superior mask estimation ability has been established when $q = 2$; thus in this work, the fuzzification q is set to 2.

The minimization problem in (23) can be solved using Lagrange multipliers and is typically implemented as an alternating optimization scheme due to the open nature of its solution (Kühne et al., 2010; Theodoridis & Koutroumbas, 2006). Initialized with a random partitioning, the cost function J_{fcm} is iteratively minimized by alternating the updates for the cluster centres and memberships

$$\boldsymbol{v_n^*} = \sum_{\forall(k,l)} \frac{u_n(k,l)^q \theta(k,l)}{\sum_{\forall(k,l)} u_n(k,l)^q} \,, \quad \forall n, \tag{25}$$

$$u_n^*(k,l) = \left[\sum_{j=1}^{N} \left(\frac{D_n(k,l)}{D_j(k,l)} \right)^{\frac{1}{q-1}} \right]^{-1} \,, \quad \forall n,k,l, \tag{26}$$

where (*) denotes the optimal value, until a suitable termination criterion is satisfied. Typically, convergence is defined as when the difference between successive partition matrices is less than some predetermined threshold ϵ (Bezdek, 1981). However, as is also the case with the k-means (Section 3.1), it is known that the alternating optimization scheme presented may converge to a local, as opposed to global, optimum; thus, it is suggested to independently implement the algorithm several times prior to selecting the most fitting result.

3.3 Gaussian mixture model clustering

To further examine the separation ability of the MENUET-FCM scheme another clustering approach, based upon GMM clustering, is presented in this study. A GMM of a multivariate distribution $\Theta(k,l)$ may be represented by a weighted sum of G component Gaussian

Summary: C-means Algorithm

1 Initialize partition matrix $\boldsymbol{U}$ randomly
2 For $j = 1, 2, \ldots$
3 Update centroids $\boldsymbol{v_n}$ according to (25)
4 Update partition matrix $\boldsymbol{U}$ with calculated memberships u_n according to (26)
5 Repeat until for some j^* the convergence threshold ϵ is met
6 Assign $u_n^*(k,l)$ and $\boldsymbol{v_n^*}$ to each TF point (k,l).

densities as given by

$$p(\boldsymbol{\Theta}|\Lambda) = \sum_i^G w_i g(\boldsymbol{\Theta}|\lambda_i) \,, \tag{27}$$

where $w_i, i = 1, \ldots, G$ are the mixture weights, $g(\boldsymbol{\Theta}|\Lambda)$ are the component Gaussian densities, and Λ is the vector of hidden parameters such that $\Lambda = \{\lambda_1, \ldots, \lambda_G\}$ of the Gaussian components. Each component density is a D-variate Gaussian function of the form

$$g(\boldsymbol{\Theta}|\boldsymbol{\mu_i}, \Sigma_i) = \frac{1}{(2\pi)^{D/2}|\Sigma_i|^{1/2}} \exp\left\{-\frac{1}{2}(\boldsymbol{\Theta} - \boldsymbol{\mu_i})'\Sigma_i^{-1}(\boldsymbol{\Theta} - \boldsymbol{\mu_i})\right\} \,, \tag{28}$$

with mean vector $\boldsymbol{\mu}_i$ and covariance matrix Σ_i. The constraint on the mixture weights is such as to satisfy the condition $\sum_{i=1}^{G} w_i = 1$.

The goal of the GMM-EM clustering is to fit the source mixture data into a Gaussian mixture model and then estimate the maximum likelihood of the hidden parameters $\Lambda = \{\lambda_1, \ldots, \lambda_G\}$, where each $\{\lambda_i\}$ has its associated mean vector $\boldsymbol{\mu}_i$ and covariance matrix Σ_i, associated with the mixture densities in the maximum likelihood of the features $\boldsymbol{\Theta}(k,l)$. The features $\boldsymbol{\Theta}(k,l)$ in this section will henceforth be denoted as $\boldsymbol{\Theta}$ for simplicity. Under the assumption of independence between the features, the likelihood of the parameters, $\mathcal{L}(\Lambda|\boldsymbol{\Theta})$ is related to $\boldsymbol{\Theta}$ by

$$p(\boldsymbol{\Theta}|\Lambda) = \prod_{t=1}^{T} p(\boldsymbol{\theta}_t|\Lambda) = \mathcal{L}(\Lambda|\boldsymbol{\Theta}) \,, \tag{29}$$

where T is the total number of TF cells per feature (i.e. $k * l$). The estimation of the optimum hidden parameter set Λ^* relies on the maximization of (29)

$$\Lambda^* = \underset{\Lambda}{\operatorname{argmax}}\ \mathcal{L}(\Lambda|\boldsymbol{\Theta}) \,. \tag{30}$$

Due to the fact that the log of $\mathcal{L}(*)$ is typically calculated in lieu of $\mathcal{L}(*)$, the function (29) is a nonlinear function of Λ. Therefore, the maximization in the G mixture components is a difficult problem. However, the maximum-likelihood (ML) estimates of these parameters may be calculated using the Expectation-Maximization (EM) algorithm (Izumi et al., 2007). The EM algorithm is iterated until a predetermined convergence threshold ϵ is reached.

The choice of the number of Gaussian mixtures for fitting the microphone array data is critical, and is typically determined by trial and error (Araki et al., 2007). In this study, the number of mixture components is set equal to the number of sources in order to facilitate the association of clusters to sources. In the case where $G > N$, the association will have an ambiguous nature.

This assumption that each resulting Gaussian cluster uniquely fits one source therefore allows the calculation of the probability that a TF cell originates from the nth source; this is because the probability is equivalent to the probability that the TF cell originates from the nth mixture component. It is assumed in this study that the probability of membership follows a normal distribution as

$$p(\boldsymbol{\theta}(k,l)|\lambda_n^*) = \frac{1}{(2\pi|\Sigma_n^*|)^{1/2}} \exp\left\{-\frac{1}{2}(\boldsymbol{\theta}(k,l) - \boldsymbol{\mu}_n^*)'\Sigma_n^{*-1}(\boldsymbol{\theta}(k,l) - \boldsymbol{\mu}_n^*)\right\}, \tag{31}$$

where $\lambda_n^* \in \Lambda^* = \{\lambda_1^*, \ldots, \lambda_N^*\}$.

Summary: GMM-EM Algorithm

1 Assume initial parameter set Λ
2 For $j = 1, 2, \ldots$
3 Calculate expectation $\mathcal{L}(\Lambda|\Theta)$ according to EM as in (Izumi et al., 2007)
4 Estimate Λ^j according to (Izumi et al., 2007)
5 Repeat until for some j^* the convergence threshold ϵ is met
6 Assign λ_n^* to each TF point (k,l).

4. Experimental evaluations

4.1 Experimental setup

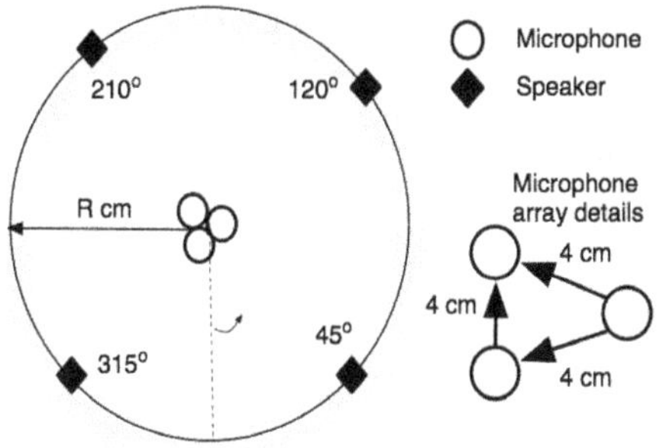

Fig. 3. The room setup for the three sensor nonlinear arrangement experimental evaluations.

The experimental setup was such as to reproduce that in (Araki et al., 2007) and (Jafari et al., 2011) for comparative purposes. Fig. 3 depicts the speaker and sensor arrangement, and Table 1 details the experimental conditions. The wall reflections of the enclosure, as well as the room impulse responses for each sensor, were simulated using the image model method for small-room acoustics (Lehmann & Johansson, 2008). The room reverberation was quantified in the measure RT_{60}, where RT_{60} is defined as the time required for reflections of a direct sound to decay by 60dB below the level of the direct sound (Lehmann & Johansson, 2008).

For the noise-robust evaluations, spatially uncorrelated white noise was added to each sensor mixture such that the overall channel SNR assumed a value as in Table 1. The SNR definition as in (Loizou, 2007) was implemented, which employs the standardized method given in

(ITU-T, 1994) to objectively measure the speech. The four speech sources, the genders of which were randomly generated, were realized with phonetically-rich utterances from the TIMIT database (Garofolo et al., 1993) , and a representative number of mixtures for evaluative purposes constructed in total. In order to avoid the spatial aliasing problem, the microphones were placed at a maximum distance of 4cm apart.

Experimental conditions	
Number of microphones	$M = 3$
Number of sources	$N = 4$
R	50cm
Source signals	6 s
Reverberation time	0 ms, 128 ms 300ms
	(450ms for clean evaluations only)
Input channel SNR	0 dB - 30 dB
Sampling rate	8 kHz
STFT window	Hann
STFT frame size	64 ms
STFT frame overlap	50%

Table 1. The parameters used in experimental evaluations.

As briefly discussed in Section 3.1 and 3.2, it is widely recognized that the performance of the clustering algorithms is largely dependent on the initialization of the algorithm. For both the MENUET and MENUET-FCM, the best of 100 runs was selected for initialization in order to minimize the possibility of finding a local, as opposed to global, optimum. In order to ensure the GMM fitting of the mixtures in the MENUET-GMM evaluations, the initial values for the mean and variance in the parameter set Λ had to be selected appropriately. The initialization of the parameters has been proven to be an imperative yet difficult task; should the selection be unsuccessful, the GMM fitting may completely fail (Araki et al., 2007). In this study, the mean and variance for each parameter set were initialized using the k-means algorithm.

4.1.1 Evaluation measures

For the purposes of speech separation performance evaluation, two versions of the publicly available MATLAB toolboxes *BSS_EVAL* were implemented (Vincent et al., 2006; 2007). This performance criteria is applicable to all source separation approaches, and no prior information of the separation algorithm is required. Separation performance was evaluated with respect to the global image-to-spatial-distortion ratio (ISR), signal-to-interference ratio (SIR), signal-to-artifact ratio (SAR) and signal-to-distortion ratio (SDR) as defined in (Vincent et al., 2007); for all instances, a higher ratio is deemed as better separation performance.

This assumes the decomposition of the estimated source $\hat{s}_n(t)$ as

$$\hat{s}_{mn}(t) = s_{mn}^{img}(t) + \hat{e}_{mn}^{spat}(t) + \hat{e}_{mn}^{int}(t) + \hat{e}_{mn}^{artif}(t)\ , \tag{32}$$

where $s_{mn}^{img}(t)$ corresponds to the true source image, and $\hat{e}_{mn}^{spat}(t)$, $\hat{e}_{mn}^{int}(t)$ and $\hat{e}_{mn}^{artif}(t)$ are the undesired error components that correlate to the spatial distortion, interferences and artifacts respectively. This decomposition is motivated by the auditory notion of distinction between sounds originating from the target source, sounds from other sound sources present, and "gurgling" noise corresponding to $s_{mn}^{img}(t) + \hat{e}_{mn}^{spat}(t)$, $\hat{e}_{mn}^{int}(t)$ and $\hat{e}_{mn}^{artif}(t)$, respectively. The decomposition of the estimated signal was executed using the function *bss_eval_images*, which computes the spatial distortion and interferences by means of a least-squares projection of the estimated source image onto the corresponding signal subspaces. As recommended in (Vincent et al., 2007), the filter length was set to the maximal tractable length of 512 (64ms).

The ISR of the nth recovered source is then calculated as

$$\mathrm{ISR}_n = 10\log_{10}\frac{\sum_{m=1}^{M}\sum_{t} s_{mn}^{img}(t)^2}{\sum_{m=1}^{M}\sum_{t}\hat{e}_{mn}^{spat}(t)^2}, \tag{33}$$

which provides a measure for the relative amount of distortion present in the recovered signal.

The SIR, given by

$$\mathrm{SIR}_n = 10\log_{10}\frac{\sum_{m=1}^{M}\sum_{t}(s_{mn}^{img}(t) + \hat{e}_{mn}^{spat}(t))^2}{\sum_{m=1}^{M}\sum_{t}\hat{e}_{mn}^{int}(t)^2}, \tag{34}$$

provides an estimate of the relative amount of interference in the target source estimate. For all SIR evaluations the gain $\mathrm{SIR}_{\mathrm{gain}} = \mathrm{SIR}_{\mathrm{output}} - \mathrm{SIR}_{\mathrm{input}}$ was computed in order to quantify the improvement between the input and the output of the proposed studies.

The SAR is computed as

$$\mathrm{SAR}_n = 10\log_{10}\frac{\sum_{m=1}^{M}\sum_{t}(s_{mn}^{img}(t) + \hat{e}_{mn}^{spat}(t) + \hat{e}_{mn}^{int}(t))^2}{\sum_{m=1}^{M}\sum_{t}\hat{e}_{mn}^{artif}(t)^2}, \tag{35}$$

in order to give a quantifiable measure of the amount of artifacts present in the nth source estimate.

As an estimate of the total error in the nth recovered source (or equivalently, a measure for the separation quality), the SDR is calculated as

$$\mathrm{SDR}_n = 10\log_{10}\frac{\sum_{m=1}^{M}\sum_{t} s_{mn}^{img}(t)^2}{\sum_{m=1}^{M}\sum_{t}\left[\hat{e}_{mn}^{spat}(t) + \hat{e}_{mn}^{int}(t) + \hat{e}_{mn}^{artif}(t)\right]^2}. \tag{36}$$

Similarly, the SNR of the estimated output signal was also evaluated using the *BSS_EVAL* toolkit. The estimated source $\hat{s}_n(t)$ was assumed to follow the following decomposition (Vincent et al., 2006)

$$\hat{s_n}(t) = s_n^{target}(t) + \hat{e}_n^{noise}(t) + \hat{e}_n^{int}(t) + \hat{e}_n^{artif}(t) , \tag{37}$$

where $s_n^{target}(t)$ is an allowed distortion of the original source, and $\hat{e}_n^{noise}(t)$, $\hat{e}_n^{int}(t)$ and $\hat{e}_n^{artif}(t)$ are the noise, interferences and artifacts error terms respectively. The decomposition of the estimated signal in this instance was executed using the function *bss_decomp_filt*, which permits time-invariant filter distortions of the target source. As recommended in (Vincent et al., 2006), the filter length was set to 256 taps (32ms). The global SNR for the nth source was subsequently calculated as

$$\mathrm{SNR}_n = 10\log_{10}\frac{||s_n^{target}(t) + \hat{e}_n^{int}(t)||^2}{||\hat{e}_n^{noise}(t)||^2} . \tag{38}$$

4.2 Results

4.2.1 Initial evaluations of fuzzy *c*-means clustering

Firstly, to establish the feasibility of the *c*-means clustering as a credible approach to the TF mask estimation problem for underdetermined BSS, the algorithm was applied to a range of feature sets as defined in (Araki et al., 2007). The authors of (Araki et al., 2007) present a comprehensive review of suitable location features for BSS within the TF masking framework, and evaluate their effectiveness using the *k*-means clustering algorithm. The experimental setup for these set of evaluations was such as to replicate that in (Araki et al., 2007) to as close a degree as possible. In an enclosure of dimensions 4.55m x 3.55m x 2.5m, two omnidirectional microphones were placed a distance of 4cm apart at an elevation of 1.2m. Three speech sources, also at an elevation of 1.2m, were situated at 30°, 70° and 135°; and the distance R between the array and speakers was set to 50cm. The room reverberation was constant at 128ms. The speech sources were randomly chosen from both genders of the TIMIT database in order to emulate the investigations in (Araki et al., 2007) which utilized English utterances.

It is observed from the comparison of separation performance with respect to SIR improvement as shown in Table 2 that the *c*-means outperformed the original *k*-means clustering in all but one feature set. This firstly establishes the applicability of the *c*-means clustering in the proposed BSS framework, and secondly demonstrates the robustness of the *c*-means clustering against a variety of spatial features. The results of this investigation provide further motivation to extend the fuzzy TF masking scheme to other sensor arrangements and acoustic conditions.

4.2.2 Separation performance in reverberant conditions

Once the feasibility of the fuzzy *c*-means clustering for source separation was established, the study was extended to a nonlinear three sensor and four source arrangement as in Fig. 3. The separation results with respect to the ISR, SIR gain, SDR and SAR for a range of reverberation times are given in Fig. 4(a)-(d) respectively. Fig. 4(a) depicts the ISR results; from here it is evident that there are considerable improvements in the MENUET-FCM over

Feature $\boldsymbol{\theta}(k,l)$	k-means (dB)	c-means (dB)
$\boldsymbol{\theta}(k,l)=\left[\frac{\lvert X_2(k,l)\rvert}{\lvert X_1(k,l)\rvert},\frac{1}{2\pi f}\arg\left[\frac{X_2(k,l)}{X_1(k,l)}\right]\right]^T$	1.8	**2.1**
$\boldsymbol{\theta}(k,l)=\left[\frac{\lvert X_2(k,l)\rvert}{\lvert X_1(k,l)\rvert}-\frac{1}{\frac{\lvert X_2(k,l)\rvert}{\lvert X_1(k,l)\rvert}},\frac{1}{2\pi f}\arg\left[\frac{X_2(k,l)}{X_1(k,l)}\right]\right]^T$	1.1	**1.6**
$\boldsymbol{\theta}(k,l)=\left[\frac{\lvert X_2(k,l)\rvert}{\lvert X_1(k,l)\rvert},\frac{1}{2\pi fc^{-1}d}\arg\left[\frac{X_2(k,l)}{X_1(k,l)}\right]\right]^T$	7.8	**9.2**
$\boldsymbol{\theta}(k,l)=\frac{1}{2\pi f}\arg\left[\frac{X_2(k,l)}{X_1(k,l)}\right]$	**10.2**	8.0
$\boldsymbol{\theta}(k,l)=\frac{1}{2\pi fc^{-1}d}\arg\left[\frac{X_2(k,l)}{X_1(k,l)}\right]$	10.1	**17.2**
$\boldsymbol{\theta}(k,l)=\left[\frac{\lvert X_1(k,l)\rvert}{A(k,l)},\frac{\lvert X_2(k,l)\rvert}{A(k,l)},\frac{1}{2\pi}\arg\left[\frac{X_2(k,l)}{X_1(k,l)}\right]\right]^T$	4.2	**5.4**
$\boldsymbol{\theta}(k,l)=\left[\frac{\lvert X_1(k,l)\rvert}{A(k,l)},\frac{\lvert X_2(k,l)\rvert}{A(k,l)},\frac{1}{2\pi fc^{-1}d}\arg\left[\frac{X_2(k,l)}{X_1(k,l)}\right]\right]^T$	10.4	**17.4**
$\bar{\theta}_j(k,l)=\lvert X_j(k,l)\rvert\exp\left[j\frac{\arg[X_j(k,l)/X_J(k,l)]}{\alpha_j f}\right]$, $\boldsymbol{\theta}(k,l)\leftarrow\frac{\bar{\boldsymbol{\theta}}(k,l)}{\lVert\bar{\boldsymbol{\theta}}(k,l)\rVert}$	10.2	**17.2**

Table 2. Comparison of separation performance in terms of SIR improvement in dB of typical spatial features. Separation results are evaluated with SIR_{gain} for the TF masking approach to BSS when the hard k-means and fuzzy c-means algorithms are implemented for mask estimation. The reverberation was constant at RT_{60} = 128ms.

both the MENUET and MENUET-GMM. Additionally, the MENUET-GMM demonstrates a slight improvement over the MENUET.

The SIR gain as in Fig. 4(b) clearly demonstrates the superiority in source separation with the MENUET-FCM. For example, at the high reverberation time of 450ms, the proposed MENUET-FCM outperformed both the baseline MENUET and MENUET-GMM by almost 5dB.

Similar results were noted for the SDR, with substantial improvements when fuzzy masks are used. As the SDR provides a measure of the total error in the algorithm, this suggests that the fuzzy TF masking approach to BSS is more robust against algorithmic error than the other algorithms.

The superiority of the fuzzy masking scheme is further established in the SAR values depicted in Fig. 4(d). A consistently high value is achieved across all reverberation times, unlike the other approaches which fail to attain such values. This indicates that the fuzzy TF masking scheme yields source estimates with fewer artifacts present. This is in accordance with the study as in (Araki et al., 2006a) which demonstrated that soft TF masks bear the ability to significantly reduce the musical noise in recovered signals as a result of the inherent characteristic of the fuzzy mask to prevent excess zero padding in the recovered source signals.

It is additionally observed that there is a significantly reduced standard deviation resulting from the FCM algorithm which further implies consistency in the algorithm's source separation ability.

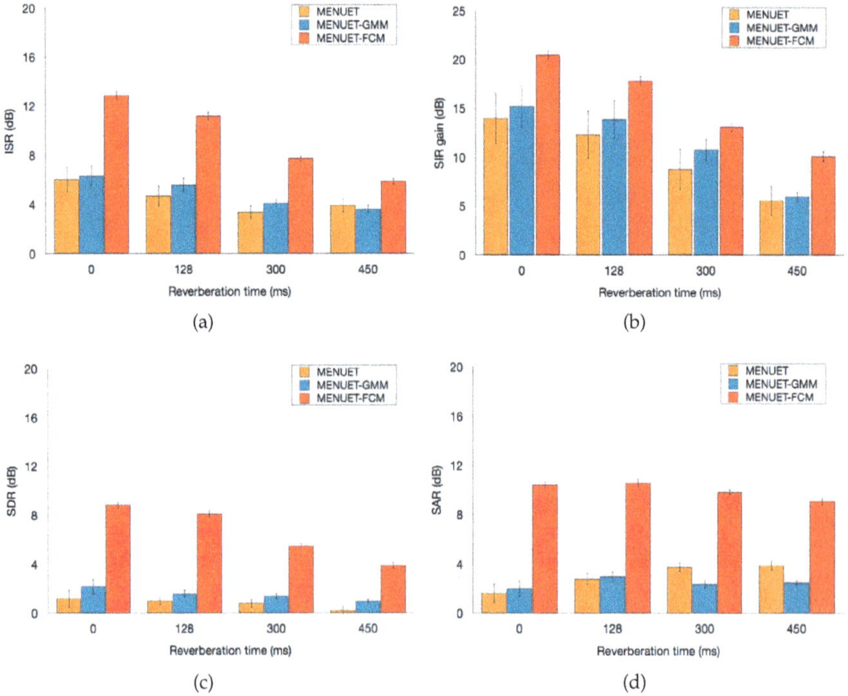

Fig. 4. Source separation results in reverberant conditions using three separation approaches: MENUET, MENUET-GMM and MENUET-FCM. Performance results given in terms of (a) ISR, (b) SIR gain, (c) SDR and (d) SAR for all RT_{60} values. The error bars denote the standard deviation over all evaluations.

4.2.3 Separation performance in reverberant conditions with additive noise

The impact of additive white channel noise on separation quality was evaluated next. The reverberation was varied from 0ms to 300ms, and the SNR at the sensors of the microphone array was varied from 0dB to 30dB in 5dB increments.

Tables 3(a)-(d) depicts the separation results of the evaluations with respect to the measured ISR, SIR gain, SDR and SAR respectively. It is clear from the table that the proposed MENUET-FCM algorithm has significantly increased separation ability over all tested conditions and for all performance criteria. In particular, the MENUET-FCM scenario demonstrates excellent separation ability even in the higher 300ms reverberation condition.

Conditions	ISR (dB)		
SNR_{in} (dB)	HKM	GMM	FCM
RT_{60} = 0ms			
0	**4.92**	3.68	4.52
5	5.13	4.07	**5.83**
10	**6.93**	4.61	6.53
15	7.18	6.09	**8.37**
20	7.81	6.21	**11.81**
25	7.96	7.15	**11.98**
30	6.87	7.48	**12.62**
RT_{60} = 128ms			
0	3.18	3.21	**4.15**
5	4.05	4.16	**5.03**
10	4.34	4.59	**5.91**
15	5.13	4.77	**7.91**
20	5.71	4.89	**10.41**
25	6.24	5.67	**10.85**
30	5.24	6.04	**11.08**
RT_{60} = 300ms			
0	3.49	2.84	**3.86**
5	3.05	3.00	**4.12**
10	3.42	4.04	**5.05**
15	3.55	4.11	**5.91**
20	3.64	4.13	**7.05**
25	4.10	4.59	**7.99**
30	3.55	4.66	**8.21**

(a)

Conditions	SIR gain (dB)		
SNR_{in} (dB)	HKM	GMM	FCM
RT_{60} = 0ms			
0	5.01	3.49	**4.95**
5	6.21	4.89	**7.01**
10	7.83	5.34	**8.86**
15	8.01	6.00	**17.89**
20	8.22	6.64	**19.15**
25	8.56	7.12	**19.08**
30	7.16	9.65	**19.4**
RT_{60} = 128ms			
0	2.78	2.84	**4.46**
5	3.08	3.27	**5.91**
10	3.46	3.86	**7.50**
15	5.17	5.03	**13.04**
20	6.72	5.48	**16.90**
25	7.01	7.58	**16.78**
30	5.17	8.36	**17.61**
RT_{60} = 300ms			
0	2.96	1.79	**3.8**
5	2.95	3.05	**4.12**
10	3.02	3.97	**6.11**
15	4.28	4.49	**8.53**
20	4.99	5.24	**10.78**
25	5.32	6.65	**11.53**
30	4.12	7.54	**13.81**

(b)

Conditions	SDR (dB)		
SNR_{in} (dB)	HKM	GMM	FCM
RT_{60} = 0ms			
0	-1.88	-2.41	**-0.20**
5	0.15	-1.14	**1.76**
10	0.88	-0.24	**3.15**
15	1.03	0.16	**6.01**
20	1.19	0.45	**8.20**
25	1.37	1.29	**8.39**
30	0.76	1.39	**8.50**
RT_{60} = 128ms			
0	-2.22	-2.41	**-0.29**
5	-0.76	-0.71	**1.64**
10	-0.50	-0.32	**2.94**
15	0.57	-0.08	**6.19**
20	0.68	0.09	**7.37**
25	0.98	1.13	**7.56**
30	-0.70	1.51	**7.98**
RT_{60} = 300ms			
0	-1.41	-2.6	**-0.36**
5	-1.07	-1.98	**1.23**
10	-0.78	-0.31	**2.13**
15	-0.35	-0.10	**3.24**
20	-0.41	-0.09	**4.35**
25	0.15	0.27	**4.93**
30	-0.41	-0.61	**5.97**

(c)

Conditions	SAR (dB)		
SNR_{in} (dB)	HKM	GMM	FCM
RT_{60} = 0ms			
0	-4.83	-5.44	**-2.47**
5	-2.44	-2.65	**1.85**
10	0.08	-0.76	**4.62**
15	0.59	0.17	**7.84**
20	1.83	0.74	**10.17**
25	1.91	1.78	**10.19**
30	2.18	2.23	**10.22**
RT_{60} = 128ms			
0	-4.42	-4.14	**-1.30**
5	-1.19	-1.01	**2.55**
10	-0.80	-0.04	**5.60**
15	1.65	1.61	**8.78**
20	2.58	1.87	**10.39**
25	2.93	2.98	**10.71**
30	2.71	3.38	**10.85**
RT_{60} = 300ms			
0	-3.51	-4.14	**-1.29**
5	-1.64	-1.91	**1.82**
10	-0.71	-0.07	**4.53**
15	2.02	1.69	**7.37**
20	2.73	1.85	**8.24**
25	3.62	2.87	**9.02**
30	3.43	3.03	**10.48**

(d)

Table 3. Source separation results for reverberant noisy mixtures using three separation approaches: MENUET, MENUET-GMM and MENUET-FCM. Performance results are given in terms of (a) ISR, (b) SIR gain, (c) SDR and (d) SAR for all RT_{60} and SNR values. The highest achieved ratios per acoustic scenario are denoted in boldface.

4.2.4 SNR evaluations

For the purposes of speech quality assessment, the SNR of each recovered speech signal was calculated with the definition as in (Vincent et al., 2006) and averaged across all evaluations, with the results shown in Table 4. The MENUET-FCM approach is again observed to be more robust against additive channel noise at the recovered output. However, a remarkable improvement in SNR values for the recovered speech sources for all clustering techniques is also observed. This suggests that the original MENUET, MENUET-GMM and MENUET-FCM have implementations beyond that of simply BSS and in fact maybe useful in applications that also require speech enhancement capabilities. This has important repercussions as it demonstrates that these approaches are able to withstand additive noise without significant degradations in performance, and thus bear the potential to additionally be utilized as a speech enhancement stage in a BSS system.

5. Discussion

The experimental results presented have demonstrated that the implementation of the fuzzy c-means clustering with the nonlinear microphone array setup as in the MENUET renders superior separation performance in conditions where reverberation and/or additive channel noise exist.

The feasibility of the fuzzy c-means clustering was firstly tested on a range of spatial feature vectors in an underdetermined setting using a stereo microphone array, and compared against the original baseline k-means clustering of the MENUET algorithm. The successful outcome of this prompted further investigation, with a natural extension to a nonlinear microphone array. The GMM-EM clustering algorithm was also implemented as a second baseline to further assess the quality of the c-means against alternative binary masking schemes other than the k-means. Evaluations confirmed the superiority of c-means clustering with positive improvements recorded for the average performance in all acoustic settings. In addition to this, the consistent performance even in increased reverberation establishes the potential of fuzzy c-means clustering for the TF masking approach.

However, rather than solely focus upon the reverberant BSS problem, this study refreshingly extended it to be inclusive of additive channel noise. It was suggested that due to the fuzzy c-means' documented robustness in reverberant environments, the extension to the noisy reverberant case would demonstrate similar abilities. Evaluations confirmed this hypothesis with especially noteworthy improvements in the measured SIR gain and SDR. Furthermore, the MENUET, MENUET-GMM and MENUET-FCM approaches were all proven to possess inherent speech enhancement abilities, with higher SNRs measured at the recovered signals.

However, a possible hindrance in the MENUET-GMM clustering was discussed previously regarding the correct selection of the number of fitted Gaussians (Section 3.3). Should the number of Gaussians be increased in a bid to improve the performance, an appropriate clustering approach should then be applied in order to group the Gaussians originating from the same speaker together; for example, a nearest neighbour or correlative clustering algorithm may be used.

Ultimately, the goal of any speech processing system is to mimic the auditory and cognitive ability of humans to as close a degree as possible, and the appropriate implementation of a BSS

Conditions	SNR (dB)		
SNR_{in} (dB)	HKM	GMM	FCM
$RT_{60} = 0ms$			
0	15.41	14.40	**17.05**
5	18.10	17.19	**21.96**
10	21.25	19.90	**25.04**
15	21.91	21.18	**28.89**
20	23.50	22.50	**32.61**
25	23.29	23.97	**32.68**
30	23.62	24.50	**32.91**
$RT_{60} = 128ms$			
0	14.25	14.04	**17.68**
5	18.25	18.98	**21.87**
10	18.50	19.65	**25.37**
15	22.16	22.87	**28.93**
20	23.17	23.46	**32.22**
25	23.58	24.96	**31.99**
30	23.40	25.10	**33.00**
$RT_{60} = 300ms$			
0	15.11	13.31	**16.95**
5	16.96	17.11	**20.83**
10	18.35	19.31	**23.54**
15	22.08	22.10	**26.92**
20	22.50	22.45	**28.01**
25	23.44	23.27	**29.10**
30	24.16	23.71	**30.70**

Table 4. Results for the measured SNR at the BSS output averaged over all the recovered signals. Results given for all RT_{60} and input channel SNR values. The highest achieved ratio per acoustic scenario is denoted in boldface.

scheme is an encouraging step towards reaching this goal. This study has demonstrated that with the use of suitable time-frequency masking techniques, robust blind source separation can be achieved in the presence of both reverberation and additive channel noise. The success of the MENUET-FCM suggests that future work into this subject is highly feasible for real-life speech processing systems.

6. Conclusions

This chapter has presented an introduction into advancements in the time-frequency approach to multichannel BSS. A non-exhaustive review of mask estimation techniques was discussed with insight into the shortcomings affiliated with such existing masking techniques. In a bid to overcome such shortcomings, the novel amalgamation of two existing BSS approaches was proposed and thus evaluated in (simulated) realistic multisource environments.

It was suggested that a binary masking scheme for the TF masking approach to BSS is inadequate at encapsulating the inevitable reverberation present in any acoustic setup, and thus a more suitable means for clustering the observation data, such as the fuzzy c-means, should be considered. The presented MENUET-FCM algorithm integrated the fuzzy c-means clustering with the established MENUET technique for automatic TF mask estimation.

In a number of experiments designed to evaluate the feasibility and performance of the c-means in the BSS context, the MENUET-FCM was found to outperform both the original MENUET and MENUET-GMM in source separation performance. The experiments varied in conditions from a stereo (linear) microphone array setup to a nonlinear arrangement, in both anechoic and reverberant conditions. Furthermore, additive white channel noise was also included in the evaluations in order to better reflect the conditions of realistic acoustic environments.

Future work should endeavor upon the refinement of the robustness of the feature extraction/mask estimation stage, and on the betterment of the clustering technique in order to propel the MENUET-FCM to a sincerely blind system. Details are presented in the following section. Furthermore, the evaluation of the BSS performance in alternative contexts such as automatic speech recognition should also be considered in order to gain greater perspective on its potential for implementation in real-life speech processing systems.

7. Future directions

Future work should focus upon the improvement of the robustness of the mask estimation (clustering) stage of the algorithm. For example, an alternative distance measure in the FCM can be considered: it has been shown (Hathaway et al., 2000) that the Euclidean distance metric as employed in the c-means distance calculation may not be robust to the outliers due to undesired interferences in the acoustic environment. A measure such as the l_1-norm could be implemented in a bid to reduce error (Kühne et al., 2010).

Additionally, the authors of (Kühne et al., 2010) also considered the implementation of observation weights and contextual information in an effort to emphasize the reliable features whilst simultaneously attenuating the unreliable features. In such a study, a suitable metric is required to determine such reliability: in the formulation of such a metric, consideration may be given to the behavior of proximate TF cells through a property such as variance (Kühne et al., 2009).

Alternatively, the robustness in the feature extraction stage can also be investigated. As described in Section 2.2, the inevitable conditions of reverberation and nonideal channels interfere with the reliability of the extracted features. A robust approach to the feature extraction would further ensure the accuracy of the TF mask estimation. The authors of (Reju et al., 2010) employ a feature extraction scheme based upon the Hermitian angle between the observation vector and a reference vector; and in a spirit similar to the MENUET-FCM, the features were clustered using the FCM and encouraging separation results were reported.

Furthermore, in a bid to move the MENUET-FCM BSS algorithm to that of a truly blind and autonomous nature, a modification to the FCM is suggested. The automatic detection of the number of clusters may prove to be of significance as all three of the clustering techniques in this chapter have required *a priori* knowledge of the number of sources. The authors of (Sun et al., 2004) describe two possible algorithms which employ a validation technique to automatically detect the optimum number of clusters to suit the data. Successful results of this technique have been reported in the BSS framework (Reju et al., 2010).

In the current investigation evaluations were limited to artificial corruption provided by a simulated room environment, as such extensions for source separation in more realistic noise scenarios (e.g. as in the CHiME data (Christensen et al., 2010), or the SiSEC data (Araki & Nesta, 2011)) will be a subject of focus in future research.

Finally, as a further evaluation measure, the separation quality of the MENUET-FCM can be evaluated in an alternative context. A natural application of the BSS scheme presented in this chapter is as a front-end to a complete speech processing system; for example, one which incorporates automatic speech recognition. The application of the MENUET-FCM to such a discipline would truly determine its functionality and relevance to modern speech systems.

8. Acknowledgements

This research is partly funded by the Australian Research Council Grant No. DP1096348.

9. References

Abrard, F. & Deville, Y. (2005). A time-frequency blind signal separation method applicable to underdetermined mixtures of dependent sources, *Signal Processing* 85: 1389–1403.

Araki, S., Makino, S., Blin, A., Mukai, R. & Sawada, H. (2004). Underdetermined blind separation for speech in real environments with sparseness and ica, *Acoustics, Speech, and Signal Processing, 2004. Proceedings. (ICASSP '04). IEEE International Conference on*, Vol. 3, pp. iii – 881–4 vol.3.

Araki, S., Nakatani, T., Sawada, H. & Makino, S. (2009). Blind sparse source separation for unknown number of sources using gaussian mixture model fitting with dirichlet prior, *Proceedings of the IEEE International Conference on Acoustics, Speech and Signal Processing*, Taipei, Taiwan, pp. 33 –36.

Araki, S. & Nesta, F. (2011). Signal separation evaluation campaign (sisec 2011). URL: *http://sisec.wiki.irisa.fr/tiki-index.php*

Araki, S., Sawada, H., Mukai, R. & Makino, S. (2005). A novel blind source separation method with observation vector clustering, *International Workshop on Acoustic Echo and Noise Control*, pp. 117–120.

Araki, S., Sawada, H., Mukai, R. & Makino, S. (2006a). Blind sparse source separation with spatially smoothed time-frequency masking, *Proceedings of the International Workshop on Acoustic Echo and Noise Control*, Paris, France.

Araki, S., Sawada, H., Mukai, R. & Makino, S. (2006b). Doa estimation for multiple sparse sources with normalized observation vector clustering, *Proceedings of the IEEE International Conference on Acoustics, Speech and Signal Processing*, Vol. 5, Toulouse, France.

Araki, S., Sawada, H., Mukai, R. & Makino, S. (2007). Underdetermined blind sparse source separation for arbitrarily arranged multiple sensors, *Signal Processing* 87: 1833–1847.

Bezdek, J. (1981). *Pattern recognition with fuzzy objective function algorithms*, Plenum Press, New York.

Cherry, E. C. (1953). Some experiments on the recognition of speech, with one and with two ears, *Journal of the Acoustical Society of America* 25(5): 975–979.

Christensen, H., J., B., N., M. & Green, P. (2010). The chime corpus: a resource and a challenge for computational hearing in multisource environments, *Proceedings of Interspeech*, Makuhari, Japan.

Cichocki, A., Kasprzak, W. & Amari, S.-I. (1996). Adaptive approach to blind source separation with cancellation of additive and convolutional noise, *Proceedings of International Conference on Signal Processing*, Beijing, China, pp. 412–415.

Coviello, C. & Sibul, L. (2004). Blind source separation and beamforming: algebraic technique analysis, *IEEE Transactions on Aerospace and Electronic Systems* 40(1): 221 – 235.

Duda, R., Hart, P. & Stork, D. (2000). *Pattern Classification*, 2nd edn, Wiley Interscience.

Garofolo, J. S., Lamel, L. F., Fisher, W. M., Fiscus, J. G., Pallett, D. S., Dahlgren, N. L. & Zue, V. (1993). Timit acoustic-phonetic continuous speech corpus.

Georgiev, P., Theis, F. & Cichocki, A. (2005). Sparse component analysis and blind source separation of underdetermined mixtures, *IEEE Transactions onNeural Networks* 16(4): 992 –996.

Godsill, S., Rayner, P. & Cappé, O. (1997). *Digital Audio Restoration*, Kluwer Academic Publishers.

Hathaway, R., Bezdek, J. & Hu, Y. (2000). Generalized fuzzy c-means clustering strategies using lp norm distances, *IEEE Transactions on Fuzzy Systems* 8(5): 576 –582.

Hyvarinen, H., Karhunen, J. & Oja, E. (2001). *Independent Component Analysis*, John Wiley & Sons, Inc.

ITU-T (1994). Objective measurement of active speech level, *Technical report*, International Telecommunication Union.

Izumi, Y., Ono, N. & Sagayama, S. (2007). Sparseness-based 2ch bss using the em algorithm in reverberant environment, *Proceedings of the IEEE Workshop on Applications of Signal Processing to Audio and Acoustics*, New Paltz, New York, pp. 147 –150.

Jafari, I., Haque, S., Togneri, R. & Nordholm, S. (2011). Underdetermined blind source separation with fuzzy clustering for arbitrarily arranged sensors, *Proceedings of Interspeech, 2011*, Florence, Italy.

Kühne, M., Togneri, R. & Nordholm, S. (2009). Robust source localization in reverberant environments based on weighted fuzzy clustering, *IEEE Signal Processing Letters* 16(2): 85.

Kühne, M., Togneri, R. & Nordholm, S. (2010). A novel fuzzy clustering algorithm using observation weighting and context information for reverberant blind speech separation, *Signal Processing* 90: 653–669.

Lehmann, E. A. & Johansson, A. M. (2008). Prediction of energy decay in room impulse responses simulated with an image-source model, *Journal of the Acoustical Society of America* 124(1): 269–277.

Li, G. & Lutman, M. (2006). Sparseness and speech perception in noise, *Proceedings of the International Conference on Spoken Language Processing*, Pittsburgh, Pennsylvania.

Li, H., Wang, H. & Xiao, B. (2006). Blind separation of noisy mixed speech signals based on wavelet transform and independent component analysis, *Proceedings of the International Conference on Signal Processing*, Vol. 1, Guilin, China.

Lippmann, R. (1997). Speech recognition by humans and machines, *Speech Communication* 22(1): 1–15.

Loizou, P. C. (2007). *Speech Enhancement: Theory and Practice*, CRC Press, Boca Raton.

Mandel, M., Ellis, D. & Jebara, T. (2006). An em algorithm for localizing multiple sound sources in reverberant environments, *Proceedings of Annual Conference on Neural Information Processing Systems*, Vancouver, Canada.

Melia, T. & Rickard, S. (2007). Underdetermined blind source separation in echoic environments using desprit, *EURASIP Journal on Advances in Signal Processing* 2007.

Mitianoudis, N. & Davies, M. (2003). Audio source separation of convolutive mixtures, *IEEE Transactions of Speech and Audio Processing* 11(5): 489–497.

Rabiner, L. (1978). *Digital Processing of Speech Signals*, Signal Processing Series, Prentice-Hall, New Jersey.

Reju, V., Koh, S. N. & Soon, I. Y. (2010). Underdetermined convolutive blind source separation via time-frequency masking, *Audio, Speech, and Language Processing, IEEE Transactions on* 18(1): 101 –116.

Roy, R. & Kailath, T. (1989). Esprit - estimation of signal parameters via rotational invariance techniques, *IEEE Transactions on Acoustics, Speech and Signal Processing* 37(7).

Sawada, H., Araki, S. & Makino, S. (2007). A two-stage frequency-domain blind source separation method for underdetermined convolutive mixtures, *Proceedings of IEEE Workshop on Applications of Signal Processing to Audio and Acoustics*, New Paltz, NY.

Sawada, H., Araki, S. & Makino, S. (2011). Underdetermined convolutive blind source separation via frequency bin-wise clustering and permutation alignment, *IEEE Transactions on Audio, Speech, and Language Processing* 19(3): 516 –527.

Shi, Z., Tan, X., Jiang, Z., Zhang, H. & Guo, C. (2010). Noisy blind source separation by nonlinear autocorrelation, *Proceedings of International Congress on Image and Signal Processing*, Vol. 7, Yantai, China, pp. 3152 –3156.

Smaragdis, P. (1998). Blind separation of convolved mixtures in the frequency domain, *Neurocomputing* 22: 21–34.

Sun, H., Wang, W., Zhang, X. & Li, Y. (2004). Fcm–based model selection algorithms for determining the number of clusters, *Pattern Recognition* 37: 2027–2037.

Theodoridis, S. & Koutroumbas, K. (2006). *Pattern Recognition*, 3rd edn, Academic Press, New York.

Vincent, E., Gribonval, R. & Fevotte, C. (2006). Performance measurement in blind audio source separation, *IEEE Transactions on Audio, Speech, and Language Processing* 14(4): 1462 –1469.

Vincent, E., Sawada, H., Bofill, P., Makino, S. & Rosca, J. (2007). First stereo audio source separation evaluation campaign: data, algorithms and results, *Proceedings of International Conference on Independent Component Analysis and Signal Separation*, London, England.

Yilmaz, O. & Rickard, S. (2004). Blind separation of speech mixtures via time-frequency masking, *IEEE Transactions on Signal Processing* 52(7): 1830–1847.

16

Blind Implicit Source Separation – A New Concept in BSS Theory

Fernando J. Mato-Méndez and Manuel A. Sobreira-Seoane
University of Vigo
Spain

1. Introduction

The *Blind Source Separation* (BSS) problem was first introduced (Herault et al., 1985; Jutten & Herault, 1988) in the context of biological problems (Ans et al., 1983; Herault & Ans, 1984) with the aim of being able to separate a set of signals generated by the central nervous system. A few years later, several methods based on BSS were applied to other fields of industry and research (Deville, 1999). The BSS problem arises from the need to recover the original sources from a blindly mixture. This extraction is characterised as a blind process because the lack of information about the following topics: the characterisation of the sources, the number of sources present at the time of the mixture, and the way that this mixture is performed. Although this kind of information is unknown, the problem described can be solved if the input signals to the mixture process are statistically independent. Related literature provides several methods, most of which have been classified according to the context in which the mixture is performed: linear mixture model, convolutive mixture model, and non-linear mixture model. The first part of this chapter is devoted to describe the most relevant existing works in applying these methods to the audio field. Many of the real problems, however, do not support this simplification, so this part stresses the need for full characterisation of the problem, mainly about the mixing process and the nature of the sources involved.

Typically, the goal of the BSS theory is to extract a set of variables matching the sources involved in the mixture. We have detected, however, the existence of other research fields where the goal is to extract from the mixture another set of variables which appear as implicit functions of the hidden sources. Extracting these variables brings a new challenge for the BSS theory, becoming particularly complex when the sources have a noisy nature. In the second part of this chapter, a complete definition of this new problem is introduced, for which the BSS problem in its classical form must be reformulated. Used by first time in (Mato-Méndez & Sobreira-Seoane, 2011), within a pattern recognition context, the *Blind Implicit Source Separation* (BISS) concept opens an interesting research field. The BSS-PCA algorithm proposed in the research work referenced above solves with success the problem of classification of traffic noise. Within this algorithm, the BISS problem is handled in an embedded way. Motivated by the promising results achieved, a new compact expression for the BISS solution is now proposed. The new BISS-PCA method introduced here robustly solves the feature extraction process for the problem described. The conclusions of this research can be generalized to other application fields, so we believe that this chapter will be of special interest for the readers.

2. Blind audio source separation

The aim of BSS theory is to extract p unknown sources, from m mixtures acquired through a sensors network. To solve this problem, the literature provides a wide set of methods, most collected in (Comon & Jutten, 2010; Hyvärinen et al., 2001). In this sense, many algorithms have been applied to the context of audio signals, and they can be classified according to the solution of three different problems. First, the denoising process from an undesired mixture provided by both, the channel noise and the sensors network noise. Second, the separation of musical sources from an audio mixture. Finally, the problem created by the "cocktail party" effect (Cherry, 1953), generated when several speakers talk at the same time in reverberant field conditions. Other problems appearing in the state of the art can be analysed as a combination of the above.

2.1 Mixture models

The study of the solution becomes very complex taking into account the existence of different types of problems and application contexts. For many years, however, they have been addressed according to how the mixing process is performed. A generic mixture model for the BSS problem can be written as

$$\mathbf{x}(n) = \mathcal{H}(\mathbf{s}(n) + \epsilon(n)), \tag{1}$$

where $\mathcal{H}$ is a function of both, the channel and the sensor network, and ϵ is a Gaussian additive noise signal, independent of the p sources of $\mathbf{s}$. Thus, existing methods can be classified according to this criterion (see (Comon & Jutten, 2010; Mansour & Kawamoto, 2003; Pedersen et al., 2007; Puntonet G., 2003) for more detail) into the categories that are described below.

2.1.1 Instantaneous mixtures

Source separation from instantaneous mixtures has been one of the first applications of BSS in the audio field. For signals acquired into a recording studio, the mixing process can be considered instantaneous: first, the signals associated with each of the sources can be considered independent because being acquired at different times and at different spatial locations. Second, the multipath contributions associated with both, the sources and sensors, can be neglected thanks to the acquisition process of the mixture and third, studios can be considered as "noise free" controlled environments. So the signals recorded under these conditions does not contains neither relevant undesired reflections or significant noise contributions. Thus, many authors approach this problem by means of an instantaneous mixture model. For this situation, the channel is characterised by have no memory, for which the mixture acquired by the $j-th$ sensor can be modelled as

$$x_j(n) = \sum_{i=1}^{p} h_{ij} s_i(n). \tag{2}$$

In this context, the function $\mathcal{H}$ in (1) can be identified with a real matrix verifying that

$$\mathbf{x}(n) = \mathbf{H}\mathbf{s}(n), \tag{3}$$

where the vector $\mathbf{x}$ contains the contributions of the m sensors in the array. So, the separation problem is reduced to solve the system of Eq. (3). In this case, the solution can be achieved by applying ICA on this equation. Before proceed, it is necessary to have at least the same number of mixtures than sources. Besides, at most, only one source can show a Gaussian distribution. Under these conditions, the separation is performed by calculating an estimation of the mixing matrix that minimises the statistical dependence between components of the original signals.

The contribution of the sensor array and the channel makes not possible to neglect the noise effect in most applications. Therefore, the signal acquired by the $j-th$ sensor can be expressed as

$$x_j(n) = \sum_{i=1}^{p} h_{ij} s_i(n) + \epsilon_j^s(n) + \epsilon_j^c(n), \tag{4}$$

where $\epsilon_j^s(n)$ is the noise signal acquired by the $j-th$ sensor, and $\epsilon_j^c(n)$ is the noise signal provided by the channel. The last signal is typically characterised as wide-band noise, with $\mathcal{N}(\mu_{\epsilon_j^c}, \sigma_{\epsilon_j^c})$ distribution for that sensor. It is usual to express the sum of these two noise signals as

$$\epsilon_j(n) = \epsilon_j^s(n) + \epsilon_j^c(n). \tag{5}$$

Taking into account this undesired effect, Eq. (3) must be rewritten as

$$\mathbf{x}(n) = \mathbf{H}\mathbf{s}(n) + \epsilon(n), \tag{6}$$

where the vector ϵ contains the values of the noise signals associated with the m sensors. There are a large number of algorithms that apply ICA on instantaneous mixing problems, which are deeply studied in (Comon & Jutten, 2010). These algorithms show a reasonable separation quality, even when applied on noisy mixtures. According to the criteria used in the application of ICA, the literature provides research contributions based on: second order statistics (Mansour & Ohnishi, 2000; Matsuoka et al., 1995), higher order statistics (Ihm & Park, 1999; Jutten et al., 1991a; Mansour & Ohnishi, 1999; Moreau, 2001), the probability density function (Amari & Cichocki, 1998; Bofill & Zibulevsky, 2000; Cichocki et al., 1997; 1998; Diamantaras & Chassioti, 2000; Hild et al., 2001; Lappalainen, 1999; Lee et al., 1999; Pham & Cardoso, 2001) and geometric models (Mansour et al., 2002; Prieto et al., 1998; 1999; Puntonet et al., 1995; 2000).

2.1.2 Convolutive mixtures

When the mixture is not instantaneous, the channel has memory, so the signal acquired by the $j-th$ sensor can be expressed as

$$x_j(n) = \sum_{i=1}^{p} \sum_{l=0}^{r-1} h_{ij}^l s_i(n-l) + \epsilon_j(n), \tag{7}$$

where r is the order of the FIR filter that models the mixture. Thus, this mixture can be modelled by means of the expression

$$\mathbf{x}(n) = [\mathbf{H}(z)]\mathbf{s}(n) + \epsilon(n) = \sum_l \mathbf{H}(l)\mathbf{s}(n-l) + \epsilon(n). \tag{8}$$

This is the convolutive model, where $\mathbf{H}(l)$ is the matrix that models the channel and $\mathbf{H}(z)$ the matrix that models the effects of sources on the observations. Therefore, this last matrix can be written by means of the $\mathcal{Z}$ transform as

$$[\mathbf{H}(z)] = \mathcal{Z}[\mathbf{H}(n)] = \sum_{l} \mathbf{H}(l) z^{-l}. \quad (9)$$

Several ICA-based algorithms can be applied in this case to carry the separation process out. In the context of audio, the convolutive problem is classically analysed by means of second order statistics (Ehlers & Schuster, 1997; Ikram & Morgan, 2001; Kawamoto et al., 1999; Rahbar & Reilly, 2001; Sahlin & Broman, 1998; Weinstein et al., 1993), higher order statistics (Charkani & Deville, 1999; Jutten et al., 1991b; Nguyen et al., 1992; Nguyen & Jutten, 1995; Van Gerven et al., 1994) and probability density function (Bell & Sejnowski, 1995; Koutras et al., 1999; 2000; Lee et al., 1997a;b; Torkkola, 1996).

2.1.3 Nonlinear mixtures

In a more general approach, the $\mathcal{H}$ function in Eq. (1) does not support a linear form. This is the case for the separation problem of traffic noise sources in a general context. In this problem, the original sources can not be observed and it is unknown how their signals have been mixed. So, if possible, the extraction of the signals that make up the resulting mixture can be *a priori* characterised as a blind separation process.

For nonlinear mixtures it is usual to simplify the problem by using a post-nonlinear mixture model as

$$\mathbf{x}(n) = \mathcal{H}_1[\mathbf{H}_2 \mathbf{s}(n)] + \epsilon(n), \quad (10)$$

being $\mathbf{H}_2$ a real matrix and $\mathcal{H}_1$ a nonlinear function. To solve it, research works based on second order statistics (Molgedey & Schuster, 1994) and based on the probability density function (Solazzi et al., 2001; Valpola et al., 2001) can be consulted.

2.2 Full problem approach

The usual procedure in BSS is to analyse the problem by means of identifying its mixing model. A proper application of the methods described, however, requires an additional knowledge about both, the mixing process and the nature of the sources involved. Thus, to set an accurate strategy of separation it is necessary to add other informations.

The BSS problem for those situations in which the number of observations is higher than the number of sources (over-determined problem), or equal (determined problem), is well studied. For other situations (underdetermined problem), much remains to be done. This new approach leads to research works focused on solving underdetermined problems (Nion et al., 2010; Rickard et al., 2005; Sawada et al., 2011; Zhang et al., 1999a), and focused on optimising the solution for over-determined problems (Joho et al., 2000; Yonggang & Chambers, 2011; Zhang et al., 1999a;b).

In addition, a prior knowledge about both, the statistical and spectral characterisation of the sources, will lead to more efficient separation methods. Thus, the information can be extracted by means of BSS algorithms that exploit the study of second order statistics for non-stationarity sources (Kawamoto et al., 1999; Mansour & Ohnishi, 2000; Matsuoka et al., 1995; Pham & Cardoso, 2001; Weinstein et al., 1993) and cyclo-stationarity sources (Knaak et al., 2002; 2003). These will also be suitable for the separation of whiteness sources (Mansour et al., 1996; 2000). Some information, however, contained in wide-band sources can not be extracted only using second order statistics. In this case, algorithms based on higher order statistics must be applied.

Finally, many of the algorithms show an excellent performance working on synthetic mixtures. However, a significant degradation in the results is detected when they are applied on real mixtures. In addition, a distinction between both, master-recorded and live-recorded mixtures, must be done. Research works carried out to solve audio signals separation in real conditions can be found in (Kawamoto et al., 1999; Koutras et al., 2000; Lee et al., 1997a; Nguyen et al., 1992; Sahlin & Broman, 1998).

The BSS problem applied to extract signals from a noisy mixture is well studied. The residual signal in this case is typically characterised as white noise. A particularly complex problem occurs, however, when the signals to extract are noise signals. Besides, these are in general characterised as coloured noise, as it occurs for traffic noise sources. In this sense, the research carried out by us regarding the application of BSS to traffic noise real mixtures may be consider a pioneer work. The more closest researches can be found in the study of mechanical fault diagnosis in combustion engines. This is a less complex problem because the signal acquisition process is performed by isolating the engine. The research is focused in applying BSS for the study of its vibrational behaviour. Existing papers (Antoni, 2005; Gelle et al., 2000; Knaak & Filbert, 2001; Knaak et al., 2002; Wang et al., 2009; Wu et al., 2002; Ypma et al., 2002) show the difficulty in the search for satisfactory solutions. The complexity of application of BSS theory will become higher by incorporating other sources for the generation of the traffic noise signal. The next section is devoted to the study of this problem in the context of pattern recognition, for which the BSS problem needs to be reformulated.

3. Blind implicit source separation

This new concept is related to the classical definition of sources into a BSS problem and it has been detected by us in classification problems of noise signals. In a generic classification problem, the main goal is to assign an unknown pattern $\boldsymbol{\varphi}$ to a given class $\mathscr{C}_i$. This class belongs to the set $\mathscr{C}$ of c classes previously determined. The starting condition is that each pattern shall be represented by a single vector of features, and it can not belong to more than one class. Under these hypotesis, this pattern may be uniquely represented by $\boldsymbol{\varphi} = [\varphi_1, \varphi_2, \ldots, \varphi_d]^T$, where d is the number of the extracted features and the dimensionality of the classification problem. For a better understanding of the new BSS concept, the following two examples of application may be considered:

- *Mechanical fault diagnosis in combustion engines*
  For the context described, the fault diagnosis can be seen as the combination of two problems to be solved: a classification problem, and a source separation problem. Thus, the BSS application has two purposes: the first task, being able to separate relevant information from the wide-band noise associated with the vibration of the structure. This relevant information is contained within the spectral lines associated with the combustion noise, so that the first task may be characterised as a denoising process. The second task is focused in extracting the information contained within the set of spectral lines and assign it to one of the engine phases. Thus, the strategy followed seeks to improve the identification of possible faults associated with one of the engine phases. This identification task can be

viewed as a classification problem. The prior application of BSS results therefore in a better definition of the boundaries that separates the two classes previously established (faulty, non-faulty).

- *Classification of traffic noise*
  Although, in a colloquial sense, being able to separate two sources of traffic noise might seem synonymous with being able to classify them, both concepts differ in practice because the processing methods applied. There appears, however, a clear correlation between both, the difficulty in applying blind separation algorithms on specific classes of sources and the difficulty in applying classification algorithms on them. To compare the problem with the above, it must be simplified by considering only the combustion noise. In this case, the classification problem consists in assigning an unknown pattern with a predetermined class of vehicles regarding its noise emitted level. In this case, a single engine can belong to two categories of vehicles. Unlike the previous case, the features vector does not provide discriminative information, so an extraction of information from extra sources is needed. The trouble, as the reader may guess, is the lack of uniqueness for the solution. This issue occurs for other sources considered, so the problem is not successfully solved by adding them into the feature extraction process.

As it will be shown, the problem of classification of traffic noise is much more complex than the one described in the example. The signal acquired by means of a sensors network is a combination of a large number of noise sources. Thus, the associated BSS problem becomes into an extremely complex problem to solve:

- For a isolated pass-by, the vibration behaviour of the engine becomes more complex due to the change of the mechanical model handled. This model is now in motion, and it is affected by its interaction with the other parts of the structure. The information associated with the spectral lines, located at low frequencies, is now altered by energy from other systems such as suspension or brakes. The resulting signal is thus combined with noise induced by the exhaust system.
- The turbulences created by the vehicle in motion (aerodynamic noise) spread energy at high frequency on the acquired signal. Both, the distribution and intensity of this energy, will depend on the geometry and speed of the vehicle. For a given geometry, the higher the speed of the vehicle, the higher the emission at high frequencies will be.
- Once exceeded 50 km/h, for motorcycles and cars, and 70 km/h for trucks, most of the energy in the acquired signal is now associated with rolling noise. This noise is generated by the contact of the wheel with the pavement surface. Thus, a masking of information associated with the three sources of noise described above is produced.
- The consideration of other features modifies the resulting signal: directivity pattern of the vehicle, vehicle maintenance/age, road conservation, ground effect, Doppler effect, type of pavement, distance from source to the sensor network, atmospheric conditions and reflexions of the signal on different surfaces close to the road (buildings, noise barriers, ...).
- The traffic noise signal results from a combined pass-by of vehicles. This combination adds both, an interfering pattern and a masking effect, into the mixing process of the signals associated with each of the sources.

Several calculation methods have been developed to predict the noise levels emitted by the traffic road. These are based on mathematical models trying to find the best approximation to the real model described above. This real model is too complex to be implemented, so an approach is carried out by simplifying the number of sources to be considered. Thus, part of the information needed to carry out this prediction is obtained by means of indirect methods. Regarding the European prediction model (CNOSSOS-EU, 2010), information about the average speed of the road, the pavement type and the traffic road intensity is then needed. This information must be collected according to the vehicle type categorisation performed. Thus, we decided to address the design of a portable device capable to provide such information in real time. For this purpose, the more complex trouble lies in the classifier design. Within this, the incorporation of BSS techniques was proposed with the hope to improve the feature extraction process. To address this task into an intercity context, the mixing process can be modelled according to the scheme of Fig. 1, where $s_i(n)$ is the signal associated with the vehicle to be classified.

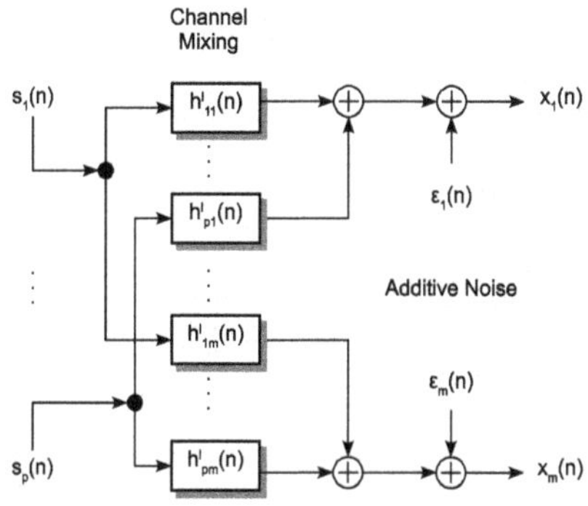

Fig. 1. BSS model of convolutive mixture for the problem of classification of traffic noise on intercity roads.

The goal will be to extract the feature vector of an event whose signal is hidden in mixture where overlapping events are present. The extraction of the signal s_i itself does not help, because this signal carries information associated to other events. It is therefore necessary to find another way to extract this features vector by means of the discriminative information associated with the event to be classified. So, it is proposed to express this information through the acquired mixture as $s_\Gamma^i(n) = \Gamma_i(\mathbf{x}(n))$. Thus, the problem to be solved consists in finding $\boldsymbol{\varphi}_i = [\varphi_{i1}, \ldots, \varphi_{id}]^T$, by means of

$$s_\Gamma^i(n) \xrightarrow{\varphi} (\varphi_{i1}(n), \ldots, \varphi_{id}(n))^T, \tag{11}$$

for wich the BSS problem can be expressed as

$$\boldsymbol{\varphi}_\mathrm{i}(n) = \varphi(s_\Gamma^i(n)) = \varphi(\Gamma_i(\mathbf{x}(n))) = \varphi(\Gamma_i([\mathbf{H}(z)]\mathbf{s}(n) + \boldsymbol{\epsilon}(n))). \tag{12}$$

As the reader can see, the BSS sources in its classical form remain hidden. For this reason, we have named this new BSS problem as *Blind Implicit Source Separation* (BISS). To solve it, the sources definition handled in Fig. 1 is thus no longer valid.

3.1 Dimensionality reduction

One of the typical problems that appear in pattern recognition is the need to reduce the dimensionality of the feature space. For this task both, *Principal Component Analysis* (PCA) and *Independent Component Analysis* (ICA), are the most usual techniques employed. The performance obtained, however, may be different according to the problem to be solved. As it will be seen through this section, the explanation lies in the way that both techniques are applied.

An overview of the original feature space shows in general the existence of values that do not efficiently contribute to the extraction of discriminative information for classification purposes. Under this assumption, for years a large number of techniques has been developed (Fodor, 2002). The goal is to reduce the dimensionality of the original problem, while minimising the possible loss of information related with this process. Most are based on the search of subspaces with better discriminative directions to project the data (Friedman & Tukey, 1974). This projection process involves a loss of information. So a compromise solution is achieved by means of a cost function. There are research works (Huber, 1985), however, which prove that the new subspaces show a higher noise immunity. Furthermore, it is achieved a better capability to filter features with a low discriminative power. So, it results in a better estimation of the density functions (Friedman et al., 1984).

But there are two issues that must be taken into account and that are closely related to the transformations to be used at this stage. First, outliers will be added due to the high variability of the patterns to classify, so an increase of between-class overlap inevitably will occur. Thus, this issue leads to a degradation in the classifier performance. Furthermore, the choice of a suitable rotation of the original data will allow a better view of the discriminative information, as it is shown in Fig. 2. So, it will be very important to find those transformations that contribute to both, a best definition of the between-class boundaries and a best clustering of the within-class information.

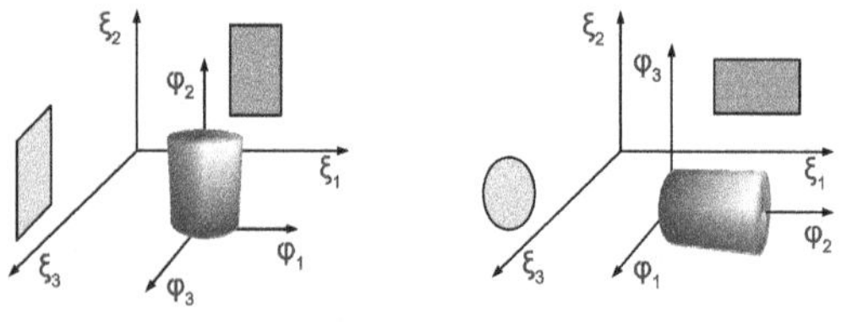

(a) (b)

Fig. 2. Example of projection pursuit. Set of projection directions achieved by means of a wrong geometric rotation (a). Set of projection directions achieved by means of an accurate geometric rotation (b).

Most of the techniques developed for dimensionality reduction are based on the assumption of normality of the original data. For these, it is also shown that most of the projections in problems of high dimensionality allow to achieve transformed data with a statistical distribution that can be considered approximately normal. Among them, a technique of proven effectiveness is PCA (Fukunaga, 1990).

In certain cases, however, PCA may not provide the best directions for projecting the data, as it is shown in Fig. 3. (b). Moreover, this technique limits the analysis to second order statistics so that, for features with a certain degree of statistical dependence between them, ICA (Hyvärinen et al., 2001) will be more suitable. In this technique, the search of independence between components is the basis for the projection directions pursuit, so it can be considered as a dimensionality reduction technique, and therefore an alternative to PCA.

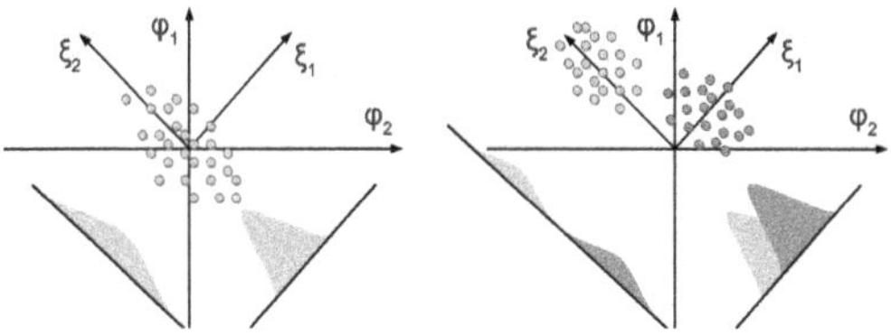

Fig. 3. Example of PCA projection. One class (a). Two classes (b): accurate direction (left), and wrong direction (right).

ICA application, however, is subject to two major restrictions:

1. The assumption of independence of the data is a stronger condition than the assumption of incorrelation, so the conditions for ICA application are more restrictive compared with PCA.
2. The data must show a non-Gaussian distribution, so ICA is not applicable to normal populations, as it occurs with the space of features studied here.

The traffic noise signal verifies the two previous hypotheses: the samples may be considered independent, because being acquired at different times and have different sources spatial location. Furthermore, these samples follow a non-Gaussian distribution, as it is shown in (Mato-Méndez & Sobreira-Seoane, 2008b). Although the extraction of features can be made by using only one sensor, the assumptions handled are the following:

1. For isolated pass-bys, the acquired signal is the result of the combination of both, the signal associated with the vehicle and the background noise.
2. For combined pass-bys, the problem becomes more complex because adding energy associated with other vehicles to the signal associated with the vehicle intended to be classified.
3. The removal of this residual information by source separation techniques would improve the extraction process.

So why not apply this technique to the acquired signal?.

3.2 ICA approach

Taking in mind the ideas described into the previous section, the application of ICA is proposed within a first research work (Mato-Méndez & Sobreira-Seoane, 2008a). This application is carried out by transforming the convolutive problem, which follows the model in Fig. 1, into a linear problem. This transformation is achieved by performing a set of synthetic mixtures by means of the signal acquired. At this point, the reader must remember that the goal is to obtain a higher separability degree of the extracted features. It is not therefore the extraction of the signals associated with the vehicles involved within the mixture process. From this point of view, the transformation carried out is accurate. Thus, the problem to solve now is to find $\hat{\boldsymbol{\varphi}}_i = [\hat{\varphi}_{i1}, \ldots, \hat{\varphi}_{id}]^T$, an estimation of $\boldsymbol{\varphi}_i$, by applying ICA to the new mixture performed $\tilde{\mathbf{x}}$.

In ICA, the separation is conducted by estimating the mixing matrix which minimises the statistical dependence between components of the original signals. To apply it, at most only one source can show a Gaussian distribution. Besides, once the number of sources is known, it is necessary to get at least an equal number of mixtures. For the linear case, the process of extracting the independent components match with solving the blind source separation problem. Under these hypothesis, the mathematical approach of the mixture can be expressed as

$$[\tilde{x}_1(n), \ldots, \tilde{x}_m(n)]^T \approx \begin{pmatrix} a_{11} & \cdots & a_{1p} \\ \vdots & \ddots & \vdots \\ a_{m1} & \cdots & a_{mp} \end{pmatrix} [s_1(n), \ldots, s_p(n)]^T. \quad (13)$$

The convolutive problem can be therefore expressed by means of a linear system of m mixture equations with p unknowns, $\tilde{\mathbf{X}} \approx \mathbf{A} \cdot \mathbf{S}$, where $\mathbf{A}$ represents the mixing matrix, and $\mathbf{S}$ and $\tilde{\mathbf{X}}$ are the vectors of sources and observations respectively. The solution for the linear problem is then conducted by finding the separation matrix $\mathbf{B}$, which is an estimate of the inverse of the mixture matrix $\mathbf{A}$. Although the uniqueness for the solution does not exist from a strict mathematical approach, regarding the independence of the extracted signals this uniqueness can be achieved (Cao & Liu, 1996). In this sense, to ensure the separability of the sources it is sufficient with applying a set of conditions before proceed:

1. The separation process is feasible if the linear function associated with the mixture is bijective, i.e., the regularity of the mixing matrix is needed to be able of estimate $\mathbf{B}$.
2. Regarding the independence of the sources, if $p - 1$ sources shows a non-Gaussian distribution, the independence of pairs of the extracted components is ensured. As result, the possibility of separating the original sources is also ensured.
3. The combined presence of Gaussian and non-Gaussian sources at the time of the mixture will allow the separation of the last ones. This separation will be impossible, however, for the first ones.

Under the above assumptions, an estimation of both unknowns, the coefficients of the matrix $\mathbf{A}$ and the values of the vector $\mathbf{s}$, can therefore be achieved. Although the independence between the recovered sources is ensured in this way, there still exist two unsolved problems in calculating the solution: the uncertainty associated with the energy of the signals obtained, and the uncertainty on the order that they appear. Despite these two uncertainties, ICA proves the existence of uniqueness solving the BSS problem. Furthermore, the existence of these two uncertainties is not an inconvenience for classification purposes.

The process is conducted in two steps. In a first stage, the orthogonal projection of the input mixtures is performed by means of a decorrrelation process. This stage therefore simplifies the solution to a data rotation. Thus, the separation matrix can be factorized as $\mathbf{B} = \mathbf{R} \cdot \mathbf{W}$, being $\mathbf{W}$ a whitening matrix and $\mathbf{R}$ a rotation matrix. The whitening process is started by subtracting the mean from the samples. After this, it concludes by applying an orthonormalization process on the centred samples by means of the *Singular Value Decomposition* (SVD). Proceeding as above, the covariance matrix $\mathbf{\Sigma} = E[\mathbf{s}(n) \cdot \mathbf{s}^T(n)]$ match with the identity matrix. It is true that the study of second order statistics, and more specifically the analysis provided by the decorrelation, allows to carry out a whitening of the samples. This is, however, a necessary but not sufficient condition to ensure the independence of the samples. The difficulty lies in the uncertainty introduced by their possible rotation. This is the reason why, at most, only one of the original sources may show a Gaussian distribution. If this condition is not ensured, the separation of two Gaussian sources is not possible. It is due because the joint distribution for these sources will show a circular symmetry.

Among the wide set of ICA-based algorithms, the developed by Aapo Hyvärinen (Hyvärinen, 1999) is used in (Mato-Méndez & Sobreira-Seoane, 2008a;b) due to its excellent relationship between quality and computational cost. Also known as FastICA, this algorithm in fixed point use both statistics, the kurtosis and negentropy, as non-gaussianity criteria. The decorrelation process is performed by applying on $\tilde{\mathbf{X}}$ the SVD decomposition, widely used in data mining. The idea of this decomposition method was first raised by Carl Eckart and Gale Young in 1936 (Eckart & Young, 1936), by approximating a rectangular matrix by another of lower rank. It was not until 1980, however, that a computational version was proposed by Virginia C. Klema and Alan J. Laub (Klema & Laub, 1980). This new version allowed to discover its performance in solving complex problems. SVD decomposition makes possible to detect and to sort the projection directions that contain the values of higher variance, by means of the use of two square matrices containing the singular vectors. Thus, the dimensionality reduction can be achieved by means of SVD, allowing to find subspaces that best approximate the original data. By applying SVD on $\tilde{\mathbf{X}}$, this matrix can be expressed as $\tilde{\mathbf{X}} \approx \mathbf{U}\mathbf{\Lambda}^{\frac{1}{2}}\mathbf{V}^T$, i.e.,

$$\begin{pmatrix} \tilde{x}_1^1 & \cdots & \tilde{x}_1^n \\ \vdots & \ddots & \vdots \\ \tilde{x}_m^1 & \cdots & \tilde{x}_m^n \end{pmatrix} \approx \begin{pmatrix} u_1^1 & \cdots & u_1^m \\ \vdots & \ddots & \vdots \\ u_m^1 & \cdots & u_m^m \end{pmatrix} \begin{pmatrix} \sqrt{\mathbf{\Lambda}} & 0 \\ 0 & 0 \end{pmatrix} \begin{pmatrix} v_1^1 & \cdots & v_1^n \\ \vdots & \ddots & \vdots \\ v_n^1 & \cdots & v_n^n \end{pmatrix}, \tag{14}$$

where

$$\sqrt{\mathbf{\Lambda}} = \begin{pmatrix} \sqrt{\lambda_1} & & 0 \\ & \ddots & \\ 0 & & \sqrt{\lambda_r} \end{pmatrix}. \tag{15}$$

Fig. 4 graphically shows the changes that take place for a two-dimensional case. The left-multiplication by $\mathbf{V}^T$ allows to transform both vectors, $\mathbf{v}_1$ and $\mathbf{v}_2$ showed in Fig. 4 (a), to the unit vectors of Fig. 4 (b). After this step, these vectors are scaled by the product of the covariance matrix $\mathbf{\Sigma}$, by transforming the unit circle into an ellipse of axes $\sigma_1 \Gamma_1$ and $\sigma_2 \Gamma_2$, as it is showed in Fig. 4 (c). Finally, the right-multiplication by the matrix $\mathbf{U}$ leads to a new rotation of the axes and the consequent rotation of the resulting ellipse of Fig. 4 (c) to its final position showed in Fig. 4 (d).

Thus, the whitening matrix can be expressed as

$$\mathbf{W} = \mathbf{V}^T \approx \mathbf{\Lambda}^{\frac{1}{2}^{-1}}\mathbf{U}^T\tilde{\mathbf{X}}. \tag{16}$$

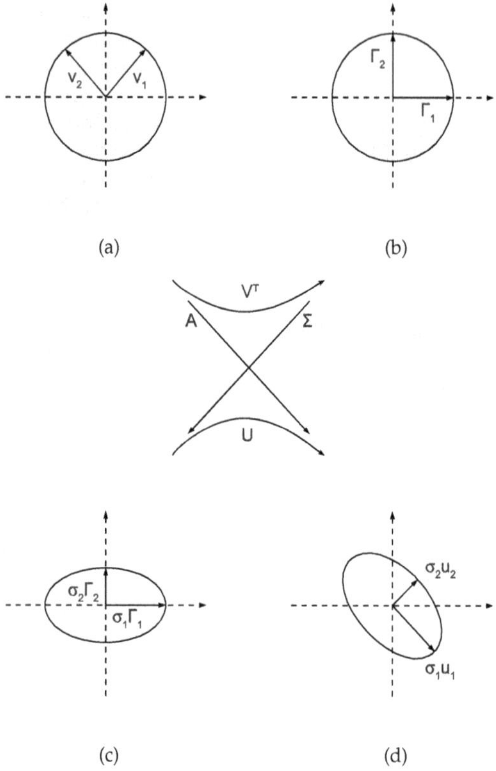

Fig. 4. Graphical evolution for the four steps involved in the SVD decomposition for a two-dimensional case.

Finally, after obtaining the matrix $\mathbf{R}$ by finding a non-normal orthogonal projection, a estimation of the sources can be achieved by means of $\hat{\mathbf{S}} = \mathbf{RW}$. Taking into account that both, $\mathbf{U}$ and $\mathbf{V}$, are unitary matrices, and that the remaining $m - r$ eigenvalues are null, the singular value decomposition of the matrix $\tilde{\mathbf{X}}$ allows to express Eq. (14) as

$$\begin{pmatrix} \tilde{x}_1^1 & \cdots & \tilde{x}_1^n \\ \vdots & \ddots & \vdots \\ \tilde{x}_m^1 & \cdots & \tilde{x}_m^n \end{pmatrix} \approx \sum_{k=1}^{r} \sqrt{\lambda_k} \mathbf{u_k} \mathbf{v_k^T}, \tag{17}$$

where $\{\lambda_1, \ldots, \lambda_r\}$ is the set of singular eigenvalues of $\tilde{\mathbf{X}}$. A suitable approximation for this matrix can be achieved therefore by means of

$$\hat{\tilde{\mathbf{X}}} = \sum_{k=1}^{b} \sqrt{\lambda_k} \mathbf{u_k} \mathbf{v_k^T}, \tag{18}$$

after removal the $r - b$ values, whose contribution can be neglected. This approximation is optimal for the Frobenius norm (Srebro, 2004), being equivalent to the Euclidean norm for this case. The error is thus limited to

$$E\left[\left\|\mathbf{X} - \hat{\mathbf{X}}\right\|^2\right]_{\mathcal{F}} = \sum_{i=b+1}^{m} \lambda_i. \tag{19}$$

3.3 Discussion

The method applied allows to improve the classification results. This improvement is due to the previous remotion of energy that is not related with the event that is being processed. The separability degree of the extracted features, however, is suboptimal because of various causes analysed by us, and which are summarised as follows:

- Under ICA assumptions, its application on the acquired signal will always result in a set of independent components. But, are these components related with the event to be classified?. For isolated pass-bys, the generated signal follows a sources model much more complex that the used in Fig. 1. In this case, the traffic signal is generated from a set $\{o_1, \ldots, o_q\}$ of q sources of noise, by combining the signals associated with each one of them. Discriminative information associated with each of these sources is therefore masked within this process. This situation is worst when considering combined pass-bys generated from a set $\{s_1, \ldots, s_p\}$ of p isolated pass-bys. Regarding discriminative information, the goal is to obtain a features vector that maximises the between-class separation, while minimising the within-class dispersion. In this sense, the features vector obtained by applying ICA on the acquired signal is not optimal. The trouble lies in that the extracted features contain a mix of information generated by several sources within the set $\{o_1, \ldots, o_q\}$. The reader should notice how the extraction of this information from the resulting coloured noise signal becomes a much more complex task for BSS theory. The situation becomes more complicated if a feature selection process is incorporated. The added complexity lies in how the extracted components are selected to be a part of the new calculated subspaces.
- On one hand, ICA is highly dependent on the values of skewness and kurtosis shown by the distributions associated with the signals to be separated. In this sense, PCA is most suitable to address the problem of dimensionality reduction of the feature space. By other hand, although ICA and PCA provide similar benefits for this purpose, PCA used alone can not be considered as a sources separation technique. Therefore, PCA must be combined with BSS for both purposes.
- From a classification point of view both, the distances and angles of the input values, are altered because the whitening process carried out by ICA. This fact contributes to increase the within-class dispersion resulting in a greater uncertainty on the separation boundaries. This dispersion will become even greater with the presence of outliers, for which ICA is fully vulnerable.
- The acquired signal can be considered approximately stationary for short time intervals, lower than 180 *ms* (Cevher et al., 2009). To process these type of signals, it is usual to use a HMM model, as in speech processing occurs. Thus, HMM provides a suitable model to extract hidden temporal information. This model is not supported by ICA, because the time dependence is removed by considering the matrix $\tilde{\mathbf{X}}$ as a set of *iid* random variables. Moreover, some discriminant information remains hidden in frequency.

Therefore, because these two reasons, a T-F domain is most suitable for the BSS process to apply. Finally, the linear model used to solve this BISS problem is suboptimal. The application of BSS on a convolutive mixture model can better exploit the information acquired by the sensor network.

The search for a successful solution that supports these ideas leads to the BISS-PCA method described below.

3.4 BISS-PCA method

To better address the solution, therefore, the first step is to express the mixture model as a function of the noise sources $\{o_1, \ldots, o_q\}$. This new expression can be achieved by reformulating Eq. (12) by means of the mixture model of Fig. 5.

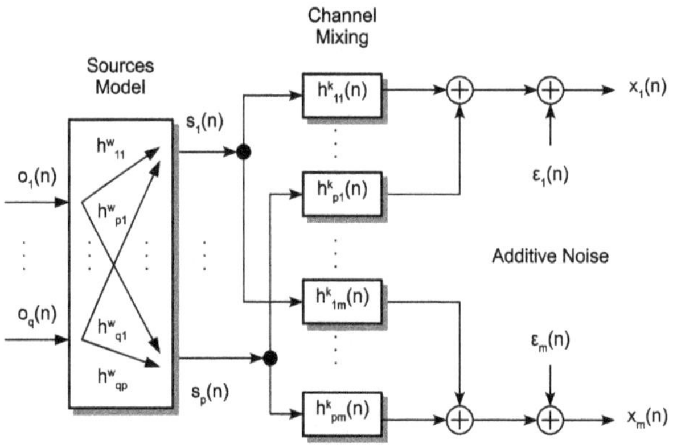

Fig. 5. Suitable BSS model of convolutive mixture for the problem of classification of traffic noise on intercity roads.

For this more suitable model, the signal provided by the $j-th$ sensor can be expressed in terms of the sources set $\{s_1, \ldots, s_p\}$ as

$$x_j(n) = \sum_{i=1}^{p} \sum_{k=0}^{r-1} h_{ij}^k s_i(n-k) + \epsilon_j(n), \tag{20}$$

where r is the order of the FIR filter that models the mixture. The signal s_i is in turn generated by the noise sources set $\{o_1, \ldots, o_q\}$. This last one can be characterised as an instantaneous mixture, after applying a far-field approximation. This is a valid approximation, given that the distances between the sources $\{o_1, \ldots, o_q\}$ are much smaller than the distance from this set to the sensors network. So, Eq. (20) can be expressed as

$$x_j(n) = \sum_{i=1}^{p} \sum_{k=0}^{r-1} h_{ij}^k \sum_{b=1}^{q} h_{ib}^w o_b(n-k) + \epsilon_j(n), \tag{21}$$

where h_{ib}^{w} indicates the contributions of the noise source o_b on the signal s_i. Thus, the above expression can be reordered as

$$x_j(n) = \sum_{b=1}^{q} \sum_{i=1}^{p} h_{ib}^{w} \sum_{k=0}^{r-1} h_{ij}^{k} o_b(n-k) + \epsilon_j(n). \tag{22}$$

This last equation already allows to express the BISS problem as a function of $\{o_1, \ldots, o_q\}$. To do this, since the goal is to extract a feature vector closest to the noise sources related with the event to be classified, this vector will be different from Eq. (11). With this consideration, the BISS problem consists in finding $\zeta_i = [\zeta_{i1}, \cdots, \zeta_{id}]^T$, by solving

$$\zeta_{\mathrm{i}}(n) = \zeta(\mathbf{o}_{\Gamma}^{\mathrm{i}}(n)) = \zeta(\Gamma_i(\mathbf{x}(n))) = \zeta(\Gamma_i(\mathbf{H}^{\mathbf{W}}[\mathbf{H}(z)]\mathbf{o}(n) + \epsilon(n))). \tag{23}$$

To achieve a better solution, it is proposed to carry out the features projection on subspaces closer to the sources $\{o_1, \ldots, o_q\}$, by means of a three-stage strategy (see (Mato-Méndez & Sobreira-Seoane, 2011) for more detail). The first stage deals with the segmentation of the acquired signal, by selecting a fragment of signal centred in the event to classify. For finding discriminative information nearest to these sources, an abstracted features vector $\boldsymbol{\psi}_i = [\psi_{i1}, \psi_{i2}, \ldots, \psi_{if}]^T$ is extracted after removing energy unrelated to the event into a T-F domain by adapting the technique proposed in (Rickard et al., 2001). The last step deals with the suppression of possible correlation between the components of $\boldsymbol{\psi}_i$ by projecting them on the directions of maximum variance. This goal can be efficiently achieved by means of the *Karhunen-Loeve Transformation* (KLT) transformation. It was originally proposed by Kari Karhunen y Michel Loeve (Karhunen, 1946; Loeve, 1945) as a method of development in series for continuous random processes. Widely used in signal processing, it is commonly applied in pattern recognition by means of the linear transformation $\boldsymbol{\zeta}_i = \mathbf{A}_{\mathrm{i}}{}^T \boldsymbol{\psi}_i$. The goal is to obtain the values of the matrix $\mathbf{A}_{\mathrm{i}}$ verifying that $\mathbf{R}_{\zeta_i}$ is diagonal. Thus,

$$\mathbf{R}_{\zeta_i} = E[\boldsymbol{\zeta}_i \boldsymbol{\zeta}_i{}^T] = E[\mathbf{A}_{\mathrm{i}} \boldsymbol{\psi}_i (\mathbf{A}_{\mathrm{i}} \boldsymbol{\psi}_i)^T] = \mathbf{A}_{\mathrm{i}} E[\boldsymbol{\psi}_i \boldsymbol{\psi}_i{}^T] \mathbf{A}_{\mathrm{i}}{}^T = \mathbf{A}_{\mathrm{i}} \mathbf{R}_{\psi_i} \mathbf{A}_{\mathrm{i}}{}^T. \tag{24}$$

It is sufficient with assign to the columns of the matrix $\mathbf{A}_{\mathrm{i}}$ the eigenvectors of the matrix $\mathbf{R}_{\psi_i}$. So that an orthogonal basis can be achieved by means of them, because $\mathbf{R}_{\psi_i}$ is a symmetric matrix. It is achieved thus that $\mathbf{R}_{\zeta_i} = \boldsymbol{\Lambda}_i$, diagonal matrix formed by the eigenvalues [1] of $\mathbf{R}_{\psi_i}$.

Although PCA (Fukunaga, 1990; Jackson, 1991) is usually identified as the same technique, it differs in the calculation components of the matrix $\mathbf{A}_{\mathrm{i}}$ when applying the transform KLT. In this case, columns of the matrix $\mathbf{A}_{\mathrm{i}}$ are matched with the eigenvectors of the covariance matrix of $\boldsymbol{\psi}_i$. The calculation is performed by obtaining each component so as to maximise the variance of the dataset

$$\zeta_{il} = \sum_{k=1}^{f} a_{il}^{k} \psi_i^k, \forall\, l = 1, \ldots, f, \tag{25}$$

under the restriction $\sum_{k=1}^{f} {a_{il}^{k}}^2 = 1, \forall\, l = 1, \ldots, f$. Before proceed it is necessary to achieve a set of data having zero mean. So that centring the data by means of a mean estimator is previously needed. After this adjust, the estimation of the covariance matrix will match the autocorrelation matrix, so that $\boldsymbol{\Sigma}_{\psi_i} = \mathbf{R}_{\psi_i} = E\{\boldsymbol{\psi}_i \boldsymbol{\psi}_i{}^T\}$. Thus both, the set of eigenvalues

[1] The set of eigenvalues of the matrix $\boldsymbol{\Lambda}_i$ also be positive because $\mathbf{R}_{\psi_i}$ is a positive definite matrix.

$\{\lambda_{i1}, \ldots, \lambda_{if}\}$ and the set of associated eigenvectors $\{a_{i1}, \ldots, a_{if}\}$ can be easily calculated. In this way we achieve to project the original data on the new subspace obtained, by means of $\zeta_{il} = a_{il}^T \psi_{il}, \forall\, l = 1, \ldots, f$. Its variance will then given by $\sigma_{\zeta_{il}}^2 = E[\zeta_{il}^2] - E^2[\zeta_{il}] = E[\zeta_{il}^2] = \lambda_{il}$, being also verified that

$$\sum_{l=1}^{f} E[\psi_{il}^2] = \sum_{l=1}^{f} \lambda_{il}. \tag{26}$$

Once the eigenvalues are sorted in descending order of weight, the d eigenvectors corresponding with the d major eigenvalues are chosen. These eigenvectors are the ones which define the set of *"Principal Components"*.

This strategy allows to reduce the dimensionality of the features space by projecting the original data on the directions of maximum variance, as it is shown in Fig. (3). (a). This is made while minimising the cost in loss of information associated with the process: taking into account that $\mathbf{A_i}$ is an orthogonal matrix, $\boldsymbol{\psi}_i$ can be expressed as

$$\boldsymbol{\psi}_i = \mathbf{A_i}\zeta_i = \sum_{l=1}^{f} \zeta_{il} a_{il}, \tag{27}$$

and $\hat{\boldsymbol{\psi}}_i$ as

$$\hat{\boldsymbol{\psi}}_\mathrm{i} = \sum_{l=1}^{d} \zeta_{il} a_{il}. \tag{28}$$

The error is limited to

$$E\left[\left\|\boldsymbol{\psi}_i - \hat{\boldsymbol{\psi}}_\mathrm{i}\right\|^2\right] = E\left[\left\|\sum_{l=1}^{f} \zeta_{il} a_{il} - \sum_{l=1}^{d} \zeta_{il} a_{il}\right\|^2\right] = E\left[\left\|\sum_{l=d+1}^{f} \zeta_{l} a_{il}\right\|^2\right]. \tag{29}$$

Substituting the values of ζ_{il} by $\zeta_{il} = a_{il}^T \psi_{il}, \forall\, l = d+1, \ldots, f$, it is easily obtained that

$$E\left[\left\|\boldsymbol{\psi}_i - \hat{\boldsymbol{\psi}}_i\right\|^2\right] = \sum_{l=d+1}^{f} a_{il}^T E[\boldsymbol{\psi}_i \boldsymbol{\psi}_i{}^T] a_{il} = \sum_{l=d+1}^{f} a_{il}^T \lambda_{il} a_{il} = \sum_{l=d+1}^{f} \lambda_{il}. \tag{30}$$

Then it follows from the above expression how the loss of residual information is minimised, in an optimal way, according to the least squares criterion.

4. Advances

The BSS-PCA algorithm summarises the concepts addressed through this chapter. This algorithm shows an accuracy of 94.83 % in traffic noise classification, drastically improving results achieved before. In addition, BSS-PCA allows to obtain a substantial reduction in uncertainty assigned by CNOSSOS-EU to this task for the prediction of the noise level emitted by traffic road. This uncertainty is calculated by considering most usual methods in vehicle counts. A full analysis on the benefits of this classifier can be found in (Mato-Méndez & Sobreira-Seoane, 2011).

The BISS-PCA method has been recently extended into a new research work. A new technique has been developed, achieving greater discriminative capability for a different set of features that the one used by BISS-PCA. Fig. 6 shows an example of the discriminative capability

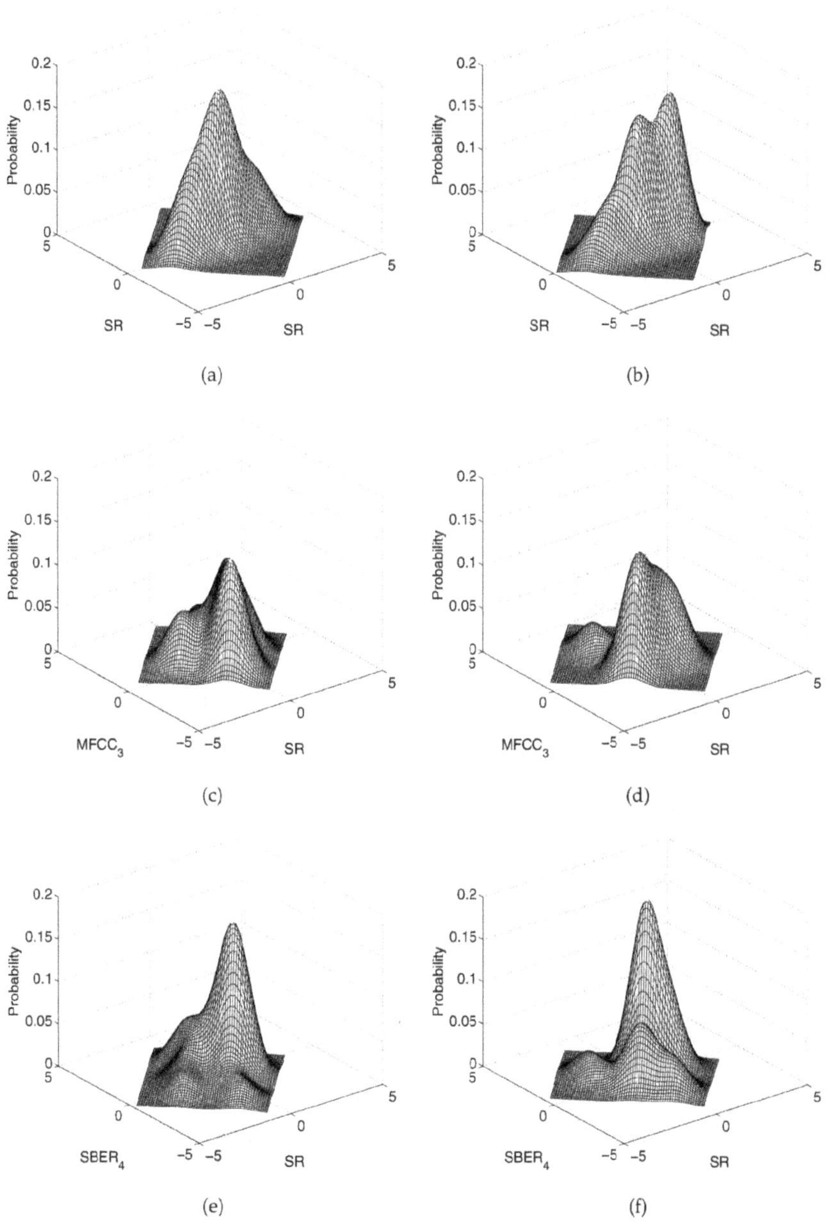

Fig. 6. Improvements (figures (b), (d), and (f)) in the separation boundaries for following vehicle classes: motorcycles (figures (a) and (b)), and cars (figures (c)-(f)).

analysed before (figures (a), (c) and (e)) and after (figures (b), (d) and (f)) applying this new technique. By means of this example, we want to show the effect of this technique over one feature (SR) working alone and combined with another feature ($MFCC_3$ or $SBER_4$). These three features are fully described in the work cited above. It can be observed (figure (a)) how SR shows no between-class discriminative capability for the motorcycle class. After applying the new technique, however, a decision boundary appears. This fact allows now be able to discriminate between two classes (figure (b)). By other hand, the discriminative capability of an isolated feature is generally lower than shown by one subset of the features vector. Figures (c) and (d) correspond to cars class, for which SR is applied in combination with $MFCC_3$. It can be observed how the new technique improves the separability degree for this combination of features. Finally, a suitable selection (SR combined with $SBER_4$) leads to a better discrimination of all classes considered (motorcycles, cars and trucks). An example of this is shown in figures (e) and (f) for cars class. The separability between this class and both, motorcycle class and truck class, is clearly improved after applying this new technique (figure (f)).

5. Conclusions

The application of the existing BSS techniques requires a thorough study of the problem to solve. In many cases, however, the BSS problem is simplified by identifying its mixture model. The first part of this chapter has been devoted to review this issue, which has allowed better understand the need for additional information about the problem to be solved. After it, a new BSS problem has been introduced and discussed. This problem appears in situations for which the variables to extract are presented as implicit functions of the original sources. For this reason, we have named this new problem as *Blind Implicit Source Separation* (BISS). Achieving a solution becomes a specially complex task when the original sources are identified with noise sources. In these cases, the sources models used in BSS are no longer valid and the separation problem needs to be reformulated. Throughout this chapter, a full characterisation for the BISS problem has been presented.

An example of BISS problem occurs for the classification of traffic noise. Through the chapter, a detailed description about it within an intercity context has been given. To solve it, a first approximation has been proposed, by applying ICA to synthetic mixtures obtained from the signal acquired by a sensor network. After a results analysis, however, it has been shown how ICA does not optimally solves this problem.

After this, a thorough study on how better solve the BISS problem is conducted. As result, a novel feature extraction technique has been then introduced. This technique is used in embedded form by the BSS-PCA classifier developed (Mato-Méndez & Sobreira-Seoane, 2011). Its excellent performance lies in its conception, robustly solving the BISS problem. Unlike other methods described in the state of the art in pattern recognition, this algorithm combines the use of both, an abstracted features vector and the application of BSS on the acquired signal. The compact design of this technique gives rise to the BISS-PCA method that has been introduced in this chapter. It has been explained how this method allows the extraction of discriminative information from the set of original noise sources. Unlike ICA, for which this information remains masked, this new technique allows emerge it. The features space therefore wins in resolution while a dimensionality reduction is performed.

Detected by us in pattern recognition problems, the new BISS concept opens an interesting multidisciplinary research field. This new approach allows to optimise the extraction of discriminative information that otherwise remains hidden. For classification purposes, the

BISS-PCA method introduced in this chapter can be extended to other application contexts. This work has been addressed in a recent research. As a result, a new technique solving the BISS problem has been achieved, allowing a highest resolution on the between-class boundaries for a different set of features that the one used by BISS-PCA. An example of the improvements has been shown at the end of this chapter. The results of this new research work are expected to appear soon published, so the reader is invited from this moment to have a look.

6. Acknowledgments

This work has been partially financed by the Spanish MCYT, ref. TEC2009-14414-C03-03, under the project *Research on Combined Techniques of Multichannel Sound Source Separation for Noise Reduction*. The authors would like to express their sincere gratitude to Ph.D. Ganesh R. Naik for his kind invitation to participate in this book.

7. References

Antoni, J. (2005). Blind separation of vibration components: Principles and demonstrations. *Mechanical Systems and Signal Processing*, Vol. 19, No. 6, November 2005, pp. 1166-1180, ISSN 0888-3270.

Amari, S. & Cichocki, A. (1998). Adaptive blind signal processing - Neural network approaches, *Proceedings of the IEEE*, Vol. 86, No. 10, pp. 2026–2048, 1998.

Ans, B.; Gilhodes, J. C. & Herault, J. (1983). Simulation de reseaux neuronaux (SIRENE). II. Hypothese de decodage du message de mouvement porte par les afférences fusoriales IA et II par un mecanisme de plasticite synaptique, *Comptes Rendus de l'Academie des Sciences Paris-Serie III (Sciences Naturelles)*, Vol. 297, pp. 419-422, 1983.

Bell, A.J. & Sejnowski, T.J. (1995). An information-maximization approach to blind separation and blind deconvolution. *Neural Computation*, Vol. 7, No. 6, November 1995, 1129-1159, ISSN 0899-7667.

Bofill P. & Zibulevsky M. (2000). Blind separation of more sources than mixtures using sparsity of their short-time fourier transform, *Proceedings of the International Conference on Independent Component Analysis and Blind Source Separation*, pp. 87-92, June 2000, Helsinki, Finland.

Cao, X. R. & Liu, R. W. (1996). General approach to blind source separation. *IEEE Transactions on Signal Processing*, Vol. 44, No. 3, March 1996, pp. 562–571. ISSN 1053-587X.

Cevher, V.; Chellappa, R. & McClellan, J. H. (2009). Vehicle speed estimation using acoustic wave patterns. *IEEE Transactions on Signal Processing*, Vol. 57, No. 1, January 2009, pp. 30–47. ISSN 1053-587X.

Charkani, N. & Deville, Y. (1999). Self-adaptive separation of convolutively mixed signals with a recursive structure, Part I: Stability analysis and optimization of asymptotic behaviour. *Signal Processing*, Vol. 73, No. 3, January 1999, pp. 225–254. ISSN 0165-1684.

Cherry, E. C. (1953). Some experiments on the recognition of speech, with one and with two ears. *Journal of Acoustic Society of America*, Vol. 25, No. 5, September 1953, pp. 975–979. ISSN 0001-4966.

Cichocki, A.; Sabala, I; Choi, S.; Orsier, B. & and Szupiluk R. (1997). Self adaptive independent component analysis for sub-Gaussian and super-Gaussian mixtures with unknown number of sources and additive noise, *Proceedings of the International Symposium on*

Nonlinear Theory and its Applications, Vol. 2, pp. 731-734, December 1997, Hawaii, USA.

Cichocki, A.; Sabala, I & Amari, S. (1998). Intelligent neural networks for blind signal separation with unknown number of sources, *Proceedings of Conference Engineering of Intelligent Systems*, pp. 148-154, February 1998, Tenerife, Spain.

CNOSSOS-EU Group (2010). *Draft JRC Reference Report 1 on Common Noise Assessment Methods in EU (CNOSSOS-EU)*, Version 2d.

Comon, P. & Jutten, C. (2010). *Handbook of Blind Source Separation. Independent Component Analysis and Applications*, Elsevier, ISBN 978-0-12-374726-6, USA.

Deville, Y. Towards industrial applications of blind source separation and independent component analysis, *Proceedings of International Conference on Independent Component Analysis and Blind Source Separation*, pp. 19-24, January 1999, Aussois, France.

Diamantaras, K. I. & Chassioti E. (2000). Blind separation of N binary sources from one observation: a deterministic approach, *Proceedings of the International Conference on Independent Component Analysis and Blind Source Separation*, pp. 93-98, June 2000, Helsinki, Finland.

Ehlers, F. & Schuster, H. G. (1997). Blind separation of convolutive mixtures and an application in automatic speech recognition in a noisy environment. *IEEE Transactions on Signal Processing*, Vol. 45, No. 10, October 1997, pp. 2608-2612, ISSN 1053-587X.

Eckart, C. & Young, G. (1936). The approximation of one matrix by another of lower rank. *Psychometrika*, Vol. 1, September 1936, pp. 211–218, ISSN 0033-3123.

Friedman, J. H. & Tukey, J. W. (1974). A projection pursuit algorithm for exploratory data analysis. *IEEE Transactions on Computers*, Vol. c-23, No. 9, September 1974, pp. 881–890, ISSN 0018-9340.

Friedman, J. H.; Stuetzle, W & Schroeder, A. (1984). Projection pursuit density estimation. *Journal of the American Statistical Association*, Vol. 79, No. 387, September 1984, pp. 599–608. ISSN 0162-1459.

Fodor, I. K. (2002). *A Survey of Dimension Reduction Techniques*, Center for Applied Scientific Computing, Lawrence Livermore National Laboratory.

Fukunaga, K. (1990). *Introduction to Statistical Pattern Recognition*, Academic Press, ISBN 0-12-269851-7, San Diego, CA, USA.

Gelle, G.; Colas, M. & Delaunay, G. (2000). Blind sources separation applied to rotating machines monitoring by acoustical and vibrations analysis. *Mechanical Systems and Signal Processing*, Vol. 14, No. 3, May 2000, pp. 427-442, ISSN 0888-3270.

Herault, J. & Ans, B. (1984). Neural network with modifiable synapses : Decoding of composite sensory messages under unsupervised and permanent learning. *Comptes Rendus de l'Academie des Sciences Paris-Serie III (Sciences Naturelles)*, Vol. 299, pp. 525-528, 1984.

Herault, J.; Jutten, C. & Ans, B. (1985). Detection de grandeurs primitives dans un message composite par une architecture de calcul neuromimetique en apprentissage non supervise, *Proceedings of the X GRETSI Symposium on Signal and Image Processing*, pp. 1017-1022, May 1985, Nice, France.

Hild, K. E.; Erdogmus, D. & Principe J. C. (2001). Blind source separation using Renyi's mutual information. *IEEE Signal Processing Letters*, Vol. 8, No. 6, June 2001, pp. 174-176, ISSN 1070-9908.

Huber, P. J. (1985). Projection pursuit. *The Annals of Statistics*, Vol. 13, No. 2, June 1985, pp. 435–475. ISSN 0090-5364.

Hyvärinen, A. (1999). Fast and robust fixed-point algorithms for independent component analysis. *IEEE Transactions on Neural Networks*, Vol. 10, No. 3, May 1999, pp. 626–634. ISSN 1045-9227.

Hyvärinen A.; Karhunen, J. & Oja, E. (2001). *Independent Component Analysis*, John Wiley & Sons, ISBN 978-0-471-40540-5, New York, USA.

Ihm B. C. & Park D. J. (1999). Blind separation of sources using higher-order cumulants. *Signal Processing*, Vol. 73, No. 3, January 1999, pp. 267-276, ISSN 0165-1684.

Ikram, M.Z. & Morgan, D.R. (2001). A multiresolution approach to blind separation of speech signals in a reverberant environment, *Proceedings of the IEEE International Conference on Acoustics, Speech and Signal Processing*, Vol. 5, pp. 2757-2760, May 2001, Utah, USA.

Jackson, J. E. (1991). *A User's Guide to Principal Components*, Wiley Series in Probability and Statistics, John Wiley & Sons, ISBN 9780471622673, New York, USA.

Joho, M.; Mathis, H. & Lambert, R. H. Overdetermined blind source separation: using more sensors than source signals in a noisy mixture, *Proceedings of the International Conference on Independent Component Analysis and Blind Source Separation*, pp. 81–86, June 2000, Helsinki, Finland.

Jutten, C & Herault, J. (1988). Independent components analysis versus principal components analysis, In: *Signal Processing IV, Theories and Applications*, Lacoume J. L.; Chehikian A.; Martin N. & Malbos, J., (Ed.), pp. 643-646, Elsevier Science Publishers, Grenoble, France.

Jutten, C.; Nguyen H. L.; Dijkstra, E.; Vittoz, E. & Caelen, J. (1991a). Blind separation of sources: an algorithm for separation of convolutive mixtures, *Proceedings of the International Workshop on High Order Statistics*, pp. 273-276, July 1991, Chamrousse, France.

Jutten, C.; Guérin, A. & Nguyen, H. L. (1991b). Adaptive optimization of neural algorithms, *Proceedings of the International Workshop on Neural Networks*, pp. 54-61, September 1991, Granada, Spain.

Karhunen, K. (1946). Zur spektraltheorie stochastischer prozesse. *Annales Academiae Scientiarum Fennicae. Mathematica-Physica*, Vol. 34, 1946, pp. 1–7, ISSN 1239-629X.

Kawamoto, M.; Barros, A. K.; Mansour, A.; Matsuoka, K. & Ohnishi N. (1999). Real world blind separation of convolved non-stationary signals, *Proceedings of the International Conference on Independent Component Analysis and Blind Source Separation*, pp. 347-352, January 1999, Aussois, France.

Klema, C. & Laub, A. J. (1980). The singular value decomposition: its computation and some applications. *IEEE Transactions on Automatic Control*, Vol. AC-25, No. 2, April 1980, pp. 164–176, ISSN 0018-9286.

Knaak, M. & Filbert, D. (2009). Acoustical semi-blind source separation for machine monitoring, *Proceedings of the International Conference on Independent Component Analysis and Blind Source Separation*, pp. 361-366, December 2001, San Diego, USA.

Knaak, M; Kunter, M; & Filberi, D. (2002). Blind Source Separation for Acoustical Machine Diagnosis, *Proceedings of the International Conference on Digital Signal Processing*, pp. 159-162, July 2002, Santorini, Greece.

Knaak, M.; Araki, S. & Makino, S. (2003). Geometrically constrained ICA for robust separation of sound mixtures, *Proceedings of the International Conference on Independent Component Analysis and Blind Source Separation*, pp. 951–956, April 2003, Nara, Japan.

Koutras, A.; Dermatas, E. & Kokkinakis, G. (1999). Blind signal separation and speech recognition in the frequency domain, *Proceedings of the IEEE International Conference on Electronics, Circuits and Systems*, Vol. 1, pp. 427-430, September 1999, Pafos, Cyprus.

Koutras, A.; Dermatas, E. & Kokkinakis, G. (2000). Blind speech separation of moving speakers in real reverberant environments, *Proceedings of the IEEE International Conference on Acoustics, Speech and Signal Processing*, Vol. 2, pp. 1133-1136, June 2000, Istambul, Turkey.

H. Lappalainen, H. (1999). Ensemble learning for independent component analysis, *Proceedings of the International Conference on Independent Component Analysis and Blind Source Separation*, pp. 7-12, January 1999, Aussois, France.

Lee, T., Bell, A. J. & Orglmeister, R. (1997a). Blind source separation of real world signals, *Proceedings of the International Conference on Neural Networks*, Vol. 4, pp. 2129-2134, June 1997, Houston, USA.

Lee, T.; Bell, A. J. & Lambert, R. H. (1997b). Blind separation of delayed and convolved signals, In: *Advances in Neural Information Processing Systems 9*, pp. 758-764, MIT Press.

Lee, T.W.; Lewicki, M.S.; Girolami, M. & Sejnowski T.J. (1999). Blind source separation of more sources than mixtures using overcomplete representations. *IEEE Signal Processing Letters*, Vol. 6, No. 4, April 1999, pp. 87-90, ISSN 1070-9908.

Loeve, M. (1945). Sur les fonctions aleatoires stationnaires du second ordre. *Revue Scientifique*, Vol. 83, 1945, pp. 297-310, ISSN 0370-4556.

Mansour, A.; Jutten, C. & Loubaton. P. (1996). Subspace method for blind separation of sources and for a convolutive mixture model. In: *Signal Processing VIII, Theories and Applications*, September 1996, pp. 2081–2084, Elsevier, Triest, Italy.

Mansour, A. & Ohnishi, N. (1999). Multichannel blind separation of sources algorithm based on cross-cumulant and the Levenberg-Marquardt method. *IEEE Transactions on Signal Processing*, Vol. 47, No. 11, November 1999, pp. 3172-3175, ISSN 1053-587X.

Mansour, A. & Ohnishi, N. (2000a). Discussion of simple algorithms and methods to separate non-stationary signals, *Proceedings of the IASTED International Conference on Signal Processing and Communications*, pp. 78-85, September 2000, Marbella, Spain.

Mansour, A.; Jutten, C. & Loubaton, P. (2000b). Adaptive subspace algorithm for blind separation of independent sources in convolutive mixture. *IEEE Transactions on Signal Processing*, Vol. 48, No. 2, February 2000, pp. 583–586, ISSN 1053-587X.

Mansour, A.; Ohnishi, N. & Puntonet C. G. (2002). Blind multiuser separation of instantaneous mixture algorithm based on geometrical concepts. *Signal Processing*, Vol. 82, No. 8, August 2002, pp. 1155-1175, ISSN 0165-1684.

Mansour, A. & Kawamoto, M. (2003). ICA papers classified according to their applications and performances. *IEICE Transactions on Fundamentals of Electronics, Communications and Computer Sciences*, Vol. E86-A, No. 3, March 2003, pp. 620-633, ISSN 0916-8508.

Mato-Méndez, F. J. & Sobreira-Seoane, M. A. (2008a). Automatic segmentation of traffic noise, *Proceedings of the International Congress on Acoustics*, pp. 5867-5872, June 2008, Paris, France. *Journal of the Acoustical Society of America*, Vol. 123, No. 5, June 2008, p. 3818, ISSN 0001-4966.

Mato-Méndez, F. J. & Sobreira-Seoane, M. A. (2008b). Sustracción espectral de ruido en separación ciega de fuentes de ruido de tráfico, *Proceedings of the International Acoustics European Symposium - Congresso Ibérico de Acústica - Congreso Espanol de Acústica (TecniAcústica)*, October 2008, Coimbra, Portugal.

Mato-Méndez, F. J. & Sobreira-Seoane, M. A. (2011). Blind separation to improve classification of traffic noise. *Applied Acoustics*, Vol. 72, No. 8 (Special Issue on Noise Mapping), July 2011, pp. 590-598, ISSN 0003-682X.

Matsuoka, K.; Ohoya, M. & Kawamoto M. (1995). A neural net for blind separation of non-stationary signals. *Neural Networks*, Vol. 8, No. 3, 1995, pp. 411–419, ISSN 0893-6080.

Molgedey, L. & Schuster, H. G. (1994). Separation of a mixture of independent signals using time delayed correlations. *Physical Review Letters*, Vol. 72, No. 23, June 1994, pp. 3634-3637.

Moreau, E. (2001). A generalization of joint-diagonalization criteria for source separation. *IEEE Transactions on Signal Processing*, Vol. 49, No. 3, March 2001, pp. 530–541, ISSN 1053-587X.

Nguyen, H. L.; Jutten, C. & Caelen J. (1992). Speech enhancement: analysis and comparison of methods on various real situations, In: *Signal Processing VI, Theories and Applications*, Vandewalle, J.; Boite, R.; Moonen, M. & Oosterlinck, A. (Ed.), pp. 303–306, Elsevier.

Nguyen, H. L. & Jutten, C. (1995). Blind sources separation for convolutive mixtures. *Signal Processing*, Vol. 45, No. 2, August 1995, pp. 209-229, ISSN 0165-1684.

Nion, D.; Mokios, K. N.; Sidiropoulos, N. D. & Potamianos, A. (2010). Batch and adaptive PARAFAC-based blind separation of convolutive speech mixtures. *IEEE Transactions on Audio, Speech, and Language Processing*, Vol. 18, No. 6, August 2010, pp. 1193-1207, ISSN 1558-7916.

Pedersen, M. S.; Larsen, J.; Kjems, U. & Parra L. C. (2007). A survey of convolutive blind source separation methods, In: *Multichannel Speech Processing Handbook*, Benesty, J. & Huang, A. (Ed.), pp. 1065-1084, Springer, ISBN 978-3-540-49125-5.

Pham, D. T. and Cardoso, J. F. (2001). Blind separation of instantaneous mixtures of non stationary sources. *IEEE Transactions on Signal Processing*, Vol. 49, No. 9, September 2001, pp. 1837-1848, ISSN 1053-587X.

Prieto, A.; Puntonet, C. G. & Prieto B. (1998). Separation of sources: a neural learning algorithm for blind separation of sources based on geometric properties. *Signal Processing*, Vol. 64, No. 3, February 1998, pp. 315-331, ISSN 0165-1684.

Prieto, A.; Prieto, B.; Puntonet, C. G.; Canas, A. & Martín-Smith, P. (1999). Geometric separation of linear mixtures of sources: application to speech signals, *Proceedings of the International Conference on Independent Component Analysis and Blind Source Separation*, pp. 295-300, January 1999, Aussois, France.

Puntonet, C. G.; Prieto, A.; Jutten, C.; Rodriguez Alvarez, M. & Ortega J. (1995). Separation of sources: A geometry-based procedure for reconstruction of n-valued signals. *Signal Processing*, Vol. 46, No. 3, October 1995, pp. 267-284, ISSN 0165-1684.

Puntonet, C.G.; Bauer, C.; Lang, E.W.; Alvarez, M.R. & Prieto B. (2000). Adaptive-geometric methods: application to the separation of EEG signals, *Proceedings of the International Workshop on Independent Component Analysis and Blind Separation of Signals*, pp. 273-277, June 2000, Helsinki, Finland.

Puntonet G. (2003). Procedimientos y aplicaciones en separación de senales (BSS-ICA), *Proceedings of the XVIII URSI Symposium*, September 2003, La Coruna, Spain.

Rahbar, K. & Reilly J. (2001). Blind separation of convolved sources by joint approximate diagonalization of cross-spectral density matrices, *Proceedings of the IEEE International Conference on Acoustics, Speech, and Signal Processing*, Vol. 5, pp. 2745–2748, May 2001, Utah, USA.

Rickard, S.; Balan, R. & Rosca, J. (2001). Real-time time–frequency based blind source separation, *Proceedings of the International Conference on Independent Component Analysis and Blind Source Separation*, pp. 651-656, December 2001, San Diego, USA.

Rickard, S.; Melia, T. & Fearon, C. (2005). DESPRIT - histogram based blind source separation of more sources than sensors using subspace methods, *Proceedings of the IEEE Workshop on Applications of Signal Processing to Audio and Acoustics*, pp. 5-8, October 2005, New Paltz, New York.

Sahlin, H. & Broman H. (1998). Separation of real-world signals. *Signal Processing*, Vol. 64, No. 1, January 1998, pp. 103-113, ISSN 0165-1684.

Sawada, H.; Araki, S. & Makino, S. (2011). Underdetermined convolutive blind source separation via frequency bin-wise clustering and permutation alignment. *IEEE Transactions on Audio, Speech, and Language Processing*, Vol. 19, No. 3, March 2011, pp. 516-527, ISSN 1558-7916.

Solazzi, M.; Parisi, R. & Uncini, A. (2001). Blind source separation in nonlinear mixtures by adaptive spline neural network, *Proceedings of the IEEE International Conference on Acoustics, Speech and Signal Processing*, May 2001, Utah, USA.

Srebro, N. (2004). *Learning with Matrix Factorizations*, PhD thesis, Institute of Technology. Massachusetts.

Torkkola, K. (1996). Blind separation of convolved sources based on information maximization, *Proceedings of the IEEE Workshop on Neural Networks for Signal Processing*, pp. 423-432, September 1996, Kyoto, Japan.

Valpola, H.; Honkela, A. & Karhunen J. (2001). Nonlinear static and dynamic blind source separation using ensemble learning, *Proceedings of the International Joint Conference on Neural Networks*, Vol. 4, pp. 2750–2755, July 2001, Washington D. C., USA.

Van Gerven, S.; Van Compernolle, D.; Nguyen, H. L. & Jutten, C. (1994). Blind separation of sources: a comparative study of a 2nd and a 4th order solution, In: *Signal Processing VII, Theories and Applications*, pp. 1153–1156, Elsevier, Edinburgh, Scotland.

Wang, Y.; Chi, Y.; Wu, X. & Liu, C. (2009). Extracting acoustical impulse signal of faulty bearing using blind deconvolution method, *Proceedings of the International Conference on Intelligent Computation Technology and Automation*, pp. 590-594, October 2009, Changsa, China.

Weinstein, E.; Feder, M. & Oppenheim, A.V. (1993). Multi-channel signal separation by decorrelation. *IEEE Transactions on Speech Audio Processing*, Vol. 1, No. 4, October 1993, pp. 405-413, ISSN 1063-6676.

Wu, J. B.; Chen, J.; Zhong, Z. M. & Zhong, P. (2002). Application of blind source separation method in mechanical sound signal analysis, *Proceedings of the American Society of Mechanical Engineers International Mechanical Engineering Congress and Exposition*, pp. 785-791 , November 2002, New Orleans, USA.

Ypma, A.; Leshem, A.; & Duin. R. P. D. (2002). Blind separation of rotating machine sources: bilinear forms and convolutive mixtures. *Neurocomputing*, Vol. 49, December 2002, pp. 349-368, ISSN 0925-2312.

Yonggang, Z. & Chambers, J. A. (2011). Exploiting all combinations of microphone sensors in overdetermined frequency domain blind separation of speech signals. *International Journal of Adaptive Control and Signal Processing*, Vol. 25, No. 1, 2011, pp. 88–94, ISSN 1099-1115.

Zhang, L. Q.; Amari, S. & Cichocki, A. (1999a). Natural gradient approach to blind separation of over- and undercomplete mixtures, *Proceedings of the International Conference on Independent Component Analysis and Blind Source Separation*, pp. 455–460, January 1999, Aussois, France.

Zhang, L. Q.; Cichocki, A. and Amari, S. (1999b). Natural gradient algorithm for blind separation of overdetermined mixture with additive noise, *IEEE Signal Processing Letters*, Vol. 6, No. 11, November 1999, pp. 293–295, ISSN 1070-9908.

Permissions

The contributors of this book come from diverse backgrounds, making this book a truly international effort. This book will bring forth new frontiers with its revolutionizing research information and detailed analysis of the nascent developments around the world.

We would like to thank Dr. Ganesh R. Naik, for lending his expertise to make the book truly unique. He has played a crucial role in the development of this book. Without his invaluable contribution this book wouldn't have been possible. He has made vital efforts to compile up to date information on the varied aspects of this subject to make this book a valuable addition to the collection of many professionals and students.

This book was conceptualized with the vision of imparting up-to-date information and advanced data in this field. To ensure the same, a matchless editorial board was set up. Every individual on the board went through rigorous rounds of assessment to prove their worth. After which they invested a large part of their time researching and compiling the most relevant data for our readers. Conferences and sessions were held from time to time between the editorial board and the contributing authors to present the data in the most comprehensible form. The editorial team has worked tirelessly to provide valuable and valid information to help people across the globe.

Every chapter published in this book has been scrutinized by our experts. Their significance has been extensively debated. The topics covered herein carry significant findings which will fuel the growth of the discipline. They may even be implemented as practical applications or may be referred to as a beginning point for another development. Chapters in this book were first published by InTech; hereby published with permission under the Creative Commons Attribution License or equivalent.

The editorial board has been involved in producing this book since its inception. They have spent rigorous hours researching and exploring the diverse topics which have resulted in the successful publishing of this book. They have passed on their knowledge of decades through this book. To expedite this challenging task, the publisher supported the team at every step. A small team of assistant editors was also appointed to further simplify the editing procedure and attain best results for the readers.

Our editorial team has been hand-picked from every corner of the world. Their multi-ethnicity adds dynamic inputs to the discussions which result in innovative outcomes. These outcomes are then further discussed with the researchers and contributors who give their valuable feedback and opinion regarding the same. The feedback is then collaborated with the researches and they are edited in a comprehensive manner to aid the understanding of the subject.

Apart from the editorial board, the designing team has also invested a significant amount of their time in understanding the subject and creating the most relevant covers. They scrutinized every image to scout for the most suitable representation of the subject and create an appropriate cover for the book.

The publishing team has been involved in this book since its early stages. They were actively engaged in every process, be it collecting the data, connecting with the contributors or procuring relevant information. The team has been an ardent support to the editorial, designing and production team. Their endless efforts to recruit the best for this project, has resulted in the accomplishment of this book. They are a veteran in the field of academics and their pool of knowledge is as vast as their experience in printing. Their expertise and guidance has proved useful at every step. Their uncompromising quality standards have made this book an exceptional effort. Their encouragement from time to time has been an inspiration for everyone.

The publisher and the editorial board hope that this book will prove to be a valuable piece of knowledge for researchers, students, practitioners and scholars across the globe.

List of Contributors

Ganesh R. Naik
RMIT University, Melbourne, Australia

Anil Lal and Wenwu Wang
Department of Electronic Engineering, University of Surrey, United Kingdom

Feng Jin
Dept. of Electrical & Computer Engineering, Ryerson University, Toronto, Ontario, Canada

Farook Satta
Dept. of Electrical & Computer Engineering, University of Waterloo, Waterloo, Ontario, Canada

Bin Gao and W.L. Woo
School of Electrical and Electronic Engineering, Newcastle University, England, United Kingdom

Hiroshi Saruwatari and Yu Takahashi
Nara Institute of Science and Technology, Japan

Andrés Ortiz, Lorenzo J. Tardón, Ana M. Barbancho and Isabel Barbancho
Dept. Ingeniería de Comunicaciones, ETSI Telecomunicación-University of Malaga, Campus Universitario de Teatinos s/n, Malaga, Spain

Masoud Geravanchizadeh and Masoumeh Hesam
Faculty of Electrical and Computer Engineering, University of Tabriz, Tabriz, Iran

Farid Oveisi and Ioannis Patras
Queen Mary University of London, UK

Shahrzad Oveisi
Azad University, Iran

Abbas Efranian
Iran University of Science and Technology, Iran

Atsushi Kawaguchi
Biostatistics Center, Kurume University, Kureme, Fukuoka, Japan

Young K. Truong
Department of Biostatistics, University of North Carolina at Chapel Hill, NC, USA

Xuemei Huang
Department of Neurology, Penn State University, PA, USA

Celso Hilario, Josue-Rafael Montes, Teresa Hernández, Leonardo Barriga and Hugo Jiménez
CIDESI- Centro de Ingeniería y Desarrollo Industrial, México

Jorge I. Marin-Hurtado
Universidad del Quindio, Department of Electronics Engineering, Armenia, Q., Colombia

David V. Anderson
Georgia Institute of Technology, School of Electrical and Computer Engineering, Atlanta, GA, USA

Rubén Martín-Clemente and José Luis Camargo-Olivares
University of Seville, Spain

Evan S. Hill, Sunil K. Vasireddi and William N. Frost
Department of Cell Biology and Anatomy, USA

Angela M. Bruno
Department of Cell Biology and Anatomy, USA
Interdepartmental Neuroscience Program, Rosalind Franklin University of Medicine and Science, North Chicago, IL, USA

Auxiliadora Sarmiento, Iván Durán, Pablo Aguilera and Sergio Cruces
Department of Signal Theory and Communications, University of Seville, Seville, Spain

Ingrid Jafari and Roberto Togneri
The University of Western Australia, Australia

Sven Nordholm
Curtin University, Australia

Fernando J. Mato-Méndez and Manuel A. Sobreira-Seoane
University of Vigo, Spain

www.ingramcontent.com/pod-product-compliance
Lightning Source LLC
Chambersburg PA
CBHW070718190326
41458CB00004B/1018